UNIVERSITY OF ROCHESTER LIBRARIES
3 9087 01015483 1

D1570244

ELISA and Other Solid Phase Immunoassays

ELISA and Other Solid Phase Immunoassays
Theoretical and Practical Aspects

Edited by

D.M. Kemeny and **S.J. Challacombe**

United Medical and Dental Schools of Guy's and St Thomas's Hospitals London SE1 9RT UK

A Wiley Medical Publication

JOHN WILEY & SONS

Chichester · New York · Brisbane · Toronto · Singapore

British Library Cataloguing in Publication Data:

ELISA and other solid phase immunoassays: theoretical and practical aspects
1. Immunoassay
I. Kemeny, D.J. II. Challacombe, Stephen J.
616.07′56 RB46.5

ISBN 0 471 90982 3

Typeset by Inforum Ltd, Portsmouth
Printed and bound in Great Britain by Anchor Brendon Ltd., Tiptree, Colchester.

Contents

List of Contributors

N.F. Adkinson, Jr *Division of Allergy and Clinical Immunology, Good Samaritan Hospital Baltimore, USA.*

D.L. Bates *IQ (Bio) Limited, Cambridge.*

J.E. Butler *Department of Microbiology, University of Iowa, Iowa, USA.*

S.J. Challacombe *Department of Oral Medicine and Pathology, United Medical and Dental Schools of Guy's and St Thomas's Hospitals, London.*

S.M. Chantler *Immunodiagnostics Division, Wellcome Research Laboratories, Beckenham, Kent.*

A.-L. Clayton *Immunodiagnostics Division, Wellcome Research Laboratories, Beckenham, Kent.*

C. Czerkinsky *Department of Medical Microbiology, University of Göteborg, Sweden.*

M.E. Devey *Immunology Unit, Department of Medical Microbiology, London School of Hygiene and Tropical Medicine, London.*

R.J.S.Duncan *The Wellcome Research Laboratories, Beckenham, Kent.*

R.G. Hamilton *Department of Internal Medicine, Health Science Center University of Texas, Houston, USA.*

P. Holt *Clinical Immunology Research Unit, Princess Margaret Children's Hospital, Perth, Western Australia 6001.*

A. Johansson *IQ (Bio) Limited, Cambridge.*

D.M. Kemeny *Department of Medicine, United Medical and Dental Schools of Guy's and St Thomas's Hospitals, London.*

W.J. Koopman *Department of Medicine, University of Alabama, Birmingham, USA.*

J. Mestecky *Department of Microbiology, University of Alabama, Birmingham, USA.*

J.G. Michael *Department of Pathology and Laboratory Medicine, University of Cincinnati, Cincinnati, Ohio, USA.*

L.-A. Nilsson *Department of Medical Microbiology, University of Göteborg, Sweden.*

O. Ouchterlony *Department of Medical Microbiology, University of Göteborg, Sweden.*

A.J. Pesce *Department of Pathology and Laboratory Medicine, University of Cincinnati, Cincinnati, USA.*

D. Richards *Department of Medicine, United Medical and Dental Schools of Guy's and St Thomas's Hospitals, London.*

J.D. Sedgwick *MRC Immunology Unit, Sir William Dunn School of Pathology, University of Oxford, Oxford.*

M.W. Steward *Immunology Unit, Department of Medical Microbiology, London School of Hygiene and Tropical Medicine, London.*

A. Tarkowski *Department of Rheumatology, University of Göteborg, Sweden.*

I. Weeks *Department of Biochemistry, University of Wales College of Medicine, Heath Park, Cardiff, Wales.*

S. Woodhead *Department of Biochemistry, University of Wales College of Medicine, Heath Park, Cardiff, Wales.*

Preface

Radio- and enzyme immunoassay techniques have only been available for about 30 years but have found a wide range of applications. The increasing demands for sensitivity, speed and ease of performance of these tests is reflected by the number of new assay systems that are continually being reported. These cover a wide range of disciplines and it is difficult, if not impossible, to keep up to date with all the latest developments. One of the objectives of this book, therefore, is to bring together the latest concepts and applications of solid phase assays in a readily accessible and digestible form.

In preparing this book we have been fortunate in being able to call on the experience of scientists from universities, medicine and industry all over the world. We are grateful to them for their prompt acceptance of our invitations and the quality of their contributions. We have tried to minimize the repetition which is inevitable in a multi-author book. At the same time we wished for techniques in various chapters to be able to be followed without continuous reference to others and have therefore accepted some duplication.

While the emphasis is strongly on the practical aspects of solid phase immunoassay we feel it is essential that the principles underlying them are properly understood. We have, therefore, sought to combine theory with practice throughout the book and to explain the concepts underlying each stage of the assay systems.

We wanted the book to be both a reference book and a laboratory manual. We have therefore sectioned the book for ready reference to basic aspects such as solid phase support and the quantification of results, to applied aspects such as microbiology and virology, to newer techniques such as the analysis of individual antibody producing cells, and have included an appendix which contains details of many commonly used reagents.

We thank Clare Langlan and Sheelah Lockwood for their efficient secretarial assistance and Professor Maurice Lessof for his helpful comments and discussions.

D.M. Kemeny
S.J. Challacombe

To SHK and CBC and for CGC

ELISA and Other Solid Phase Immunoassays
Edited by D.M. Kemeny and S.J. Challacombe

Published by John Wiley & Sons Ltd

CHAPTER 1

An Introduction to ELISA

D.M. Kemeny and **S. Chantler**

UMDS, Guy's Hospital, London and Wellcome Research Laboratory, Beckenham

CONTENTS

INTRODUCTION

The development of the enzyme-linked immunosorbent assay (ELISA) is best appreciated in the context of the preceding radioimmunoassays. The earliest of these measured the analyte by competition between ^{131}I-radiolabelled and unlabelled antigen for antibody and involved a separation step to distinguish between bound and free-labelled antigen. Non-specific agents such as ammonium sulphate, polyethylene glycol, trichloracetic acid or, alternatively, specific second antibodies were used to enhance precipitation of the antigen-antibody complex. It was later found that the efficiency and speed of separation was increased by immobilizing the second antibody on a solid phase such as

Sephadex (Wide and Porath, 1966), plastic (Catt and Tregear, 1967) or cellulose (Ceska and Lundqvist, 1972). The 'sandwich-type' solid phase radioimmunoassays using labelled antibodies rather then labelled antigens, known as immunoradiometric assays (IRMA), developed from these and offered several advantages over the earlier competitive radioimmunoassays (Miles and Hales, 1968; Woodhead, Addison, and Hales, 1974).

Although enzymes had been used as antibody labels for the immunohistochemical localization of antigens in tissue sections in the mid-1960s (Nakane and Pierce, 1966; Avrameas and Uriel, 1966) and their potential adoption as markers in solid phase tube assays was discussed by Miles and Hales (1968), their practical exploitation did not occur until 1971 (Engvall and Perlmann, 1971; Van Weemen and Schuurs, 1971).

Enzyme linked immunosorbent assays (ELISA) offer a number of advantages over IRMA. The reaction can be read visually without the need for expensive apparatus. The labelled reagents used are stable and are easily stored for long periods of time without loss of activity. Multiwell microtitre plates, used as the solid phase in place of tubes, are easy to handle and wash and when used with automated readers and multiple well washers allow large numbes of samples to be assayed (Voller *et al.*, 1974). A variety of enzyme-labelled antisera of good quality can be purchased commercially now and an increasingly wide range of suitable enzyme substrates and chromogens is available. Furthermore, the potential sensitivity of ELISA is greater than radioimmunoassay as many molecules of product can be generated by single molecule of enzyme (Ekins, 1980, 1985; Jackson and Ekins, 1986). As a result, ELISA has been extensively applied both for the detection of antigens and antibodies in human and veterinary infectious diseases and in immunology (Schuurs and van Weemen, 1977; Voller, Bartlett, and Bidwell, 1981; Parratt *et al.*, 1982).

This chapter outlines the basic principles involved in setting up and performing ELISA and highlights some of the likely problem areas. Homogeneous and other non-separation enzyme immunoassays are not included and interested readers should consult appropriate reviews (Voller, Bartlett, and Bidwell, 1978; Ishikawa, Kawai, and Miyai, 1981; Yolken, 1982; Blake and Gould, 1984; Ngo and Lenhoff, 1985).

ASSAY OBJECTIVES

The type of assay should be closely tailored to the particular task for which it is required. It is clearly quite unnecessary to develop an exquisitely sensitive and quantitative assay if one simply wants to screen large numbers of samples for the presence of a particular antigen or antibody present in relatively high concentration. In such circumstances the ease with which the test can be performed and its general robustness, reliability, and reproductibility in use are likely to be more important. The quantity of analyte to be measured is of particular importance in determining the choice of test employed and is exemplified by the measurement of serum immunoglobulins.

Immunoglobulin classes other than IgE are present in milligram amounts and hence are easily and accurately measured by a variety of methods including precipitation in gel (Mancini, Carbonara, and Heremans, 1965) or by nephelometry (Blanchard and Gardner, 1980). Conversely, IgE, which is usually present in serum in nanogram quantities, cannot be detected accurately by these methods. Sensitive radio immunoassays (Wide and Porath, 1966) or enzyme immunoassays (Hoffman, 1973) are mandatory and, furthermore, modifications of conventional ELISA systems are required to measure the lowest levels of IgE (<0.1 ng/ml) present in some tissue culture supernatants (Imagawa et al., 1981; Ishikawa *et al.*, 1982). Hence, even for a single analyte, differences in the assay conditions may be required depending on the level of analyte present and this should be borne in mind when developing an assay system.

ANTIBODY AND ANTIGEN REAGENTS

The performance of any immunoassay is directly dependent on the quality of the antigen, used as target or as labelled detector, and antibody, used as capture or detector. Nevertheless, some assays can work well with crude antigen preparations (Wide, Bennich, and Johansson, 1967) provided users are aware of likely cross-reactions and include appropriate controls to distinguish between specific and non-specific interactions. For instance the use of crude cell lysates of cytomegalovirus (CMV)-infected cells as target for specific antibody detection could result in the binding of not only CMV-specific antibodies but also non-CMV immunoglobulin by non-immunological interaction with Fc receptors generated on infected cells. Both types of binding would give a positive result in a conventional sandwich ELISA using labelled anti-human immunoglobulin. In these instances, selective binding of CMV antigens, by use of specific capture antibodies or the use of purified antigens, would be the preferred approach. If, on the other hand, a labelled antigen is used to detect bound antibody, crude preparations may be employed as target antigen provided the detector reagent is adequately purified. This is important for two reasons: firstly, to ensure specificity and, secondly, to allow efficient labelling of the active component. Coupling an enzyme to a heterogeneous mixture of components precludes efficient and controlled labelling of the relevant antigen or immunoglobulin and results in the conjugation of irrelevant components. Hence the level of antigen or antibody purification required for a particular application will vary.

The parameters used to select an antibody (Steward and Lew, 1985; Chantler and Evans, 1986; Shields and Turner, 1986; see also this volume Chapters 6,

7, and 9) and the degree of processing required prior to use depend upon the particular function it is performing in the assay.

Capture antibodies

The most important requirement for capture antibodies is that once bound to the solid phase they should have a high binding capacity for relevant antigens. If a small proportion of antigen is bound, both sensitivity and reproducibility will be adversely affected and small changes in the assay conditions could lead to wide variations in what is detected. The specificity, particularly when using monoclonal antibodies, must be against epitopes common to all representative antigens to be detected. Reactivity against irrelevant antigens may or may not be critical depending upon the intended use. If the labelled antigen or antibody detection system is highly specific for the relevant analyte, low level binding of irrelevant antigens in the first step can be tolerated without sacrificing assay specificity. If, as in IgM class capture assays for infectious agents, the antigen detector, however pure, reacts with all immunoglobulin classes of specific antibody then the specificity of the capture antibody is crucial to ensure discrimination between IgG, IgM, and IgA antibodies.

Monoclonal or polyclonal antibodies

Monoclonal or polyclonal antibodies may be used and absorption of cross-reactivity removed, preferably by the use of immunoadsorbents, prior to further processing. Monoclonal antibodies offer tremendous potential as capture reagents in terms of specificity and continuity of supply but because they represent a single antibody lineage within a polyclonal immune response it is sometimes, but not always, necessary to use a mixture of antibodies of differing epitope specificity (Holmes and Parnham, 1983; Soos and Siddle, 1983; Thompson and Jackson, 1984; Siddle, 1985). Most polyclonal antibodies bind well to plastics with good retention of activity. The failure of some monoclonals to bind may be overcome by altering the pH of coating (Conradie, Govender, and Visser, 1983), by chemically coupling to bovine serum albumin (Papadea, Check, and Reimer, 1985), or by binding to anti-mouse antibody-coated wells (Mangili *et al.*, 1987). The selection of the most appropriate capture antibody will depend on its spectrum of reactivity and ability to bind antigen efficiently when immobilized; this can only be confirmed by trial and error.

Fractionation of serum or ascites

In most cases it is preferable to use an immunoglobulin fraction rather than whole serum or ascites for coating wells. This avoids competitive binding and interference by the many other proteins present in serum or ascites and gives a more reproducible result. Different batches of sera and ascites not only contain different levels of total immunoglobulin but also vary in the relative amounts of specific antibody and non-antibody immunoglobulin. We have observed considerable variation in the level of activity in ascites drawn on different days from mice inoculated with hybridoma cells, hence readers should exercise caution in indiscriminate pooling of products without adequate testing.

A variety of methods is available for globulin fractionation (Hurn and Chantler 1980; see also Appendix to this volume). In our experience affinity purification of antibody from antigen immunoadsorbents, although useful where the concentration of specific antibody is low, often results in the loss of highest affinity antibody and should therefore only be adopted if essential. Further fractionation of antibody is rarely necessary except in assays for IgG rheumatoid factor where $F(ab)_2$ must be used or in sera in which there is a high risk of rheumatoid factor being present. In practice the inclusion of wells coated with antibody of unrelated specificity provides an adequate control.

The storage and handling of antibody is important. It is good practice to store whole serum or ascites in aliquots at −20 °C. Purified antibodies can be freeze-dried. The concentration should be determined by absorption at 280 nm (OD 1.4 at 280 nm approximately 1 mg/ml), the sample concentrated to at least 1 mg/ml, preferably 5 mg/ml, and stored at −20 °C in 50 per cent glycerol.

Detector antibodies

The quality of the detector antibody used for enzyme labelling and the efficacy of conjugation is directly related to assay peformance. In our experience the requirements for specificity, affinity, and purity are often more stringent than for capture reagents. Undesirable cross-reactivity with components in the assay must be removed by pre-absorption of the unlabelled antibody or by neutralization by addition of the appropriate material in the conjugate diluent. Anti-species immunoglobulin cross-reactivity may be effectively decreased by addition of 1–2 per cent of serum from the offending species. Most conjugation methods, however carefully performed, result in some polymerization of labelled antibody and presence of free enzyme. Removal of unreacted horseradish peroxidase is easily achieved by precipitation with an equal volume of saturated ammonium sulphate followed by exhaustive dialysis or by Sephadex chromatography. This simple procedure can improve the performance of some assays without deleterious effects on biological activity. However, in other sytems where high sensitivity is essential, more sophisticated fractionation and selection of optimally active fractions are mandatory.

It is important that enzyme-labelled detection antibodies are stored in protein- and glycerol-containing buffers to minimize denaturation and loss by adsorption during storage.

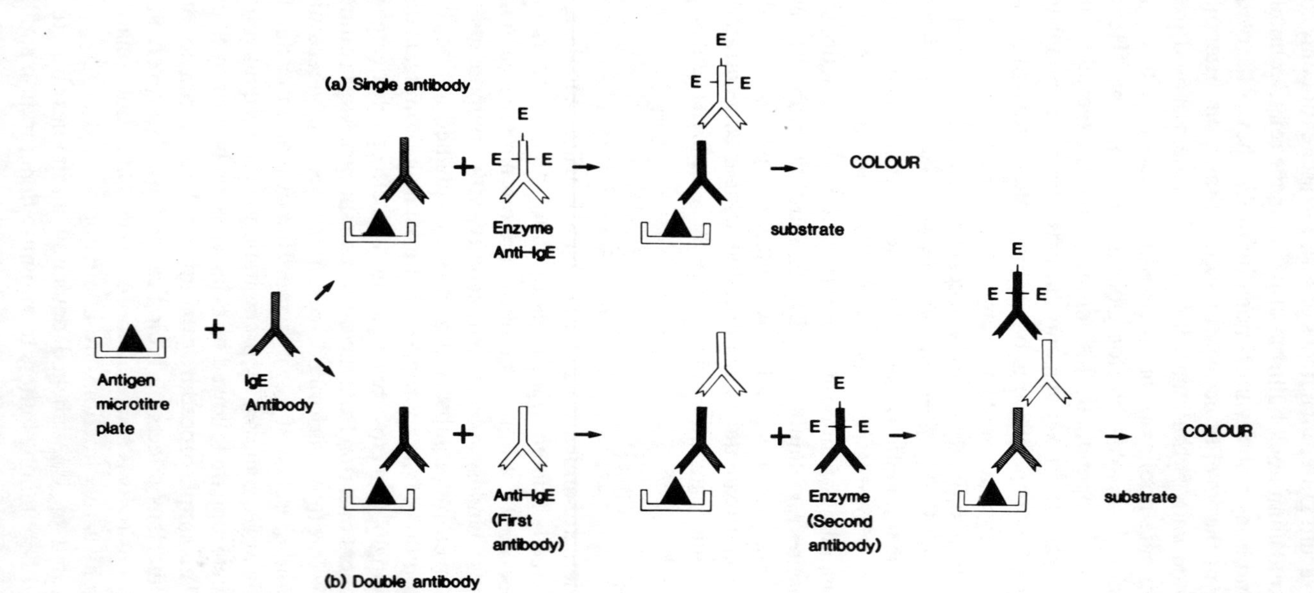

FIGURE 1 Indirect or sandwich ELISA. Antibody in the sample binds to antigen in the solid phase and is subsequently detected with (a) an enzyme-labelled antibody or (b) an unlabelled and a second, labelled, antibody

ASSAY FORMATS

We do not propose to cover all of the possible variants of the method and readers interested in procedures not covered here should consult the many excellent books and review articles published on the subject (Schuurs and Van Weemen, 1977; Voller, Bartlett, and Bidwell, 1978; Maggio, 1979; Yolken, 1982; Collins, 1985; Ngo and Lenhoff, 1985).

Indirect ELISA

Perhaps the simplest form of ELISA is the indirect or sandwich assay (Engvall and Perlmann, 1972), commonly referred to as the 'dirty plate' assay. Antigen is bound passively by incubation with the microtitre plate through an as yet ill-defined process (hence the name 'dirty plate'). The antigen solid phase is then used to bind specific antibodies in the test sample. Unbound material is removed by washing and bound antibody is detected using enzyme-labelled anti-immunoglobulin (Figure 1a). If the enzyme-labelled antibody is specific for a particular class of immunoglobulin then the class specificity of antibody can be determined.

Readers should be aware of the limitation of this approach for measurement of immunoglobulin classes present in low concentration. False-negative results due to competitive inhibition or false-positive results due to interference by IgM anti-immunoglobulin can occur (Cradock-Watson, Ridehalgh, and Chantler, 1975; Vejtorp, 1980; Chantler and Diment, 1981). Alternatively, enzyme-labelled staphylococcal protein-A can be substituted (Engvall, 1978; Yolken and Leister, 1981) but readers should be aware of the variability of protein-A binding to immunoglobulins of different species, to different immunoglobulin classes and, indeed, to immunoglobulin from different individuals of the same species, e.g. goat (Richman *et al.*, 1982).

When it is impossible or undesirable to label the antibody directly a second anti-species immunoglobulin-labelled antibody can be employed (Figure 1b). This may increase the sensitivity of the test but will also increase the likelihood of non-specific interactions due to natural antibodies present in both detector antibodies. The detection of antibody by the indirect ELISA may be limited by the ability to adsorb antigen directly to the solid phase. However, a number of approaches can be used to overcome this potential problem (Chapters 2 and 9) and these or the use of capture antibodies are usually satisfactory for most purposes.

Two-site ELISA

The concentration of antigen can be determined using a two-site ELISA. In this assay (Figure 2a) antibody bound to the microtitre wells is used to capture

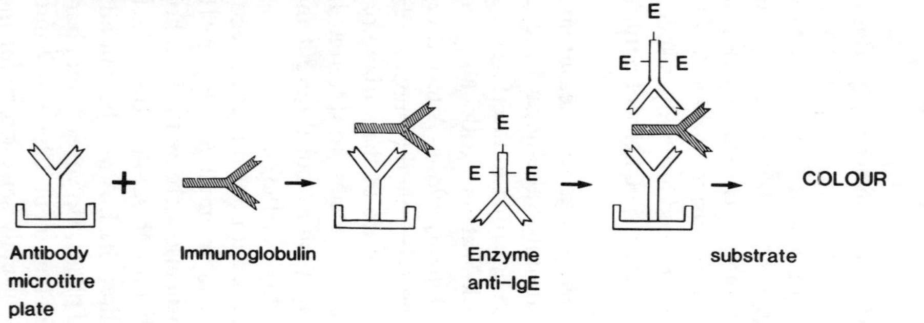

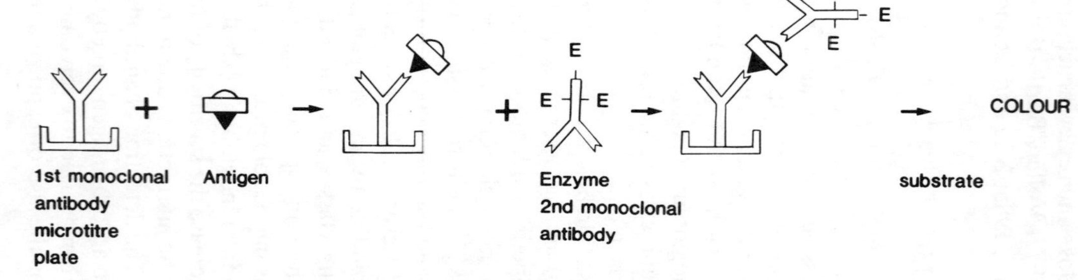

FIGURE 2 Antibody on the solid phase binds the antigen in the sample which is subsequently detected using (a) an enzyme-labelled antibody. Alternatively, (b) two monoclonal antibodies that recognize different parts of the antigen can be used

the corresponding antigen in the test sample. The bound antigen is subsequently detected using a second enzyme-labelled antibody. This provides a rapid, easy method for antigen detection, particularly if the sample and conjugate can be added simultaneously (Chapters 14 and 15). It is preferable, but certainly not essential, that the two antibodies should be from the same species to reduce the possibility of species cross-interactions.

Monoclonal antibodies to different group-common epitopes on the analyte can be used (Figure 2b) or, alternatively, combinations of monoclonal and polyclonal reagents may be used for capture and detector (Clayton *et al.*, 1986; Kemeny and Richards, 1987). As monoclonal antibodies can be produced with consistency and in quantity, they have the advantage of ensuring continuity of reagents — a source of considerable error in ELISA. The concentration of solid phase capture antibody is not usually critical (5–15 μg/ml) but enzyme-labelled antibodies must be carefully optimized. The specificity and sensitivity of the test are highly dependent upon the efficiency with which the solid phase antibody can bind the antigen in the sample and the specificity and efficient labelling of the conjugate.

Class capture assays

In certain circumstances it is desirable to separate a particular class of immunoglobulin from a sample of serum or secretion prior to measurement of antibody activity. The main reason for this is that the presence of a much larger quantity of antibody of different immunglobulin class may interefere with the detection of the clinically relevant one. (Reimer *et al.*, 1975); Chantler and Diment, 1981). Class capture assays for the detection of specific IgM antibodies to viruses (Chapter 13) and bacteria (Chapter 15), and for IgG (Diment and Pepys, 1978) and IgE antibodies to allergens (Van Loon *et al.*, 1985; Zeiss, Grammar, and Levitz, 1981) have been in existence for some years. The principle for detection of IgE antibody is shown in Figure 3.

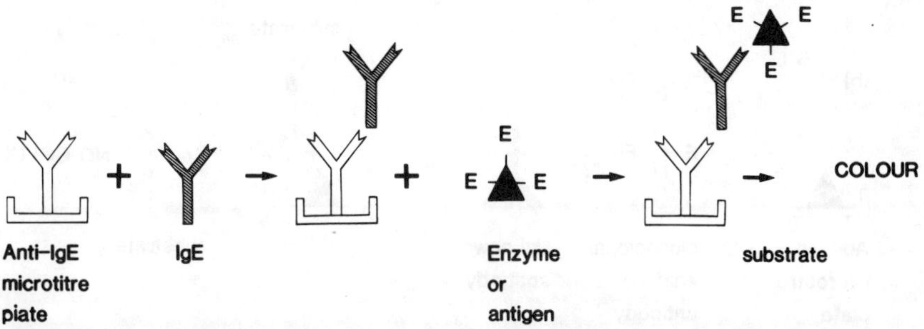

FIGURE 3 Class capture assay for IgE antibody. Immunglobulin of a particular class (e.g. IgE, IgM) is captured by the antibody on the microtitre plate and the presence of antibody activity in the bound immunoglobulin is determined by addtion of labelled antigen

The microtitre plate is coated with class-specific anti-immunoglobulin and used to capture the immunoglobulin in the sample. The antibody activity of the bound immunoglobulin is identified by subsequent addition of enzyme-labelled antigen or unlabelled antigen followed by labelled specific antibody. Alternatively, antigen can be labelled with a hapten such as TNP or with biotin and subsequently detected with enzyme-labelled anti-TNP or avidin (Moneo *et al.*, 1983).

Class capture assays of this kind have considerable advantages over conventional indirect ELISA using class-specific labelled detector antibodies principally because they avoid competitive inhibition caused by IgG antibodies of the same specificity normally present in higher concentration in the body fluid (Chantler and Diment, 1981). This is of critical importance in the measurement of IgE antibodies against food (Roundtree *et al.*, 1985) and insect venom allergens (Kemeny *et al.*, 1982). The capacity and specificity of the solid phase capture antibody and the high specificity and enzymatic activity of the detector are critically important, particularly if only relatively impure antigen preparations are available. Polyclonal, monoclonal or combinations of polyclonal and monoclonal antibodies may be used (Mortimer *et al.*, 1981; Zeiss, *et al.* 1973; Kemeny and Lessof, 1987; Chantler and Evans, 1986) but the selection of the best combination must be established by experimentation.

Competitive ELISA

Many of the theoretical advantages of solid phase immunometric assays that

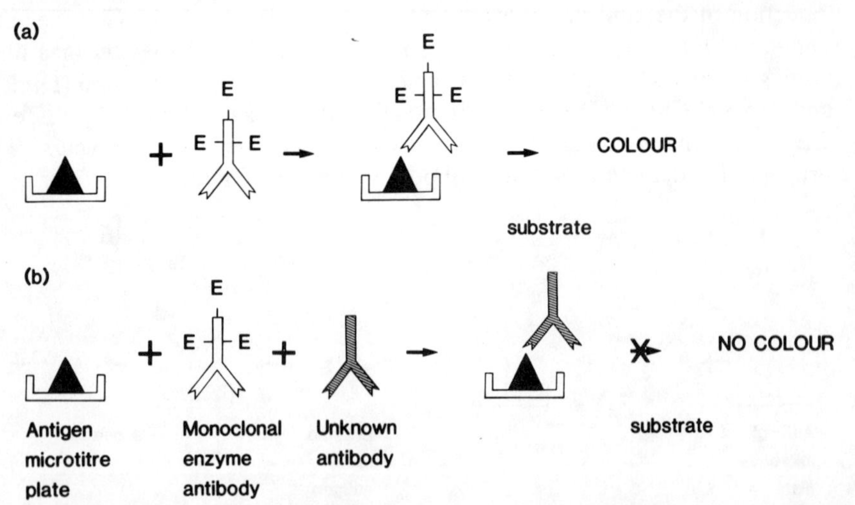

FIGURE 4 Competitive ELISA for detection of antibody. Antibody can be measured by competition with enzyme-labelled antibody for antigen on the solid phase. In the absence of antibody in the sample the enzyme-labelled antibody binds (a) but if antibody is present in the sample it is inhibited (b)

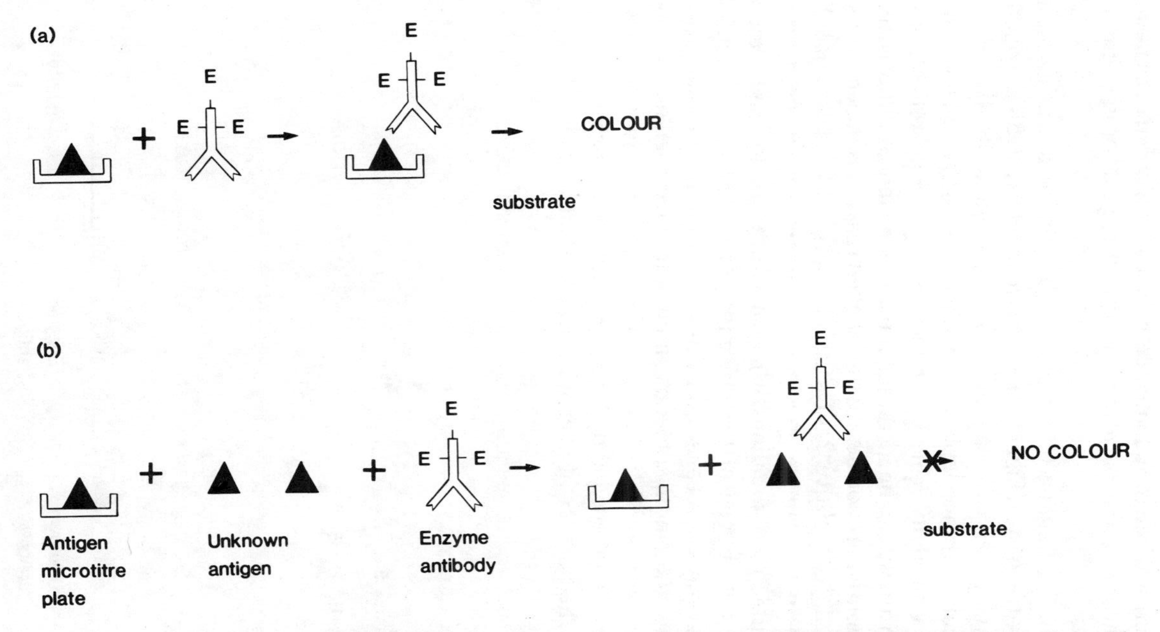

FIGURE 5 Competitve indirect ELISA for antigen detection. Antigen can be measured by competing with solid phase antigen for binding to enzyme-labelled antibody. In the absence of free antigen in the sample (a) the enzyme-labelled antibody binds. If antigen is present this will be inhibited (b)

were advanced by Miles and Hales (1968) are lost if competitive assays are carried out on a solid phase. By definition, competitive assays utilize limited concentrations of both antigen and antibody so that the amount bound to the solid phase is critical and the quantity desorbed or denatured even more so. Set against this, these techniques are technically easy to perform and offer valuable screening tests.

Competition for binding to antigen can be carried out using an antigen-coated microtitre plate (Figure 4) where a fixed level of enzyme-labelled antibody competes with varying levels of unlabelled antibody in the sample. The relative concentration, epitope specificity, and affinity of test and reagent antibodies is crucial to achieve sensitive, specific, and reliable assays for antibody. Sensitivity can be increased if the unlabelled antibody is allowed to bind in a sequential fashion before the labelled reagent is added. The signal in this and in all competitive assays is, of course, inversely related to the concentration in the sample. A similar approach can be used to measure antigen (Figure 5). In situations where it is difficult to raise experimental antibodies to the antigen or the objective is to assess the concentration of antigenic determinants recognized by human antibodies, a modification may be used. In this, the binding of human antibodies to antigen on the solid phase is inhibited by free unlabelled antigen and resultant activity is measured by adding an enzyme-labelled anti-globulin reagent. Such assays have been used to determine the potency of allergen extracts (Foucard *et al.*, 1972; Gleich *et al.*, 1974).

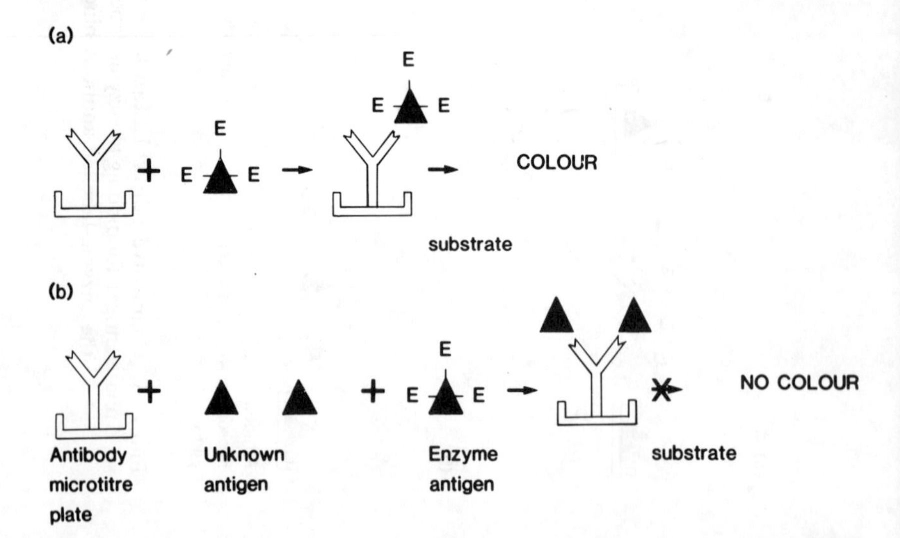

FIGURE 6 Competitive antigen capture ELiSA. Competition between (a) labelled and (b) unlabelled antigen for binding to antibody on the solid phase can be used for measurement of antigen

The presence of antigen in the sample can be determined by competition with labelled antigen for binding to antibody on a solid phase (Figure 6). This method offers an easy, rapid way of measuring antigen (Yolken and Leister, 1980). The sensitivity and specificity of the procedure is heavily reliant, however, upon having a supply of antigen which is sufficiently pure to allow efficient labelling, appropriately robust to ensure retention of antigenicity during chemical labelling, and possessing the same epitope specificity as the test antigen. This method, as with other methods for antigen detection, can be affected by the presence of antibody in the sample (Gleich and Dunnette, 1977; Clayton *et al.*, 1985). Small molecular weight markers such as haptens or biotin may be used as alternatives to enzymes. These are less likely to hinder interaction with antibody due to configurational effects.

SELECTION OF REAGENTS

Microtitre plates

The importance of the interaction between the antigen or antibody with the solid phase cannot be overemphasized (Kemeny and Challacombe, 1986). The widespread popularity of ELISA is to a large extent due to the convenience afforded by microtitre plates or strips which avoid the need to number wells individually (unlike most polystyrene tubes), and allow large numbers of samples to be processed easily and quickly. There is very little non-specific binding to some plates, which makes it possible to use an excess of the detection reagents. A limitation of most microtitre plates compared with other particulate solid phase supports is their low capacity for binding protein. Pretreatment procedures which increase their potential capacity for protein often result in increased non-specific binding. For many purposes commercially irradiated plastic microtitre plates offer the best compromise (see Appendix) and these plates seem to suffer less from protein leaching off during the assay (Urbanek, Kemeny, and Samuel, 1985). Details of the relative merits of microtitre plates and other solid phases are given in Chapter 2.

Choice of enzyme and substrate

A large variety of enzymes has been used in ELISA (Avrameas, Ternynck, and Guesdon, 1978; Voller, Bartlett, and Bidwell, 1978; Maggio, 1979; Yolken,

TABLE 1 Enzymes used in ELISA

Horseradish peroxidase
Alkaline phosphatase
β-D-Galactosidase

TABLE 2 Factors affecting the choice of enzyme

Purity
Specific activity
Sensitivity of substrate detection
Ease of conjugation
Efficacy when conjugated
Stability of conjugate

1982; Blake and Gould, 1984; Kurkstat, 1985; Schuurs and Van Weemen, 1977; Kennedy, Kricka, and Wilding, 1976; Van Weemen, 1985; Table 1). The choice of enzyme for a particular application will be determined by a number of criteria (Table 2). It is important to realize that whichever marker and product detection system is used, the optimal conditions for labelling the reagent antibody or antigen and for detecting the enzyme in an assay must be established at the earliest stage.

Horseradish peroxidase, alkaline phosphatase, and β-galactosidase have been extensively used in ELISA. Several well-established conjugation procedures (Kennedy, Kricka, and Wilding, 1976; Schuurs and Van Weemen, 1977; Ternynck and Avrameas 1977; Avrameas, Ternynck and Guesdon, 1978; Duncan, Weston and Wrigglesworth, 1983) and product detection systems are available and in our experience are suitable for most applications.

The relative merits of different enzymes, coupling procedures, and substrates are difficult to assess objectively on published information for several reasons. Comparative performance data in a single system utilizing the same antigen or antibody preparation and conjugation procedure are not, to our knowledge, available. Furthermore, the absence of information on the relative immunoreactivity and enzyme activity of the conjugates used, makes it impossible to determine whether differences in the observed sensitivity of assays using different enzymes are due to differences in the enzyme product detection system used or are due to the deleterious effect of the conjugation method on either the enzyme or antibody activity.

Horseradish peroxidase (HRP) (EC 1.11.1.7)

This has a high turnover rate, is pure, relatively cheap, and readily available. A large number of chromogens giving readily visible colours are commercially available. HRP can be readily coupled to proteins via its carbohydrate moiety by peroxidase oxidation (Nakane and Kawoi, 1974; Wilson and Nakane, 1978), by sulphydryl-maleimide conjugation (Duncan, Weston, and Wrigglesworth, 1983), or the two-step glutaraldehyde procedure (Avrameas and Ternynck, 1971).

Several chromogens may be used with the HRP substrate hydrogen peroxide. Most commonly used are orthophenylene diamine (OPD), 2,2-azino-di(3-ethylbenzothiazoline-6-sulphonate) (ABTS), 5-aminosalicylic acid (5-AS), and, more recently 3,3′, 5, 5′-tetramethylbenzidine hydrochloride (TMB) (Bos *et al.*, 1981; Porstmann *et al.*, 1985). In our experience OPD and TMB are most sensitive for the detection of low levels of enzyme and the latter has the advantage of being non-mutagenic and non-carcinogenic (Holland *et al.*, 1974; Van Weemen, 1985). For many applications where sensitivity is not the prime requirement, 5-AS is suitable. Commercially available 5-AS is poorly soluble but this can be improved by recrystallization (Ellens and Gielkins, 1980).

Whichever enzyme product detection system is used, care should be taken to ensure that the concentration of enzyme substrate and chromogen, the molarity and pH of buffers, and the temperatures used are optimized. Optimal conditions are given in the Appendix but it is prudent to check individual batches of materials to confirm the conditions for optimal sensitivity and linearity of colour development over the time period measured. It is advisable to store hydrogen peroxide in aliquots at 4 °C or, alternatively, uric oxide can be used. Most substrates for peroxidase are mutagenic (Sharpe *et al.*, 1976; Holland *et al.*, 1974; Venitt and Searle, 1976) and should be handled with appropriate care. Fluorogenic substrates have been reported for peroxidase but these are less stable than those for alkaline phosphatase and β-galactosidase (Barman, 1969). Peroxidase is sensitive to micro-organisms, to high levels of the anti-microbial agent sodium azide (Schonbaum, 1973) and to methanol (Straus, 1971) and can be inactivated by plastic — although the latter can be prevented by the addition of Tween 20 to the dilutent (Berkowitz and Webert, 1981). Peroxidase conjugates should be filtered and stored in aliquots at −20 °C in 50 per cent glycerol.

Alkaline phosphatase (AP) (EC 3.1.3.1)

Like HRP, this satisfies most of the criteria for a good enzyme label for ELISA but it is expensive. The simplest procedure for coupling AP to proteins is the one-step glutaraldehyde method (Avrameas, 1969). It is highly efficient (60–70 per cent) but the derived conjugates formed are of high molecular weight and are heterogeneous with regard to the number of enzyme molecules coupled to the antigen or antibody (Ford and Pesce, 1981). Alkaline phosphatase may also be coupled to protein by two-step glutaraldehyde or sulphydryl/maleimide procedures. In general, these methods result in a more homogeneous population of labelled proteins with regard to enzyme-protein molar ratios which allow greater sensitivity and specificity. The advantage of the latter procedure is that the number of molecules of enzyme inserted per protein molecule can be controlled for a particular application.

para-Nitrophenyl phosphate (*p*-NPP) is the substrate used for spectrophotometric measurement. It is stable, safe, and available commercially in convenient tablet form (Fernerley and Walker, 1965; Guilbaut, 1968; Garen and Levinthal, 1960). The pale yellow colour develops more slowly than with many peroxidase substrates but the signal can be improved by the use of fluorescent substrates (Rotman, Zderic, and Edelstein, 1963; Rhodes and Woltorten, 1968; Schuurs and Van Weemen, 1977).

The recent application of a cyclical enzyme-amplified system for AP detection has had a significant impact on the sensitivity that can be achieved over a fixed time of measurement (Self, 1985; see also Chapter 4). AP conjugates have good stability, are resistant to bacteriostatic agents (Belanger, 1978; Avrameas, Ternynck, and Guesdon, 1978), and may be stored at 4 °C in buffer containing 1–2 per cent w/v protein, and 50 per cent glycerol.

β-D-Galactosidase (EC 3.2.1.28)

This has a slower turnover rate than horseradish peroxidase and alkaline phosphatase. It is not normally found in plasma or other body fluids, although it can be found in some micro-organisms. β-D-Galactosidase can be coupled to proteins by one-step glutaraldehyde (Cameron and Erlanger, 1976) and maleimide procedures (Kato *et al.*, 1976; Kitagawa, 1981; Yoshitake, Imagawa, and Ishikawa, 1982). Chromogenic substrates such as *p*-nitrophenyl-β-D-galactosidase and stable fluorogenic substrates like 4-methylumbelliferyl-β-D-galactosidase (MUG) are available and using this substrate Ishkawa and Kato (1978) were able to detect 1 attomole of enzyme per hour. In another assay Labrousse *et al.* (1982) were able to measure 240 attograms (1.50×10^6 molecules) of human IgE. It is likely that the high sensitivity achieved by these workers is also influenced by the assay design and quality of antibodies used.

Purification of conjugates

The need to separate labelled and unlabelled enzyme or antibody in a conjugate will depend to a large extent on the coupling procedure used (Avrameas, 1969; Kato *et al.*, 1976), the assay design adopted, and the level of sensitivity required. Despite care in conjugate preparation, some enzyme and antibody molecules will be unlabelled and the labelled molecules will be heterogeneous with regard to the enzyme–antibody molar ratio. The extent of this heterogeneity will vary with the coupling method used. For most purposes adequate fractionation can be achieved by gel filtration using acrylamide/agarose (Boorsma and Streefkerk, 1976), Sephadex gels, or con A affinity chromatography (Arends, 1979). If highly sensitive assays are the objective then highly purified conjugates are mandatory (Ishikawa *et al.*, 1982) and we have found HPLC separation to be very effective. Chromatographic purification of conjugates invariably results in dilution hence it is essential to add protein additives such as BSA or selected sera and 50 per cent glycerol, to minimize loss of activity during storage.

ASSAY OPTIMIZATION

Despite the apparent simplicity of ELISA, the quality of assays developed by different laboratories varies tremendously and this is largely due to the level of assay optimization undertaken. The technical simplicity of ELISA makes it unacceptable for laboratories to omit the appropriate chequerboard titration of components fundamental to assay optimization. The optimal working dilution of labelled reagents, whether obtained from a commerical source or prepared in-house, will differ depending on the application and the assay format and can only be verified by trial and error. Time spent optimizing is always repaid later by the reliability of the assay and the quality of the data that can be collected.

The optimum concentration of the different components is determined by chequerboard titration, first using a wide range and subsequently small dilution increments around the optimum. Having determined the optimal antigen or

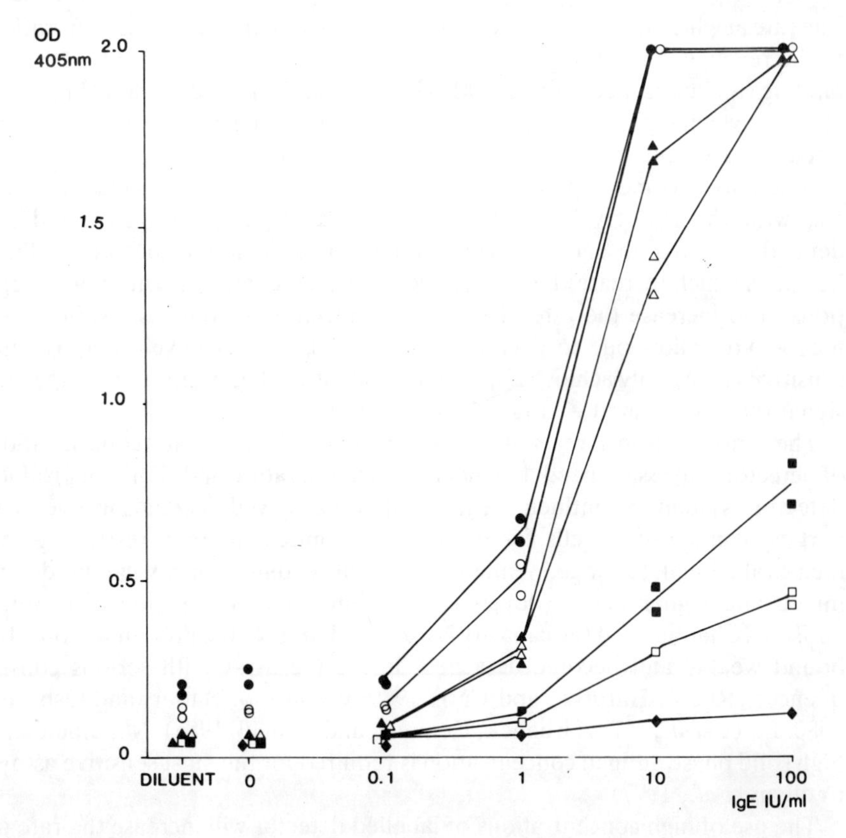

FIGURE 7 The effect of the concentration of alkaline-phosphatase-labelled rabbit anti-IgE in a two-site assay for human IgE (● 1/100, ○ 1/200, ▲ 1/500, △ 1/1000, ■ 1/2000, □ 1/5000, ♦ 1/10 00 of AP anti-IgE) (Kemeny and Richards, 1987)

antibody solid phase coating condition (i.e. the least quantity giving near-maximal binding of analyte) this is used to test the effect of varying concentrations of enzyme-labelled reagent. A typical example is shown in Figure 7 for a two-site assay for IgE. Conjugate dilutions of 1 : 500 and 1 : 1000 give good activity and low background and allow quantitation over the full range of IgE measured. Higher concentrations increase the non-specific readings and decrease the measurable analyte range. Lower conjugate levels decrease the specific absorbance measured but the non-specific binding is unchanged. Hence, in this example, a conjugate dilution of 1 : 500 would be optimal and allow for minor errors during conjugate dilution. It is prudent to re-check from time to time the optimal working dilution of reagents used over a prolonged period of study and essential to assess different batches of reagents even if prepared in an identical manner.

Speed and sensitivity of the test

The rate at which the assay proceeds is governed primarily by the concentration of the reactants, according to the Law of Mass Action (Chapter 6). Maximal binding can be reached quite quickly (Kemeny and Richards, 1987) (Figure 8) although when measuring small amounts of analyte longer incubation steps are often needed and the affinity of antibody and valency of antigen–antibody interactions become important. However, it is important to realize that the ease with which low levels of analyte are measured over a prolonged period will depend upon the absence of a concomitant increase in non-specific binding. Methods which increase the rate of specific signal detection, with few exceptions, also increase the rate of the signal detection in controls. While these manoeuvres allow one to decrease the assay time to achieve an equivalent sensitivity, they only achieve improved sensitivity if the specific to non-specific signal ratio is elevated — a fact often ignored.

The concentration of reactants, capacity of the solid phase and concentration of detector, the assay speed, the incubation temperature and choice of product detection system can influence the sensitivity achieved. The adsorptive properties of microtitre wells for proteins are limited, thereby restricting the measurable analyte range. While its practical binding capacity can be determined and higher plateau levels can be enhanced by selection of capture antisera (Chapter 9), if the capacity is exceeded some of the protein will only be bound weakly and become detached during the assay with serious consequences (Rubin, Hardtke, and Carr, 1980; Cantarero, Butler, and Osborne 1980; Pesce *et al.*, 1977; Urbanek, Kemeny, and Samuel, 1985). Nonetheless, a high solid phase antigen concentration is required for the most sensitive assays (Yolken *et al.*, 1977).

The use of high concentrations of labelled detector will increase the rate of binding (Figure 9) and allow the assay to be completed more rapidly. However,

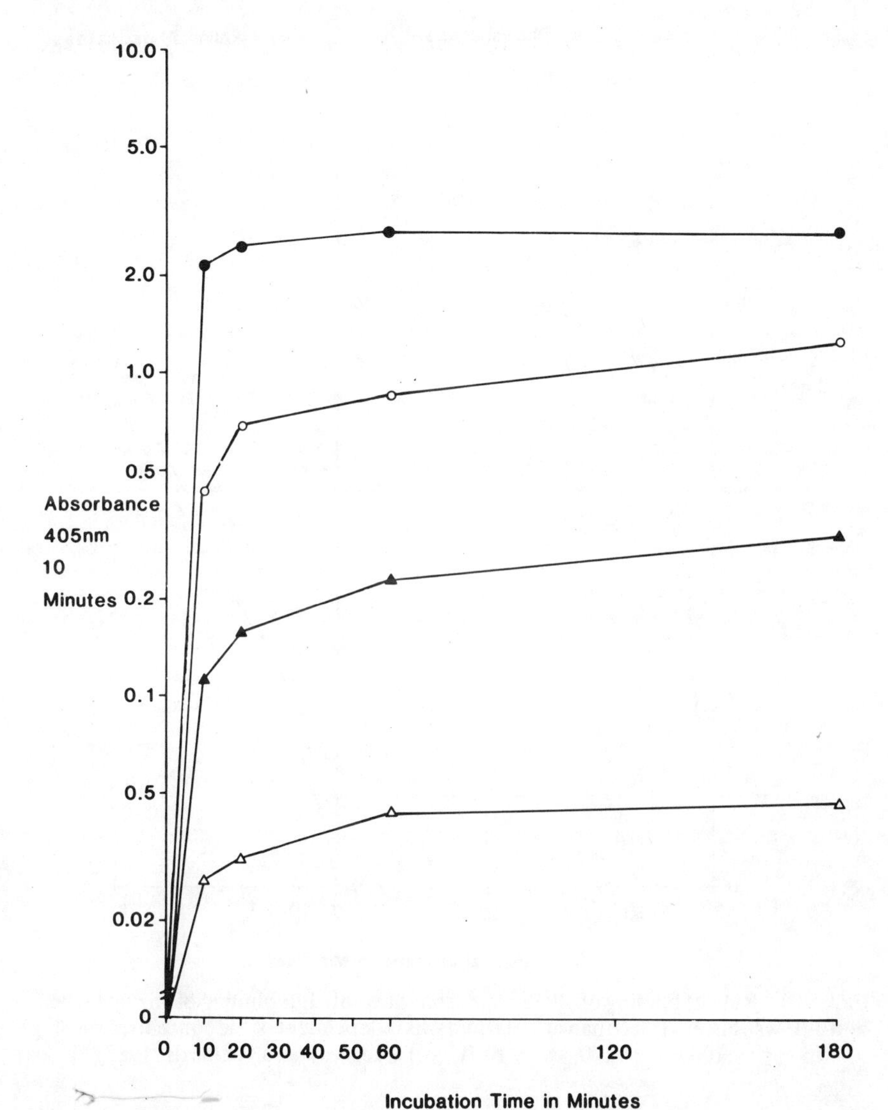

FIGURE 8 The rate of binding of IgE to the monoclonal anti-IgE-coated plate was near maximal at 3 hr. The rate of binding was higher at 1000 (●) than at 100 (○), 10 (▲) or 1 (△) IU/ml of IgE. AP anti-IgE was used at 1/50 dilution (Kemeny and Richards, 1987)

the sensitivity may not be increased due to elevated non-specific binding. Substitution of Fab′ anti-IgE-AP permits the amplification of the signal by secondary enzymes (Self, 1985; see also this volume Chapter 4) and an improvement in signal to noise ratio so that as little as 10 pg/ml of IgE can be detected (Figure 10). In some systems utilizing elevated incubation temperatures

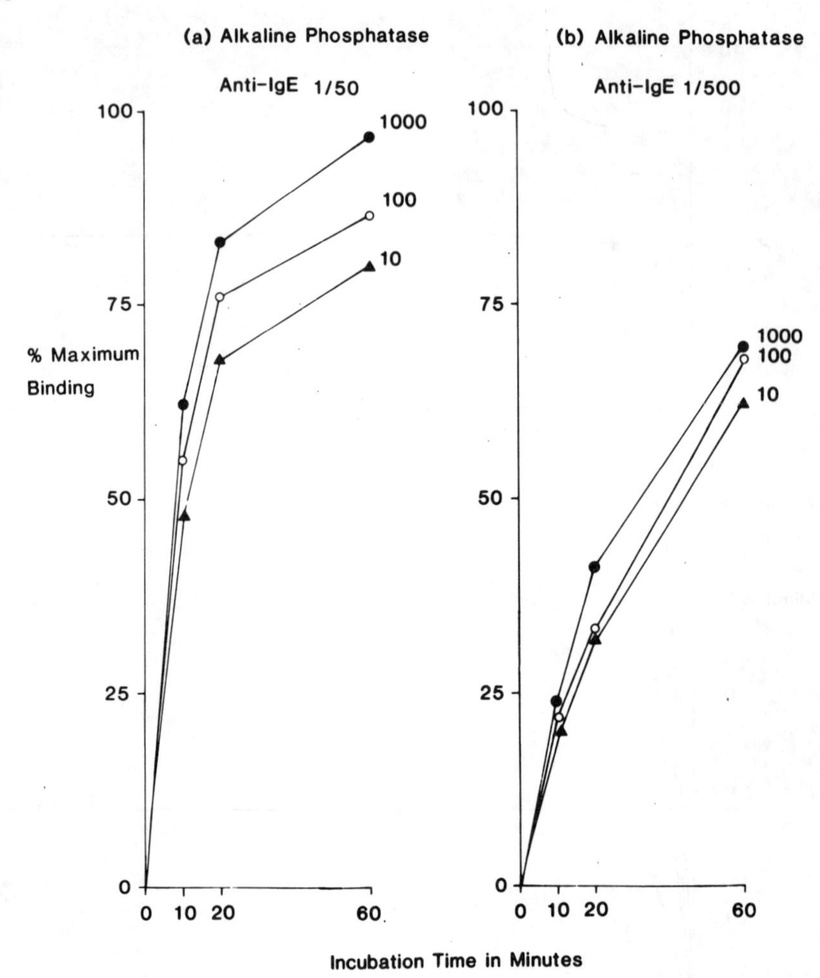

FIGURE 9 Rate of binding of alkaline phosphatase anti-IgE binding of enzyme labelled anti-IgE was faster at 1/50 than at 1/500 and was independent of the concentration of IgE (● = 1000, ○ = 100, ▲ = 10 IU/ml) (Kemeny and Richards, 1987)

assay sensitivity and speed may be improved but readers should be aware that dissociation is also higher so that for assays where sensitivity is the prime requirement, prolonged incubation times are not desirable in microtitre wells as intra-assay variation is increased (Chapter 3). Assays can be accelerated by ultrasound which is reported to increase the binding rate constant up to 500 times (Chen *et al.*, 1984).

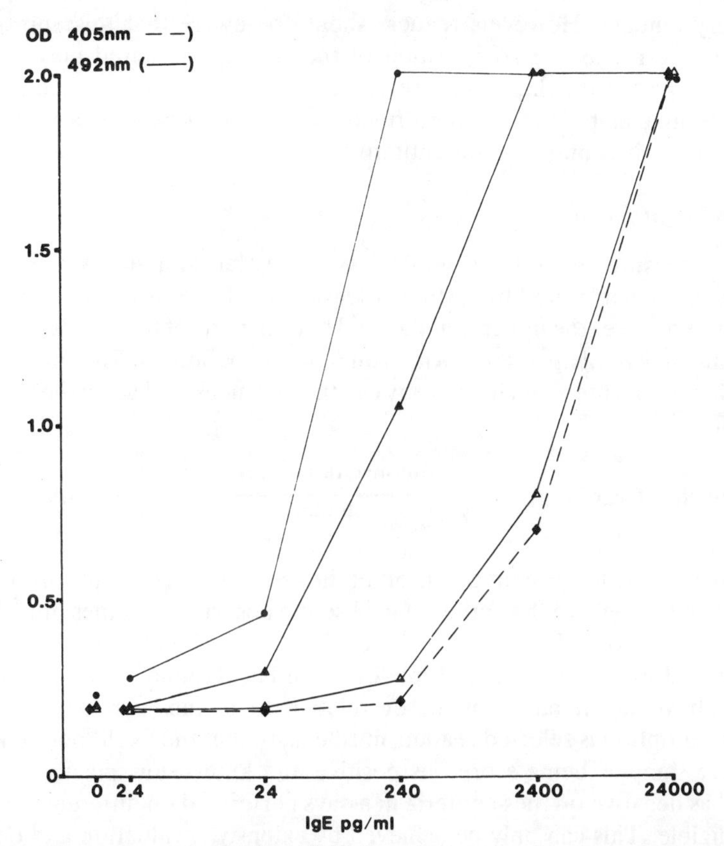

FIGURE 10 Comparison of total IgE ELISA carried out on 50 μl of sample using the intact rabbit anti-IgE-AP conjugate without (▲) and with (△) enzyme amplification (Self, 1985) and using Fab′ anti-IgE-AP conjugate with enzyme amplification at 50 μl (▲) or 300 μl (•) of sample (Kemeny and Richards, unpublished results)

Quantitation

Quantitation is discussed in detail in Chapter 3 so we will only introduce some of the more obvious, but nevertheless important, aspects here. Theoretically there should be a simple linear relationship between optical density or fluorescence and analyte concentration. As this is the exception rather than the rule, it is undesirable to express results as absolute optical density units. Where reference preparations are available, the results can be extrapolated from a standard curve run on each occasion and given a relative value. More often these reference preparations are not available. In such circumstances a pool of high positive sera or of a high concentration of antigen extract (e.g. viral or microbial) can be used as an internal reference preparation and be given an arbitrary unitage. However, readers should be aware that this approach is subject to error as the composition of the analyte measured may differ in several respects from that used as reference. While useful for assessing relative levels of an analyte in assays performed on different occasions or in different laboratories, it is only semi-quantitative.

Variability of method

For any measurement it is essential to assess the real variability of the method. This is easily performed by testing at least three dilutions of the appropriate reference to cover the upper, middle, and bottom parts of the standard curve in triplicate in several assays performed on different occasions. The coefficient of variation of the inter- and intra-assay can then be measured using the following equation:

$$\text{Coefficient of variation} = \frac{\text{standard deviation} \times 100}{\text{mean}}$$

Establishing the lower limit or cut-off of the assay always presents problems in ELISA due largely to the variation in absolute absorbance values obtained on different occasions. The cut-off has been derived in different ways by different groups and expressed as a multiple of mean background, as an absorbance value above mean background, or relative to a known reactive sample. Whatever option is selected readers must ensure that the likelihood of known negative samples being scored as positive and known low positive samples scored as negative on these criteria in assays performed on different occasions is negligible. This can only be achieved by extensive evaluation and defining the conditions within which the assay is in control. The failure to do so results in invalid data.

COMMON PROBLEMS ENCOUNTERED WITH ELISA

Despite its simplicity and convenience, in practice ELISA can be fraught with difficulties. In this section we will attempt to identify some common technical problems encountered and to offer simple ways in which they can be overcome or minimized (Table 3). Readers will appreciate that technical approaches satisfactory in some assays will not necessarily relate directly to others.

TABLE 3 Common problems with ELISA

Common problems with ELISA
Non-specific protein binding
Cross-reactivity and other unwanted interactions
Antibodies to blocking protein
Temperature variation and edge effects

Pipetting clear liquids into transparent microtitre plates can lead to error, especially if large numbers of tests are being performed. While the presence of bubbles or inadequate dispensing can usually be avoided by careful observation, the addition of coloured dyes to reagents has many practical benefits (Dr Aalberse, Red Cross, Amsterdam, personal communication). Details of these are given in the Appendix but readers should check that the additives do not interfere in their intended application.

A common problem in ELISA is non-specific, or specific, undesirable binding of different serum or other proteins. This can be reduced by the addition of protein such as BSA, animal sera with or without Tween 20, or casein hydrolysate to buffers but unfortunately the choice of additives is not universally suitable for all assays. Readers should be aware that some undesirable high backgrounds are the result of specific interactions due to cross-reactivity or low levels of natural antibodies present in the reagents used. These are best removed by the addition of a selected additive which does, of course, necessitate prior knowledge of the cause.

Serum albumin, gelatine and other proteins have been used to block non-specific binding to the solid phase. This manipulation is helpful in some situations and not in others. It should be remembered that test samples may possess antibodies to the blocking protein – many normal adults, for example, have IgG antibodies to bovine serum albumin in their blood.

As previously mentioned, optimization of the conditions for measuring bound enzyme are of paramount importance if reproducible results are to be achieved. The rate of reaction, as in all enzymatic reactions, is dependent also upon temperature and it is desirable to ensure that this is consistent between tests performed on different occasions. Temperature variations can be minimized by incubating in a controlled environment, avoiding stacking of large numbers of plates, placing on heated blocks and selecting incubation times which are of sufficient duration to allow well contents to gain the appropriate temperature.

Proper control and storage of laboratory reagents makes it possible to develop robust ELISA tests giving good reproducibility and precision. These assays are reliable and easy to use; they are increasingly being adopted in routine use and have many advantages which are likely to make them the method of choice in routine laboratories in the near future.

REFERENCES

Arends, J. (1979). Purification of peroxidase-conjugated antibody for enzyme immunoassay by affinity chromatography on concanavalin A. *J. Imm. Methods*, **25**, 171.

Avrameas, S. (1969). Coupling of enzymes to proteins with gluteraldehyde. Use of conjugates for the detection of antigens and antibodies. *Immunochemistry*, **5**, 43.

Avrameas, S., and Ternynck, T. (1971). Peroxidase labelled antibody and Fab conjugates with enhanced intracellular penetration. *Immunochemistry*, **8**, 1175.

Avrameas, S., Ternynck, T., and Guesdon, J-L. (1978). Coupling of enzymes to antibodies and antigens. *Scand. J. Immunol.*, **8** (Suppl. 7), 7–23.
Avrameas, S., and Uriel, J. (1966). Methode de Marguge d'antigenes et d'anticorps avec des enzymes et son application en immunodiffusion. *C.R. Acad.* Sci. Paris, **266**, 2543.
→Barman, T.E. (1969). *Enzyme Handbook*. Springer-Verlag, Berlin.
Belanger, L. (1978). Alternative approaches to enzyme immunoassays. *Scand. J. Immunol.*, **8** (Suppl 7), 33.
Berkowitz, D.M., and Webert, D.W. (1981). The inactivation of horseradish peroxidase by a polystyrene surface. *J. Imm. Methods*, **47**, 121.
Blake, C., and Gould, B.J. (1984). Use of enzymes in immunoassay techniques. A review. *Analyst*, **109**, 533.
Blanchard, G.C., and Gardner, R. (1980). Two nephelometric methods compared with a radial immunodiffusion method for the method for the measurement of IgG, IgA and IgM. *Clin. Biochem*, **13**, 84.
Boorsma, D.M., and Streefkerk, J.G. (1976). Peroxidase-conjugate chromatography isolation of conjugates prepared with gluteraldehyde or periodate using polyacrylamide-agarose gel. *J. Histochem. Cytochem.*, **24**, 481.
Bos, E.S., Van der Doelen, Van A.A., Rooy, N, and Schuurs, A.H. W.M. (1981). 3,3′ 5,5′ Tetramethylbenzidine as an Ames Test negative chromogen for horse-radish peroxidase in enzyme-immunoassay. *J. Immunoassay,* **2**, 187.
Cameron, D.J., and Erlanger, B.F. (1976). An enzyme-linked procedure for detection and estimation of surface receptors on cells. *J. Immunol.*, **116**, 1313.
Cantarero, L.A., Butler, J.E., and Osborne, J.W. (1980). The adsorptive characteristics of proteins for polystyrene and their significance in solid-phase immunoassays. *Anal. Biochem.*, **105**, 375.
Catt, K., and Tregear, G.W. (1967). Solid-phase radioimmunoassay in antibody-coated tubes. *Science (NY)*, **158**, 1570.
Ceska, M., Eriksson, R., and Varga, J.M. (1972). Characteristics of allergen extracts by polyacrylamide gel electrofocusing and radioimmunosorbent allergen assay. I. Distribution of allergenic components in birch pollen extracts. *Int. Arch. All. Appl. Immun.*, **42**, 430.
Ceska, M., and Lundqvist, U. (1972). A new and simple radioimmunoassay method for detection of IgE. *Immunochemistry*, **9**, 1021.
Chantler, S., and Diment, J.A. (1981). Current status of specific IgM antibody assays. In: *Immunoassays for the 80s* (eds A. Voller, A. Bartlett, and D. Bidwell.) MTP Press, Lancaster, pp. 417–30.
Chantler, S., and Evans, C.J. (1986). Selection and performance of monoclonal and polyclonal antibodies in an IgM antibody capture enzyme immunoassay. *J. Imm. Methods*, **87**, 109.
Chen, R., Weng, L., Sizto, N.C., Osonio B., Hsu, C.J., Rodjers, R., and Litman, D.J. (1984). Ultrasound accelerated immunoassay as exemplified by enzyme immunoassay of choriogonadotrophin. *Clin. Chem.*, **30**, 1446.
Clayton, A.L., Beckford, U., Roberts, C., Sutherland, S., Druce, A., Best, J., and Chantler, S. (1985). Factors influencing the sensitivity of herpes simplex virus detection in clinical specimens in a simultaneous enzyme-linked immunosorbent assay using monoclonal antibodies. *J. Med. Virol.*, **17**, 275.
Clayton, A.L., Roberts, C., Godley, M., Best, J.M., and Chantler, S. (1986). Herpes simplex virus detection by ELISA: The effect of enzyme amplification, nature of lesion sampled and specimen treatment. *J. Med. Virol.*, **20**, 89.
Collins, W.P. (1985). *Alternative Immunoassays*. John Wiley and Sons, Chichester.

Conradie, J.D., Govender, M., and Visser, L. (1983). ELISA solid-phase partial denaturation of coating antibody yields a more effective solid-phase. *J. Imm. Methods*, **59**, 289.

Cradock-Watson, J.E., Ridehalgh, M.K.S., and Chantler, S. (1975). Immunoglobulin responses after rubella infection. *Ann. NY Acad. Sci.*, **254**, 385.

Diment, J.A., and Pepys, J. (1978). Immunoglobulin separation of IgG and IgM for the radioimmunoassay of specific antibodies. In: *Affinity Chromatography* (eds O. Hoffman-Ostenhof, M. Breitenback, F. Koller, D. Kraft, and O. Scheiner. Pergamon Press, Oxford, p. 223.

Duncan, R.J.S., Weston, P.D., and Wrigglesworth, R. (1983). A new reagent which may be used to introduce sulphydryl groups into proteins, and its use in the preparation of conjugates for immunoassay. *Anal. Biochem.*, **132**, 68.

Ekins, R.P. (1980). More sensitive immunoassays. *Nature (Lond.)*, **284**, 14.

Ekins, R.P. (1985). Current concepts and future developments. In: *Alternative Immunoassays* (ed. W.P. Collins). John Wiley and Sons, Chichester, p. 219.

Ellens, D.J., and Gielkens, A.L. (1980). A simple method for the purification of 5-aminosalicylic acid — application of the product as substrate in enzyme-linked immunosorbent assay (ELISA). *J. Imm. Methods*, **37**, 325.

Engvall, E. (1978). Preparation of enzyme-labelled staphylococcal protein A and its use for detection of antibodies. *Scand. J. Immunol.*, **8** (Suppl 7), 25.

Engvall, E., and Perlmann, P. (1971). Enzyme linked immunosorbent assay (ELISA): quantitative assay of IgG. *Immunochemistry*, **8**, 871.

Engvall, E., and Perlmann, P. (1972). Enzyme linked immunosorbent assay (ELISA). III. Quantitation of specific antibodies by enzyme-linked anti-immunoglobulin in antigen coated tubes. *J. Immunol.*, **109**, 129.

Fenerley, H.N., and Walker, P.G. (1965). Kinetic behaviour of calf intestinal alkaline phosphatase with 4-methylumbelliferyl phosphate. *Biochem. J.*, **97**, 95.

Ford, D.J., and Pesce, A.J. (1981). Reaction of gluteraldehyde with proteins. In: *Enzyme Immunoassay* (eds E. Ishikawa, T. Kawai, and K. Miyai). Igaku-Shoin, Tokoyo, pp. 54–66.

Foucard, T., Johansson, S.G.O., Bennich, H., Berg, T. (1972). *In vitro* estimation of allergens by a radioimmune antiglobulin technique using human IgE antibodies. *Int. Arch. All. Appl. Immun.*, **43**, 360.

Garen, A., and Levinthal C. (1960). A fine-structure genetic and chemical study of the enzyme alkaline phosphatase from *E. coli*. I. Purification and characterization of alkaline phosphatase. *Biochim. Biophys. Acta*, **38**, 470.

Gleich, G.J., and Dunnette, S.L. (1977). Comparison of procedures for measurement of IgE protein in serum and secretions. *J. All. Clin. Immun.*, **59**, 377.

Gleich, G.J., Larson, J.B., Jones, R.T., and Baer, H. (1974). Measurement of the potency of allergy extracts by their inhibitory capacities in the radioallergosorbent test. *J. All. Clin. Immun.*, **53**, 158.

Guilbault, G.G. (1968). Use of enzymes in analytical chemistry. *Anal. Chem.*, **40**, 459.

Hoffman, D.R. (1973). Estimation of serum IgE by an enzyme linked immunosorbent assay (ELISA). *J. All. Clin. Immun.*, **51**, 303.

Holland, U.R., Saunders, B.C., Rose, F.L., and Walpole, A.L. (1974). The carcinogenic potential of *o*-toluidine. *Tetrahedron*, **30**, 3299.

Holmes, N.J., and Parnham, P. (1983). Enhancement of monoclonal antibodies against HLA-A2 is due to antibody bivalency. *J. Biol. Chem.*, **258**, 1580.

Hurn, B.A.L., and Chantler, S.M. (1980). Production of reagent antibodies. In: *Immunochemical Techniques* (eds H. Van Vunakis and J.L. Langone) *Methods in Enzymology*, **70**, 104–142. Academic Press, London.

Imagawa, M., Yoshitake, S., Ishikawa, E., Endo, Y., Ohtaki, S., Kano, E., and Tsenetoshi, Y. (1981). Highly sensitive sandwich enzyme immunoassay of human IgE with β-D-galactosidase from *Escherichia coli*. *Clin. Chim. Acta*, **117**, 199.

Ishikawa, E., and Kato, K. (1978). Ultrasensitive enzyme immunoassay. *Scand J. Immunol.* **8** (Suppl 7), 43–55.

Ishikawa, E., Kawai, T., Miyai, K. (1981). *Enzyme Immunoassay*. Igaku-Shoin, Tokyo.

Ishikawa, E., Imagawa, M., Yoshitake, S., Niitsu, Y., Urushizaki, I., Inada, N., Kanazawa, R., Tachibana, S., Nazakawa, N., and Ogawa, H. (1982). Major factors limiting sensitivity of sandwich enzyme immunoassay for ferritin, Immunoglobulin E and thyroid stimulating hormone. *Ann. Clin. Biochem.*, **19**, 379.

Jackson, T.M., and Ekins, R.P. (1986). Theoretical limitations on immunoassay sensitivity — current practices and potential advantages of fluorescent Eu^{3+} chelates on non-radiotopic tracers. *J. Imm. Methods*, **87**, 13.

Kato, K., Fukin, H., Hamaguchi, Y., and Ishikawa, E. (1976). Enzyme-linked immunoassay: conjugation of the Fab fragment of rabbit IgG with β-D-galactosidase from *E. coli* and its use for immunoassay. *J. Immunol.*, **116**, 1554.

Kemeny, D.M., and Challacombe, S.J. (1986). Advances in ELISA and other solid-phase immunoassays. *Immunology Today*, **7**, 67.

Kemeny, D.M., and Lessof, M.H. (1987). The immune response to bee venom: II. Quantitation of the absolute amounts of IgE and IgG antibody by saturation analysis. *Int. Arch. All. Appl. Immun.* **83**, 113.

Kemeny, D.M., and Richards, D. (1987). ELISA for the detection of total serum IgE: Speed and sensitivity. *Immunological techniques in microbiology,* **24**, 47.

Kemeny, D.M., Miyachi, S., Platts-Mills, T.A.E., Wilkins, S., and Lessof, M.H. (1982). The immune response to bee venom. Comparison of the antibody response to phospholipase A with the response to inhalant antigens. *Int. Arch. All. Appl.*, **68**, 268.

Kennedy, J.H., Kricka, L.J., and Wilding, P. (1976). Protein–protein coupling reaction and the applications of protein conjugates. *Clin. Chim. Acta*, **70**, 1.

Kitagawa, T. (1981). Enzyme labelling with *N*-hydroxy-succinimidyl ester of Maleimide. In: *Enzyme Immunoassay* (eds E. Ishikawa, T. Kawai, and K. Miyai). Igaku-Shoin, Tokyo, pp. 81–89.

Kurkstat, E. (1985). Progress in enzyme immunoassays: production of reagents, experimental design and interpretation. *Bull. Wld. Hlth. Org.*, **63**, 793.

Labrousse, M., Guesdon J-L., Ragimbeau, J., and Avrameas, S. (1982). Minaturization of β-galactosidase immunoassays using chromogenic and fluorogenic substrates. *J. Imm. Methods*, **48**, 133.

Maggio, T. (1979). *The Enzyme Immunoassay*. CRC Press, Boca Raton, Florida.

Mancini, G., Carbonara, A.O., and Heremans, J.F. (1965). Immunochemical quantitation of antigens by single radial immunodiffusion. *Immunochemistry*, **2**, 235.

Mangili, R., Kemeny, D.M., Li, L.K., and Viberti, G.G. (1987). Development of a sensitive enzyme-linked immunosorbent assay (ELISA) for quantification of human IgG subclass. *J. All. Clin. Immun.*, **79**, 223.

Miles, L.E.M., and Hales, C.N. (1968). Labelled antibodies and immunological assay systems. *Nature (Lond)*, **219**, 186.

Moneo, I., Cueva, M., Urena, V., Alcover, R., Bootello, A, (1983). Reverse enzyme: immunoassay for the determination of *Dermatophagoides pteronyssinus* IgE antibodies. *Int. Arch. All. Appl. Immun.* **71**, 285.

Mortimer, P.P., Tedder, R.S., Hambling, M.H., Shaji, M.H., Burkhardt, F., Schilt, U. (1981). Antibody capture radioimmunoassay for anti-rubella IgM. *J.Hyg.*, **86**, 139.

Nakane, P. and Kawaoi, A. (1974). Peroxidase-labelled antibody. A new method of conjugation. *J. Histochem. Cytochem.*, **22**, 1084.

Nakane, P., and Pierce, G.B. (1966). Enzyme-labelled antibodies. Preparation and application to localization of antigens. *J. Histochem. Cytochem.*, **14**, 929.

Ngo, T.T., and Lenhoff, H.M. (1985). *Enzyme-Mediated Immunoassay*. Plenum Press, NY.

Parratt, D., McKenzie, H., Nielsen, K.H., and Cobb, S.J. (1982). *Radioimmunoassay of Antibody and its Clinical Applications*. John Wiley & Sons, Chichester.

Papadea, C., Check, I.J., and Reimer, C.B. (1985). Monoclonal antibody based solid-phase immunoenzymometric assays for quantifying human immunoglobulin G and its subclasses in serum. *Clin. Chem.* **31**, 1940.

Pesce, A.J., Ford, D.J., Gaizutis, M., and Pollak, V.E. (1977). Binding of proteins to polystyrene in solid-phase immunoassays. *Biochim. Biophys. Acta,* **492**, 399.

Porstmann, B., Porstmann, T., Nugel, E,, and Evers, U. (1985). Which of the commonly used marker enzymes gives the best results in colorimetric and fluorimetric immunoassays: horseradish peroxidase, alkaline phosphatase or β-galactosidase? *J. Imm. Methods*, **79**, 27.

Reimer, C.B., Black, C.M., Phillips, D.J., Logan, L.C., Hunter, E.F., Pender, B.J., and McGrew, B.E. (1975). The specificity of fetal IgM: Antibody or anti-antibody. *Ann. NY Acad. Sci.*, **254**, 94.

Rhodes, M.J., and Wooltorten, L.S.C. (1968). A new fluorometric method for the determination of pyridine nucleotides in plant material and its use in following changes in pyridine nucleotides during the respiration climacteric in apples. *Phytochemistry*, **7**, 337.

Richman, D.D., Cleveland, F.H., Oxman, M.N., and Johnson, K.M. (1982). The binding of staphylococcal protein A by the sera of different animal species. *J. Immunol.*, **128**, 2300.

Rotman, B., Zderic, J.A., and Edelstein, M. (1963). Fluorogenic substrates derived from fluorescein (3,6-dihydroxy fluoran) and its monoethyl ester. *Proc. Natl. Acad. Sci. (USA)*, **50**, 1.

Roundtree, S., Cogswell, J.J., Platts-Mills, T.A.E., and Mitchell, E.B. (1985). Development of IgE and IgG antibodies to food antigens in babies at risk of allergic disease. *Arch. Dis. Childhood*, **60**, 727.

Rubin, R.L., Hardtke, M.A., Carr, R.I. (1980). The effect of high antigen density in solid-phase radioimmunoassay for antibody regardless of immunoglobulin class. *J. Imm. Methods*, **33**, 277.

Schonbaum, G.R. (1973). New complexes of peroxidase with hydroxamic acids, hydrazides and amines. *J. Biol. Chem.*, **248**, 502.

Schuurs, A.H., and Van Weemen, B.K. (1977). Enzyme immunoassay. *Clin. Chim. Acta*, **81**, 1.

Self, C.H. (1985). Enzyme amplification — a general method applied to provide an immune assisted assay for placental alkaline phosphatase. *J. Immun. Methods*, **83**, 89.

Sharpe, S.L., Cooreman, W.M., Bloome, W.J., and Lackman, G.M. (1976). Quantitative enzyme immunoassay — current status. *Clin. Chem.*, **22**, 733.

Shields, J.G., and Turner, M.W. (1986). The importance of antibody quality in sandwich ELISA systems — evaluation of selected commercial reagents. *J. Imm. Methods*, **87**, 29.

Siddle, K. (1985). Properties and application of monoclonal antibodies. In: *Alternative Immunoassays* (ed. W.P. Collins). John Wiley & Sons, Chichester, pp. 13–37.

Soos, M., and Siddle, K. (1983). Characterization of monoclonal antibodies for human luteinizing hormone and mapping of antigenic determinants on the hormone. *Clin. Chim. Acta*, **133**, 263.

Steward, M.W., and Lew, A.M. (1985). The importance of antibody affinity in the performance of immunoassays for antibody. *J. Imm. Methods*, **78**, 173.

Straus, W. (1971). Inhibition of peroxidase by methanol and by methanol–nitroferricyanide for use in immunoperoxidase procedures. *J. Histochem. Cytochem.*, **19**, 687.

Ternynck, T., and Avrameas, S. (1977). Conjugation of *p*-benzoquinone treated enzymes with antibodies and Fab fragments. *Immunochemistry*, **14**, 767.

Thompson, R.J., and Jackson, A.P. (1984). Cyclic complexes and high avidity antibodies. *Trends Biochem. Sci.*, **9**, 1.

Urbanek, R., Kemeny, D.M., and Samuel, D. (1985). Use of enzyme-linked immunosorbent assay for measurement of allergen-specific antibodies. *J. Imm. Methods*, **79**, 123.

Van Loon, A.M., Van der Logt, J.T., Heesen, F.W.A., and Van der Veen, J. (1985). Quantitation of immunoglobulin E to cytomegalovirus by an antibody capture enzyme-linked immunosorbent assay. *J. Clin. Microbiol.*, **21**, 558.

Van Weemen, B.K. (1985). ELISA: Highlights of the present state of the art. *J. Virol. Methods*, **10**, 371.

→ Van Weemen, B.K., and Schuurs. A.H.W.M. (1971). Immunoassay using antigen–enzyme conjugates. *FEBS Lett.*, **15**, 232.

→ Vejtorp, M. (1980). The interference of IgM rheumatoid factor in enzyme-linked immunosorbent assay of rubella IgM and IgG antibodies. *J. Virol. Methods*, **1**, 1.

Venitt, S., and Searle, C.E. (1976). Mutagenicity and possible carcinogenicity of hair colourants and constituents. In: *Inserm Symposium Series*. Vol 52 IARC Scientific Publication no. 13, Inserm, Paris; 263 (eds C. Rosenfeld, and W. Davis).

→ Voller, A., Bartlett, A., and Bidwell, D.E. (1978). Enzyme immunoassays with special reference to ELISA techniques. *J. Clin. Pathol.*, **31**, 507.

Voller, A., Bartlett, A., and Bidwell, D.E. (1981). *Immunoassays for the 1980s*. MTP Press, Lancaster.

Voller, A., Bidwell, D.E., Huldt, G. and Engvall, E. (1974). A microplate method of enzyme linked immunosorbent assay and its application to malaria. *Bull. Wld. Hlth. Org.* **51**, 209.

Wide, L., and Porath, J. (1966). Radioimmunoassay of proteins with the use of Sephadex-coupled antibodies. *Biochim. Biophys. Acta.* **130**, 257.

Wide, L., Bennich, H., and Johansson, S.G.O. (1967). Diagnosis of allergy by an *in vitro* test for allergen antibodies. *Lancet*, **ii**, 1105.

Wilson, M.B. and Nakane, P.K. (1978). Recent developments in the periodate method of conjugating horseradish peroxidase (HRPO) to antibodies. In: *Immunofluorescence and Related Staining Techniques* (eds W. Knapp, K. Holubar, and G. Wick), Elsevier/North Holland, pp. 215–25.

Woodhead, J.S., Addison, G.M., and Hales. C.N. (1974). Immunoradiometric assay and related techniques. *Brit. Med. Bull*, **30**, 44.

Yolken, R.H. (1982). Enzyme immunoassays for the detection of infectious antigens in body fluids: currents limitations and future prospects. *Rev. Infect. Dis.*, **4**, 35.

Yolken, R.H., and Leister, F.J. (1980). Investigation of enzyme immunoassays time courses: development of rapid assay systems. *J. Clin. Microbiol.*, **13**, 738.

Yolken, R.H., and Leister, F.J. (1981). Staphylococcal protein A–enzyme immunoglobulin conjugates: versatile tools for enzyme immunoassays. *J. Imm. Methods*, **36**, 33.

Yolken, R.H., Greeenberg, H.B., Merson, M.H., Sack, R.B., and Kapikiam, A.Z. (1977). Enzyme-linked immunosorbent assay for detection of *Escherichia coli* heat-labile enterotoxin. *J. Clin. Microbiol.*, **6**, 439.

Yoshitake, S., Imagawa, M., and Ishikawa, E. (1982). Efficient preparation of rabbit Fab′–horseradish peroxidase conjugates using maleimide components and its use for enzyme immunoassay. *Anal. Lett.*, **15**, 147.

Zeiss, C.R., Grammar, L.C., and Levitz, D. (1981). Comparison of the radioallergosorbent test and a quantitative solid-phase radioimmunoassay for the detection of ragweed-specific immunoglobulin E antibody in patients undergoing immunotherapy. *J. All. Clin. Immun.*, **67**, 105.

Zeiss, C.R., Pruzansky, J.J., Patterson, R., and Roberts, M. (1973). A solid-phase radioimmunoassay for quantitation of human reaginic antibody against ragweed antigen E. *J. Immunol.*, **110**, 414.

ELISA and Other Solid Phase Immunoassays
Edited by D.M. Kemeny and S.J. Challacombe

CHAPTER 2

Microtitre Plates and Other Solid Phase Supports

D.M. Kemeny and **S.J. Challacombe**

UMDS, Guy's Hospital, London

CONTENTS

INTRODUCTION

The first solid-phase radioimmunometric assays used antibody-coated Sephadex or plastic tubes (Wide and Porath, 1966; Catt and Tregear, 1967) and offered many advantages over the fluid phase competitive radioimmunoassay (Yalow and Berson, 1960). They were easier to perform and because only one component, the analyte, was present at a limiting concentration they were less vulnerable to pipetting and other errors (Miles and Hales, 1968). The requirements of a suitable solid phase are much the same whether the label is an enzyme or a radiosotope. The stable binding of antigen or antibody to the solid phase is of central importance, since if that fails then the rest of the assay will do so too. There are a number of general requirements of solid phases that are common to all assays. They should have a high capacity for antigen or antibody which should retain its immunological activity, once bound. They should be stable when stored and it is vital that bound material does not become detached during the course of the assay.

The different solid phase supports that are used can be divided into those which have a high capacity (agarose, Sephadex, cellulose, and nitrocellulose) and those with a low capacity (polystyrene, polyvinylchloride (PVC), nylon, glass). The capacity of different solid phase matrices can be manipulated by a number of procedures and the choice of which to use depends on the specific requirements of the assay. For example, cyanogen bromide-activated Sephadex (Wide, Bennich, and Johansson, 1967) or cellulose (Ceska, Eriksson, and Varga, 1972), both of which have a high capacity for binding a wide range of different proteins, can be used to detect IgE antibodies even when coated with impure antigen preparations. On the other hand plastic microtitre plates do not have high background binding and so are more suitable for measurement of IgG antibodies (Grant *et al.*, 1981). The properties of the solid phase can be exploited in other ways — the close proximity (< 1 micron) of the fibres that make up filter paper can be used to bring the reactants close together and so speed up the assay (Giegel, 1985) or reduce the amount of sample needed (Kemeny and Richards, 1987). On the other hand, saturation of antibody with ^{125}I or enzyme-labelled antigen in the fluid phase (Adkinson, Sobotka, and Lichtenstein, 1979) or with antibody bound to the surface of plastic (Zeiss *et al.*, 1973; Kemeny and Lessof, 1987) works better than with low capacity matrices as non-specific binding of the antigen is lower. The ease with which assays can be set up and processed is an important factor in determining the choice of matrix and there is little doubt that the widespread popularity of ELISA owes much to the convenience of the microtitre plate format (Voller *et al.*, 1974). However, the limited protein binding capacity of these plates restricts their use and much work has gone into increasing this.

In this chapter we will discuss the relative merits of a number of different solid phase supports and review some of the ingenious assay systems that have been developed. While it seems unlikely that any one system would be suitable for all uses, we will pay particular attention to the microtitre plate as this is far and away the most widely used. The reader is also directed to the many excellent books and review articles on ELISA that were referred to in Chapter 1.

LOW CAPACITY SYSTEMS

Microtitre plates

Although appreciable amounts of protein can be bound to plastic surfaces the quantity bound is invariably less than for high capacity support media such as agarose or cellulose. This is illustrated by microtitre plate assays assays for IgE antibodies (Reid, Cheung, and Lewiston, 1981), where transfer of the serum sample to a fresh antigen-coated well after incubation was found to give an equal, and sometimes greater, uptake of ^{125}I anti-IgE (Reid *et al.*, 1985)

FIGURE 1 Different microtitre plates available for ELISA

indicating that only a small proportion of the antibody had bound.

The extensive array of available microtitre plates can be bewildering (Figure 1). PVC plates are reported to have a greater capacity for protein (De Savigny and Voller, 1980) although they do not bind more than the high capacity (irradiated) plates that are now available (Urbanek, Kemeny, and Samuel, 1985). PVC plates can, of course, be cut up with scissors so that bound ^{125}I-labelled antigen or antibody can be counted. Polystyrene plates too can be cut up with a hot wire, but much less easily although many of the manufacturers also make single wells or strips of wells. Until recently the optical properties of PVC plates were inadequate for colorimetric assays but suitable plates are now available and are reported to perform well.

The process by which proteins bind to plastic or PVC is poorly understood. Charge is believed to play a part (Oreskes and Singer, 1961) but it is likely that hydrophobic interactions are more important (Standefer, 1985). Different types of plastic and PVC show widely different capacities (Figure 2a) although drying antigen on to the plate increases uptake by low capacity plastic (Figure 2b) (Urbanek, Kemeny, and Samuel, 1985).

Not only does the protein binding capacity of microtitre plates from different manufacturers vary, but there can be differences in the ability of individual proteins to bind. Comparison of the binding of four ^{125}I-radiolabelled proteins:

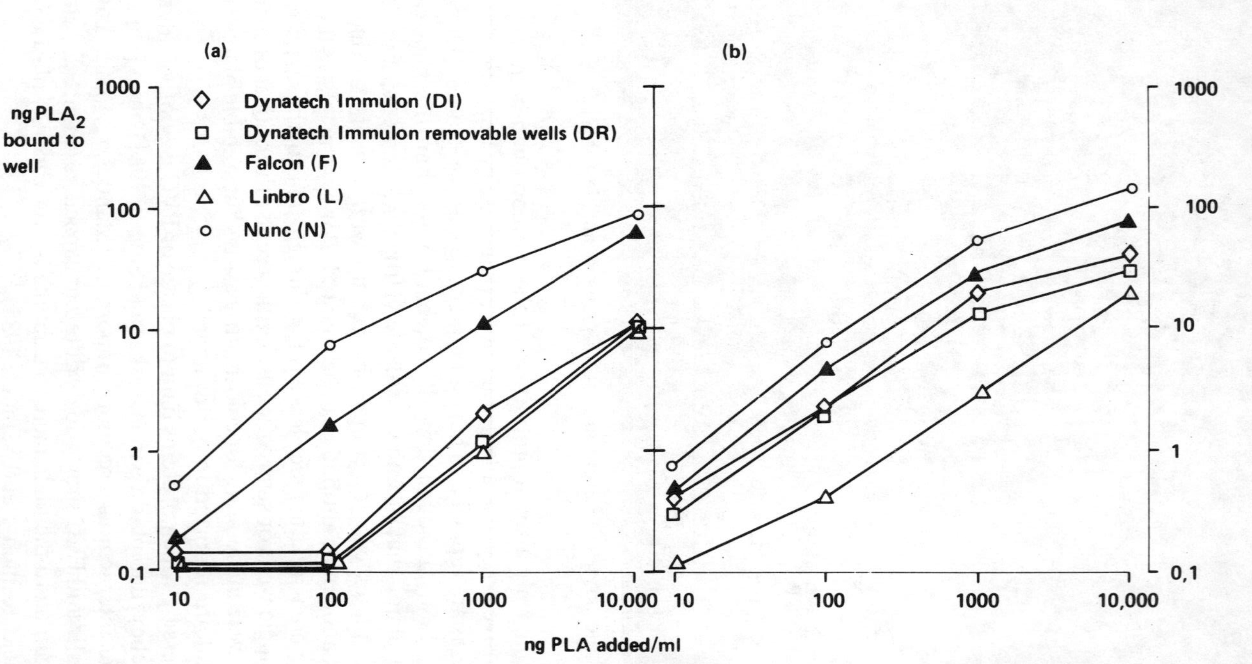

FIGURE 2 Comparison of the binding of PLA_2 to different types of microtitre plate at different concentrations (all incubation volumes 100 µl) either at 4 °C overnight (a) or under desiccating conditions at 37 °C overnight (b) (Urbanek, Kemeny and Samuel, 1985)

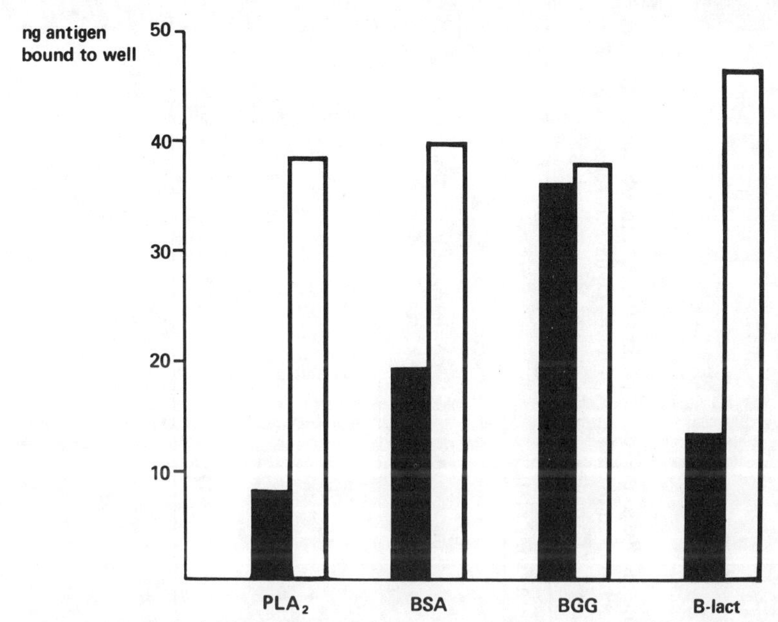

FIGURE 3 Comparison of the binding of different proteins to Dynatech M129A (▬) and NUNC immuno-1 (▭) microtitre plates. (Urbanek, Kemeny and Samuel, 1985)

bee venom phospholipase A_2 (PLA_2), bovine serum albumin (BSA), bovine gamma globulin (BGG), and β-lactoglobulin (β-LACT) to two commonly used microtitre plates, one which showed a high (Nunc immuno-1) and the other a low (Dynatech M129A) capacity for PLA_2, as shown in Figure 2 (Urbanek, Kemeny, and Samuel, 1985). The irradiated Nunc plates bound all four proteins equally well (Figure 3) but there was reduced binding of PLA_2 BSA and β-LACT to the Dynatech M129A plates. Another type of Dynatech plate (M129B) showed similar binding to the Nunc plates (Challacombe *et al.*, 1986).

The capacity of the microtitre plate to absorb protein is not the only important factor and the strength of the protein–plate interaction will determine whether any becomes detached during the assay. Any antigen or antibody thus released from the solid phase will have an adverse effect on the assay, especially if it has already bound to the analyte as such complexes are unlikely to re-attach themselves. Using ^{125}I-radiolabelled PLA_2 desorption of antigen from a number of different microtitre plates was studied (Urbanek, Kemeny, and Samuel, 1985). As much as 50 per cent of the bound PLA_2 was desorbed within 3 hr (Table 1). Similar values have been shown with BSA (Lehtonen and Viljanen, 1980) and immunoglobulin (Engvall and Perlmann, 1971). With the irradiated higher capacity plates this was reduced to 9 per cent and where the plates were pre-coated with antibodies to PLA_2, this was as low as 1 per cent in 3 days — which may in part explain the greater sensitivity of the modified

TABLE 1 The capacity for different types of microtitre plates for bee venom and PLA_2 and desorption after 3 hr incubation with 2 per cent w/v human serum albumin PBS/0.5 per cent Tween 20. DI = Dynatech immulon M129A, DR = Dynatech removawell, F = Falcon, L = Linbro, NI = Nunc Immuno-1, ND = Not determined. (Urbanek, Kemeny and Richards, 1985)

Microtitre plate type	Desorption from plate					
	Coated at 4 °C overnight			Desiccated at 37 °C		
	ng bound/ well	ng desorbed/ well	% desorbed/ well	ng bound/ well	ng desorbed/ well	% desorbed/ well
DI	5	2	40	23	4	17
DR	4	2	50	20	7	35
F	17	6	35	46	17	37
L	2	1	50	6	3	50
NI	35	3	9	59	4	8
DR + anti-PLA_2 1 µg/well	31	0.4	1.3	ND	ND	ND

sandwich ELISA (see Chapter 9). Detachment of BSA and casein occurs most rapidly during the first few days following coating (Rubin, Hardtke, and Carr, 1980) and it may be that once these loosely bound proteins are removed the remaining antigen on the solid phase will perform well. The highest concentration of most proteins that would form a monolayer is reported to be 1 μg/ml (Cantarero, Butler, and Osborne, 1980) and above this concentration some of the additional protein bound is to other proteins rather than to the plate and such interactions are weaker. At present too little information is available from the manufacturers on the binding characteristics of different ELISA plates and it is to be hoped that in the future they will carry out such tests 'in-house' or in designated laboratories.

Coating conditions

The coating of the solid-phase can be assessed in two ways. First, by measuring the binding of ^{125}I-labelled protein and, secondly, in the assay system in which it is to be used. The effect of the amount of antigen or antibody bound to the solid phase will be different according to the assay. Some proteins do not bind the analyte non-specifically even when they are used at a high concentration while others, such as immunoglobulins, can present considerable difficulties (Chapter 8). It is generally a good idea to use a slightly higher coating concentration than necessary so that any minor dilutional errors will not upset the assay.

Appropriate negative sample and diluent controls should be included. It is notalways easy to find suitable negative control sera, for example when studying the immune response to ubiquitous antigens such as cows' milk and hens' eggs. A low titre serum may not really represent true background binding and absorption of specific antibody from a positive serum is inappropriate as non-antibody factors which affect the assay may also be removed. One approach is to optimize the conditions (except for the solid phase antigen) in similar assays where negative controls are available. For example, most people do not have bee venom- or grass pollen-specific IgG antibodies so that it is possible to determine optimum conditions which can then be used in assays for milk egg-specific antibodies where negative controls are not available. For common food antigens we generally observe a 3–4 × $\log_{10}$ range of antibody concentrations in normal sera so that although it is difficult to determine whether antibodies are present or not, the magnitude of the immune response can be compared between different groups of patients or in the same individual over time. When much smaller ranges are reported it is likely that the assay has not been properly set up.

Surface area. An obvious but important factor that will affect how much protein binds to plastic is the available surface area of the solid phase. For example, if a two-site assay for human IgE is carried out at different volumes (300 – 50 μl/well) the assay is more sensitive with the larger volume (Figure 4).

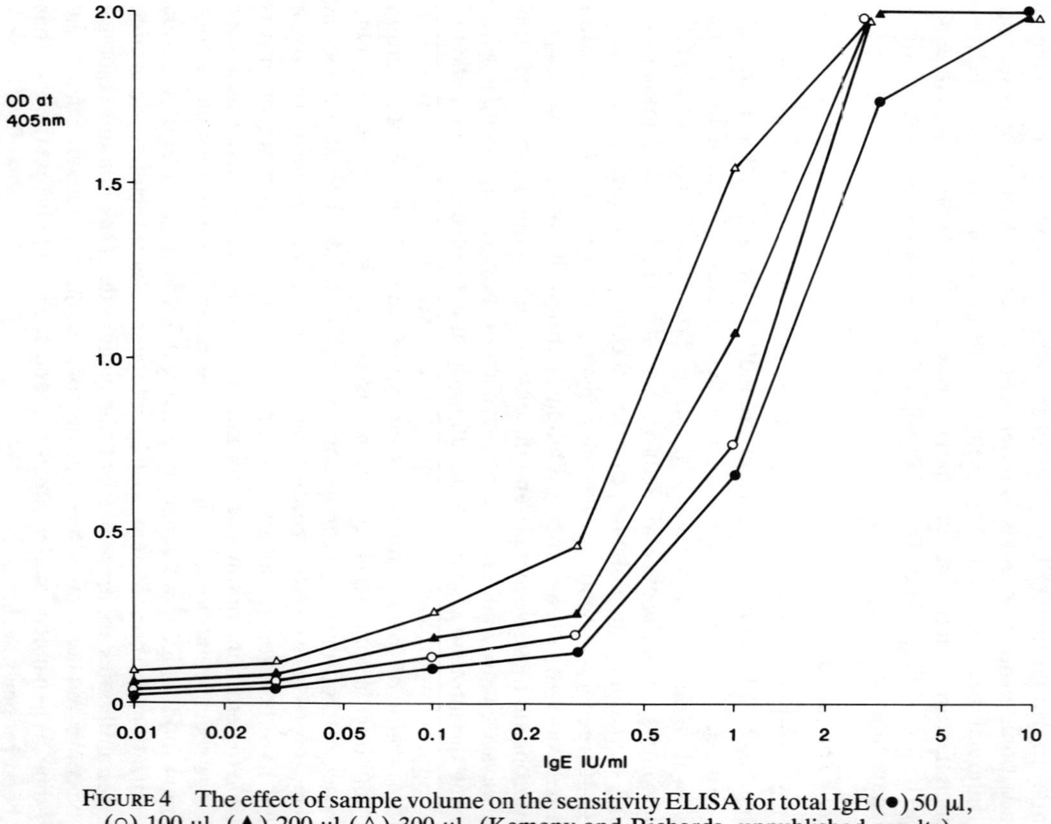

FIGURE 4 The effect of sample volume on the sensitivity ELISA for total IgE (●) 50 μl, (○) 100 μl, (▲) 200 μl (△) 300 μl. (Kemeny and Richards, unpublished results)

TABLE 2 The concentration and absolute amount of human IgE that can be detected using different incubation volumes and their relationship to surface area (Kemeny and Richards, unpublished results)

	Sample volumes in μl (internal area of well in mm^2)			
	300 (210)	200 (152.8)	100 (95.6)	50 (67.1)
Surface area/well ratio	0.70	0.76	0.96	1.34
Concentration of IgE detected in IU/ml at OD 1.0	0.56	0.8	1.2	1.4
at 1.5 × background	0.017	0.11	0.11	0.16
Absolute amount of IgE detected in IU at OD 1.0	0.16	0.16	0.12	0.07
at 1.5 × background	0.057	0.022	0.011	0.008

However, once the results are corrected for the amount of IgE added it becomes clear that the smaller volume, which has a greater surface area/volume ratio, is more sensitive (Table 2). This principle explains the rapid binding in the finned tubes described by Park (1978) and has been taken even further in the experiments of Labrousse *et al.* (1982) who found that if assays for human IgE were carried out in 0.3 μl capillaries, as little as 0.00027 IU IgE (0.65 pg) could be detected, while at 300 μl the minimum quantity of IgE that could be detected was nearly 100 times less at 0.023 IU IgE (55 pg).

Effect pH, ionic strength, and temperature. The binding of proteins to plastic is remarkably independent of the pH or ionic strength of the coating buffer although differences are greater at higher protein concentrations. For most proteins satisfactory results can be obtained at pH 9.6 which is the optimum pH for binding immunoglobulins. Binding is dependent on time (Herrmann and Collins, 1976) and temperature (Cantarero, Butler, and Osborne, 1980). For convenience we normally coat at 4 °C overnight as this leaves the whole of the following day free to carry out the test. Prolonged incubation of immunoglobulin with the solid phase can result in loss of antigenic activity (Pesce, Ford, and Gaizutis, 1978); and antibody-coated plates, which have been left for a week or more, bind less antigen. Dried antibody-coated plates, on the other hand, can be stored for a year or more if kept in air tight, waterproof packs (Voller, Bidwell, and Bartlett, 1979).

Glutaraldehyde treatment. Glutaraldehyde is commonly used to cross-link proteins and in enzyme coupling and microtitre plates treated in this way are reported to have increased capacity for protein (Howell, Nasser, and Schray, 1981; Place and Schroeder, 1982; Klasen *et al.*, 1982; Tanimori, Ishikawa, and

Kitagawa, 1983) and haptens (Suter, 1982); although this is not a universal finding (Salonen and Vaheri, 1979). The effect of glutaraldehyde on conventional plastic microtitre plates is poorly defined but it is better recognized for its ability to cross-link proteins. Multiple antigen attachment increases the stability of the solid phase antigen (Rotman and Delwel, 1983; Rotman and Scheven, 1984) and the procedure of Papadea, Check and Reimer, (1985), in which plates were first coated with BSA and subsequently treated with glutaraldehyde, is claimed to improve the binding and performance of monoclonal anti-IgG subclass antibodies compared with direct coating. Another variant of this procedure is to treat the coating protein with glutaraldehyde (Parsons, 1981) or carbodiamide (Rotman and Delwel, 1983; Rotman and Scheven, 1984).

Poly-L-lysine. The binding of carbohydrate antigens to plastic can be enhanced by prior treatment with poly-L-lysine (Gray, 1979) and pre-treatment of polyacrylic beads with poly-L-lysine provided quantitative and stable binding of ^{125}I IgM (Pachmann and Leibold, 1976).

Blocking proteins and additives. The capacity of plastic surfaces to bind added proteins firmly can work against assays as well as for them. High concentrations of proteins in the sample may cause them to bind non-specifically to the plastic and so increase the background noise of the assay. Many different approaches have been used to reduce this including addition of non-ionic detergents like Tween 20 (Berkowitz and Webert, 1981). Of course, the sample can be diluted to a point when these interactions disappear, but if the concentration of analyte is low this may not be possible. Exogenous proteins can be added to the assay and in the original description of the coated tube assay by Catt and Tregear (1967) 10 per cent aged human plasma and 0.01 per cent Merthiolate in 0.15 M NaCl were used in a final wash after coating to block any vacant binding sites. Gelatine (0.5–1 per cent) or BSA (1 per cent) are more widely used now. Treatment with blocking proteins is usually carried out after coating but it has been reported that pre-treatment with albumin can work as well (Standefer and Saunders, 1978). Blocking coated plates can reduce sensitivity by interference with antigenic determinants on the coated proteins (Butler, Peterman, and Koertge, 1985) but the benefits in terms of reduced background binding can be worthwhile in some assays. Indeed, if BSA or other common dietary proteins are used to block the plates, they may actually bind specific antibodies present in the test sample. If the same protein that is used for blocking is also included in the assay diluent this will be largely, but not completely, prevented.

Modifications of antigens used in research too may alter their behaviour and cationized proteins, which have been used to induce renal injury in experimental animals (Border *et al.*, 1982), give very high background binding in ELISA. The addition of dextran sulphate or heparin reduces this dramatically making it possible to measure the antibody response to these proteins (Pesce *et al.*, 1986; see also this volume Chapter 8).

Coloured diluents. It is possible to miss out wells when carrying out ELISA or to use a buffer of the wrong pH. We have adopted a system of coloured additives used by Dr R.C. Aalberse, Red Cross, Amsterdam, Holland in his radioimmunoassays for use in our ELISAs. Phenol red and naphthalene black (see Appendix) are added to the coating and assay buffers. These two dyes are green at neutral pH and lilac at alkaline pH. They can also be used separately (red and blue respectively) for assays which involve multiple addition of reagents.

Problems with microtitre plates

A number of other problems can occur with microtitre plates. In assays which do not reach equilibrium (probably the majority) there may be differences in binding of the sample to the outer wells (Oliver *et al.*, 1981). There can be many reasons for this but the most likely is that these wells are at a different temperature to the others. This can be reduced by placing the plates on small shelves in an incubator to improve air circulation around the plate or the duration of the sample incubation step can be increased so that differences in rates of binding are minimized. In our hands such effects occur rarely in ELISA and usually seem to indicate that one or more components are being used sub-optimally.

There may also be batch to batch variation of the plates (Wreghitt *et al.*, 1983) used which are normally only controlled for optical quality and not for protein binding. This is particularly true of lower capacity plastics and is largely solved by using the higher capacity irradiated plates. In general, it is prudent to purchase sufficient plates for a series of experiments and to re-test any new batches.

Other low capacity solid phase supports

There are a number of other solid phase supports of similar or intermediate capacity to microtitre plates. The solid phase assay of Catt and Tregear (1967) used polystyrene tubes which were also used in the earliest ELISAs by Engvall and Perlmann (1971). The capacity of these can be increased and co-polymers of styrene and maleate (Amino Dylark) (Tamimori, Ishikawa, and Kitagawa, 1983) are reported to work well, polypropylene discs (Crosignani *et al.*, 1970) and silicone rubber tubing (Ishikawa and Kato, 1978) have also been described and plastic 'dip-sticks' can be used (Felgner, 1977, 1978) although these have not been widely used.

Polystyrene balls have a higher capacity compared to microtitre plates and have been used by a number of researchers (Kalimo *et al.*, 1977; Hendry and Herrmann, 1980; Aalberse *et al.*, 1986) as well as polyaminostyrene beads which are reported to bind protein covalently (Phillips *et al.*, 1980). A variant of this is the Falcon FAST system which uses polystyrene beads on sticks, which snap into the lid of microtitre plates (Hancock and Tsang, 1986).

HIGH CAPACITY SOLID PHASE SUPPORTS

Many of the limitations of plastic or PVC are avoided with high capacity systems such as CNBr-activated agarose, Sephadex or cellulose (Axen, Porath,

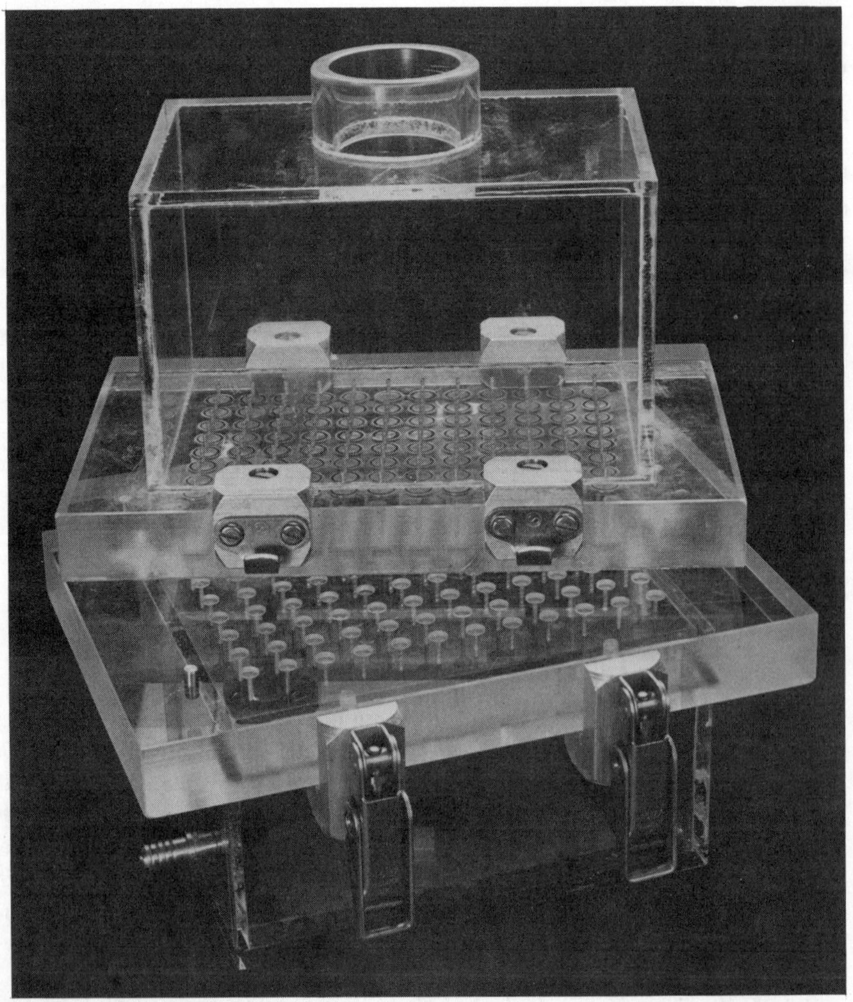

FIGURE 5 Multiple paper disc washer. (Kemeny and West, 1981)

and Ernback 1967). In their classic paper on the advantages of the solid phase immunoradiometric assay, Miles and Hales (1968) emphasized the need for the carrier to have a high capacity for antigen or antibody and indeed in the system they described 1 g of insulin-coated agarose was capable of binding 660 mg of insulin-specific antibodies.

The ease with which these matrices can be dispensed and washed differs. Paper discs are easy to handle and, using a flow-through apparatus (Figure 5), they are easy to wash (Kemeny and West, 1981). Although agarose particles require centrifugation they only need the briefest spin and using a modified centrifuge (Dr R.C. Aalberse, Red Cross, Holland and A.R. Horwell Ltd, UK) it is possible to wash 200 tubes five times in 10 minutes.

As with low capacity carriers, the efficiency with which the analyte binds is not constant for different high capacity solid phase matrices. We have found, for example, that 10 mg of castor-bean-coated agarose can remove 95 per cent of the IgE antibody from 100 μl of serum (*ca.* 20 ng) following an overnight incubation while a cellulose filter paper disc coated with the same allergen may bind less than 50 per cent of the antibody present. There are several possible explanations for this. The greater amount of antigen bound to the agarose may improve the efficienty with which it binds antibody but it is also possible that the agitation employed with agarose increases the rate at which antibody binds to antigen. Indeed, if serum samples incubated with antigen-coated discs are agitated the rate of binding can be increased (Pecoud *et al.*, 1986); in addition, ultrasound can also increase the rate of binding (Chen *et al.*, 1984). Furthermore, antigen can leach off these matrices and we have found that washing the allergen-coated paper discs immediately prior to use increases the uptake of IgE antibody in RAST (Guthrie, Kemeny and Richards, 1987).

Micro-radioallergosorbent test

An alternative approach, which we have investigated, is to make use of the absorbent quantities of filter paper discs. Of course, if the volume of serum incubated with the antigen-coated filter paper disc in RAST is reduced, a decreased amount of IgE antibody is detected when the usualy amount of ^{125}I anti-IgE is added (Figure 6a). However, if the volume of ^{125}I anti-IgE is also reduced, then sensitivity is actually increased (Figure 6b). Background binding also rises (from 0.4 to 1.2 per cent) but remains at an acceptable level. The use of a small amount of ^{125}I IgE (5 μl) gives such low counts that they are difficult to measure with a conventional gamma counter. If, however, the same amount of ^{125}I anti-IgE as is normally used is added in a smaller volume (5 μl) then the actual radioactive count bound is similar to the conventional assay (Figure 6c). Comparison of this micro-RAST (MRAST) (5 μl sample and ^{125}I anti-IgE) with the conventional RAST (CRAST) (50 μl sample and ^{125}I anti-IgE) in 48 sera from allergic patients showed a remarkably close agreement ($r = 0.96$) (Figure

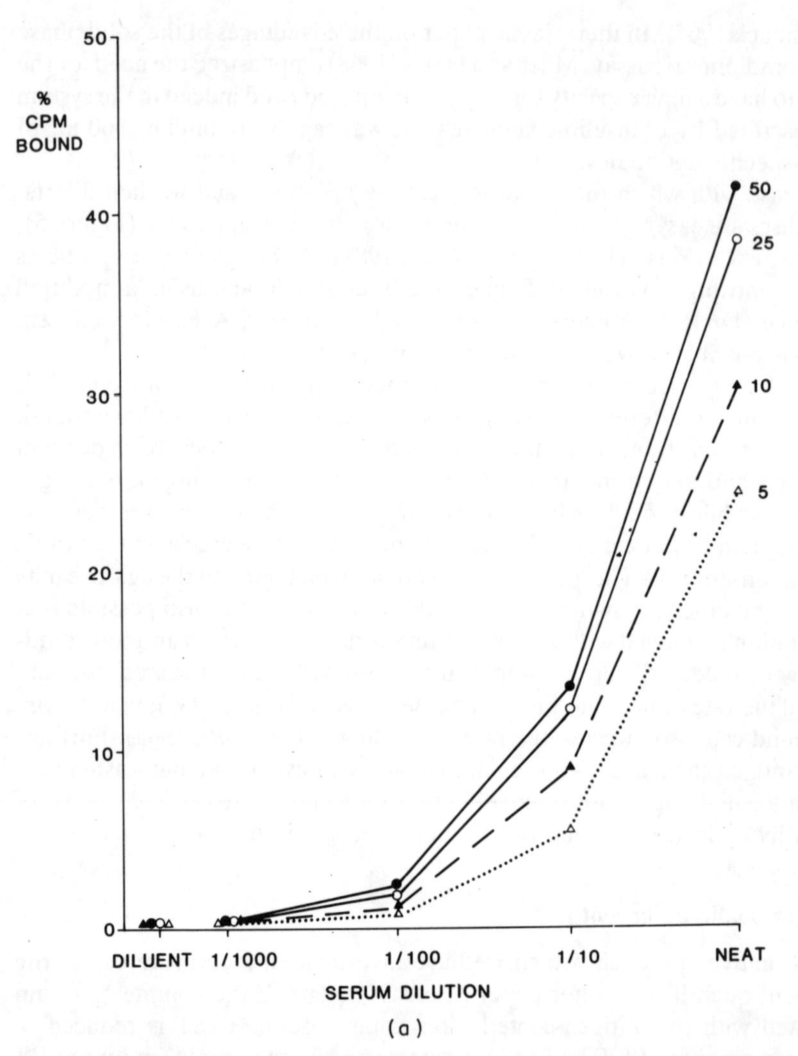

FIGURE 6 When the volume of the serum sample used in RAST (shown in μl) was reduced the percentage of ^{125}I anti-IgE bound fell (a). If the volume of ^{125}I anti-IgE was reduced in parallel (b) the percentage that bound increased at the smaller volumes but resulted in low counts. Optimum results were achieved using the same amount of ^{125}I anti-IgE that is normally used in 50 μl in a reduced volume (5 μl) — the micro-RAST (c). (Kemeny and Richards, unpublished results)

7). Not only is a similar level of sensitivity achieved with a tenfold smaller sample but the rate at which the reaction proceeds is also increased (Figure 8) compared with the conventional procedure (Figure 9). Thus by bringing the reactants into close proximity (within the fibre matrix of the filter paper disc) it

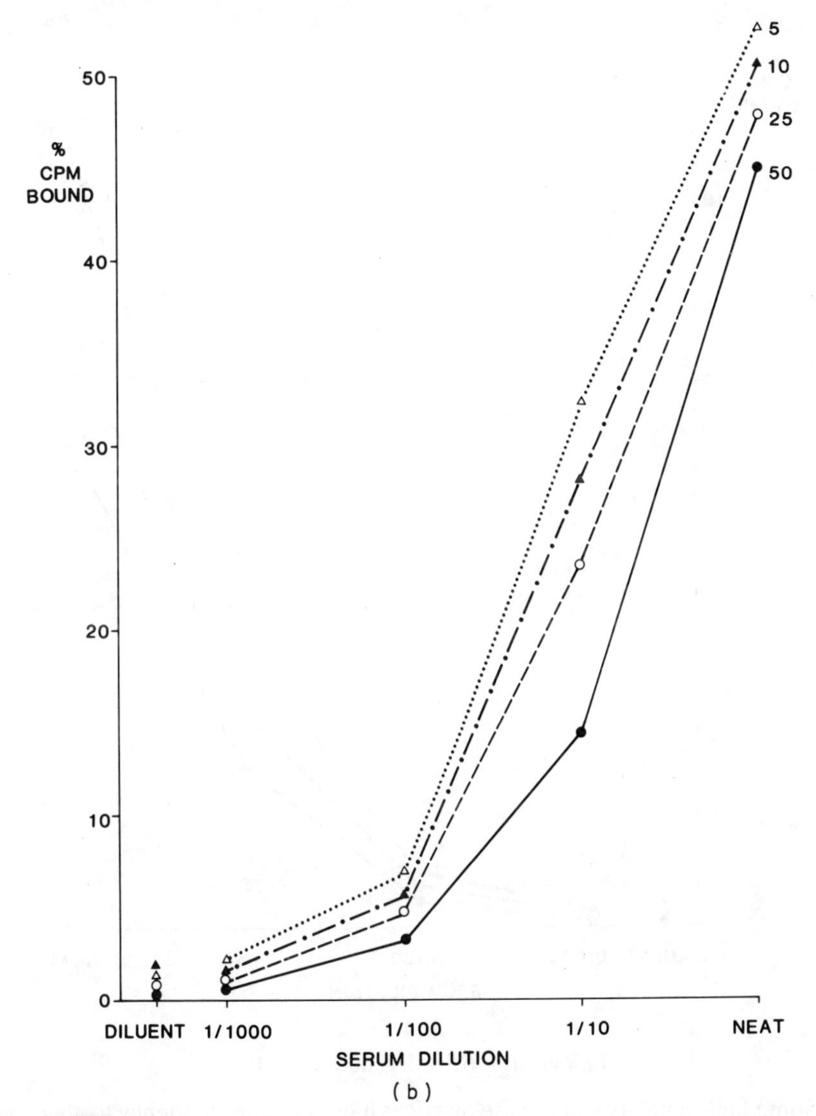

FIGURE 6(b)

is possible to accelerate the assay. A similar principle has been used in the radial partition assay Giegel *et al*. (1982) where the total reaction time can be as little as 5 min.

Apart from cyanogen-bromide-treated matrices a number of other high capacity carriers have been used. Affigel and Ag-GA from Bio Rad Ltd provide a stable solid phase support. Nitrocellulose too binds a wide range of proteins with little denaturation or leakage of bound material.

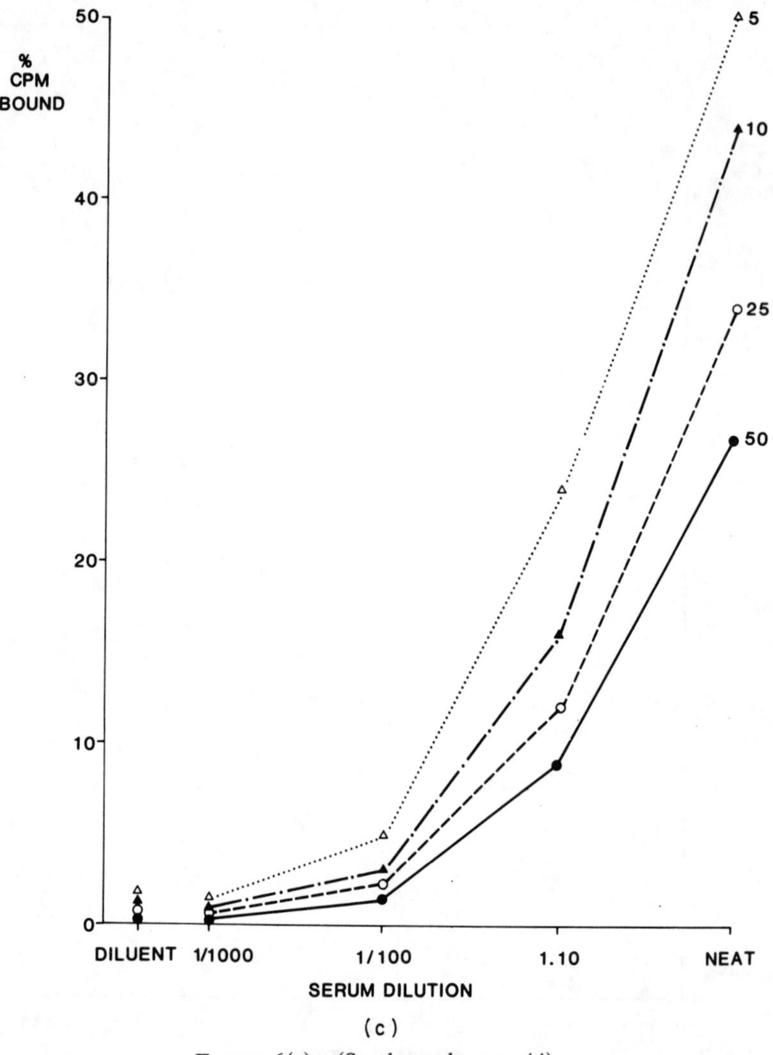

FIGURE 6(c) (See legend on p. 44).

Some high capacity solid phase matrices have a greater tendency to give high background binding than others, cyanogen-bromide (CNBr)-activated cellulose and agarose particularly tend to bind immunoglobulin non-specifically. Addition of animal serum (e.g. horse) to the assay incubation buffer can markedly reduce this (Kemeny, Lessof, and Trull, 1980) but will not prevent the binding of naturally occurring antibodies to agarose and cellulose found in some sera which have to be removed by prior absorption with the same solid phase (Hamilton and Adkinson, 1985). Some antigens such as *Staphylococcus aureus* protein A and ricinus agglutinin themselves may bind immunoglobulins

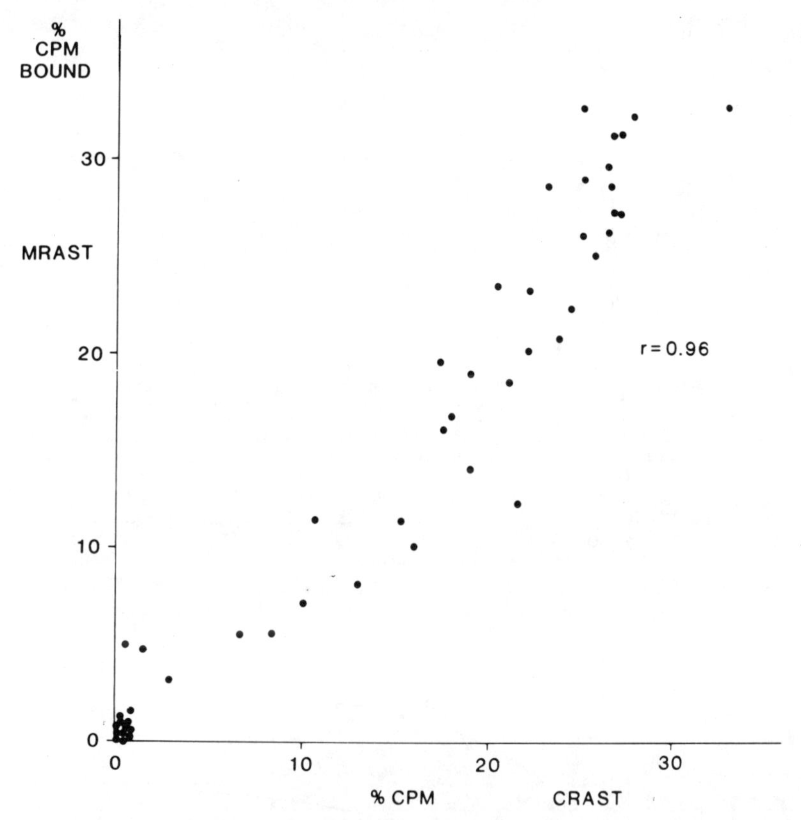

FIGURE 7 Comparison of the MRAST and the CRAST. (Kemeny and Richards, 1988)

(Saltvedt *et al.*, 1974; Johansson *et al.*, 1974), while others may possess cross-reacting carbohydrate determinants which pick up naturally occurring cross-reacting antibodies (Aalberse, Koshte, and Clemens, 1981). These may be partially inhibited by the addition of animal serum (Kemeny *et al.*, 1981) or by pre-treatment of the antigen with periodic acid (Aalberse, Kant, Koshte, 1983).

ALTERNATIVE ASSAY SYSTEMS

So far we have just talked about the use of different solid phase supports and their effect on the immunometric assay. In this section we will review some of the assay procedures that have been devised using different solid phase matrices. Some have already been discussed such as the FAST Falcon assay which utilizes coated beads that snap into the lids of microtitre plates or the radial partition assay in filter paper and these will not be discussed further here.

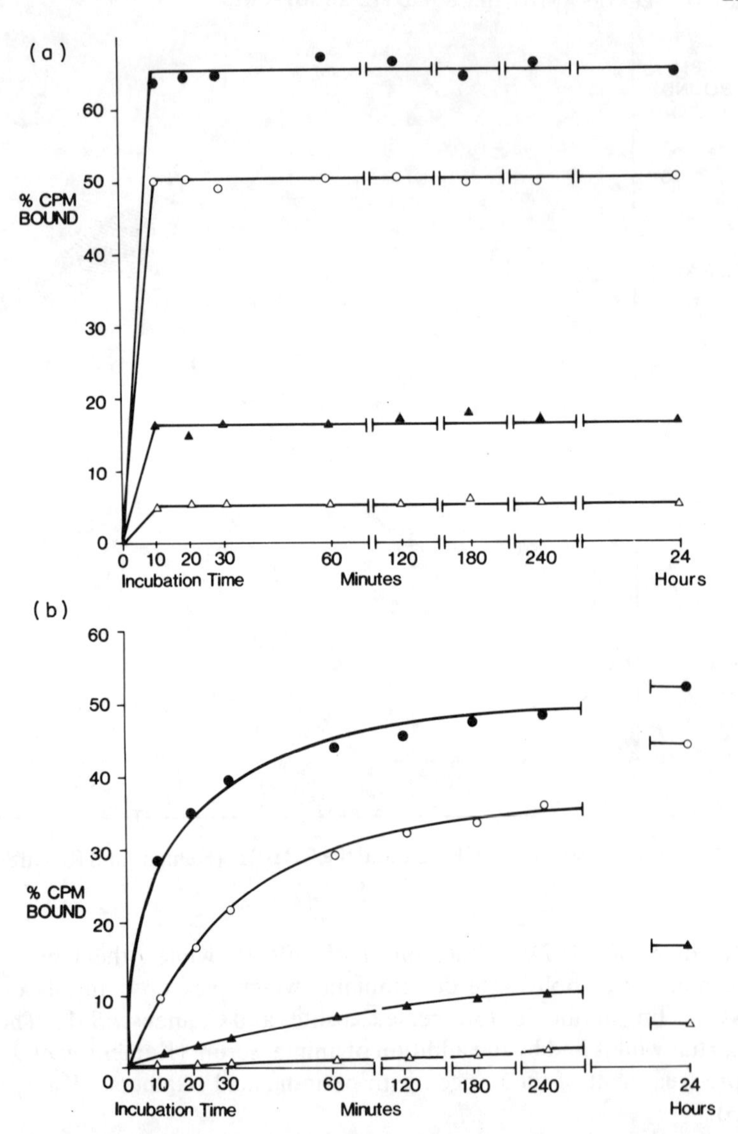

FIGURE 8 The rate of binding of (a) the IgE antibody and (b) the ^{125}I anti-IgE in the micro-RAST. Dilution of sample: ● = neat, ○ = 1/10, ▲ = 1/1000, △ = 1/1000. (Kemeny and Richards, 1988)

Many of these techniques are novel separation of bound and free antigen and antibody. The affinity-column-mediated immunoassay uses a miniature affinity chromatography column to remove unbound enzyme-labelled antibody that has been incubated with the sample (Freytag, Dickinson, and Tseng, 1984). Only enzyme-labelled antibody saturated with antigen will pass through the

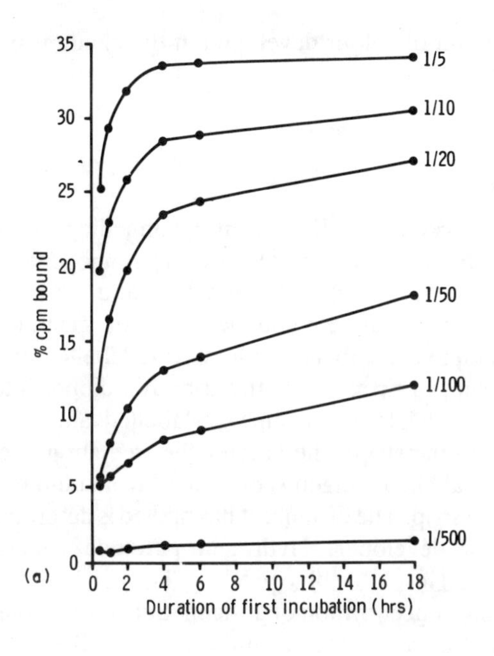

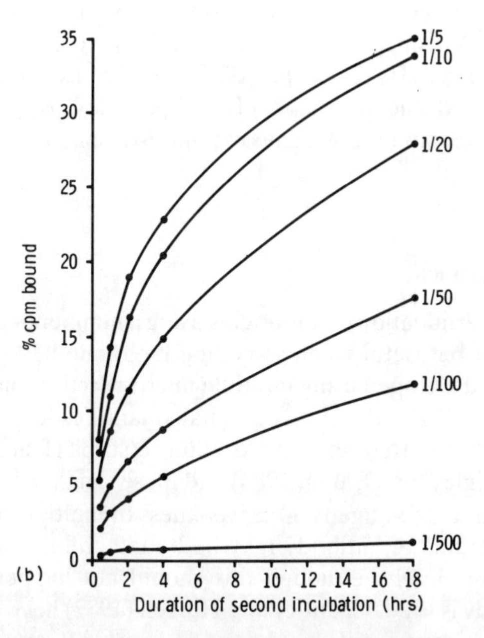

FIGURE 9 The rate of binding of (A) the IgE antibody and (B) the ^{125}I anti-IgE in the conventional RAST. (Kemeny, 1982)

column. The amount of colour developed in the eluate is proportional to the concentration of sample.

Test strip assays

Another novel procedure is the enzyme channelling immunochemical test strip. Although this is currently only used in homogeneous assays for low molecular weight analytes such as theophylline and morphine (Litman *et al.*, 1983), it seems likely that it could be used for macromolecules. In this procedure the sample is incubated with enzyme-labelled antigen and glucose oxidase. A filter paper strip coated with antibody is dipped into the mixture and by capillary action the free and enzyme-labelled antigen and the glucose oxidase migrate up the strip. The greater the free antigen concentration the more the enzyme-labelled antigen is prevented from binding and the higher it has to travel up the strip. The distance it has moved is determined by immersion of the strip in a developer. Hydrogen peroxide generated oxidizes 4-chloronaphthol to an insoluble blue product. The height of the coloured area is proportional to the concentration of antigen. Serration of the edge of the strip causes the stained area to form a uniform rocket-like appearance which is easier to read and incorporation of red cell binding agents at the bottom of the strip removes them and so makes it possible to use unseparated blood.

Guesdon and Avrameas (1981) have developed assays using magnetic particles and sucrose (Baker *et al.*, 1985) can be used for rapid and easy separation of bound and free label. Latex particles, which are counted in automated particle counters (Magnusson and Masson, 1982), have also been used.

Multiple and micro assays

Simultaneous determination of antibodies using a number of antigens bound to a solid phase can be useful when screening large numbers of samples. Such assays have been developed using multiple micro-reaction chambers (Leaback and Creme, 1980, 1981). Alternatively, small quantities (0.1–0.5 μl) of antigen or antibody can be spotted on to nitrocellulose strips (Derer *et al.*, 1984) or discs (Walsh, Wrigley, and Baldo, 1984) and these can then be used to measure antibodies to a panel of antigens. One weakness of multiple antigen systems is that variable amounts of antibody may be bound and differences in binding may not show up clearly, although this should not happen if an excess of detecting antibody is used. Chantler and Hurrell (1982) have described a novel ELISA using a series of capillary tubes coated with different snake venoms in which the sample is injected through them in series with a syringe to allow rapid identification of the offending reptile.

SUMMARY

The solid phase immunometric assay has come a long way since it was first described. The promise of increased sensitivity and ease of performance has largely been realized although many new and exciting tests are emerging at an ever-increasing rate. There is a demand for assays that can be carried out without expensive or sophisticated equipment on the one hand and for automation on the other. The widespread interest in this field makes it likely that in the future immunoassays will be more sensitive and specific and easier to carry out.

REFERENCES

Aalberse, R.C., Kant, P., and Koshte, V. (1983). Unexpected cross-reactions of human IgE antibodies. In: *Recent Developments in RAST and other Solid-Phase Immunoassay Systems* (eds D.M. Kemeny and M.H. Lessof. Excerpta Medica, Elsevier, Amsterdam, pp. 17–25.

Aalberse, R.C., Koshte, V., and Clemens, J.G.J (1981). Immunolglobulin E antibodies that cross-react with vegetable foods, pollen and hymenoptera venom. J. All. Clin. Immun., **68**, 356.

Aalberse, R.C., Van Zoonen, M., Clemens, J.G.J., and Winkel, I. (1986). The use of hapten-modified antigens instead of solid-phase-coupled antigens in a RAST-type assay. *J. Imm. Methods*, **87**, 51.

Adkinson, N.F., Sobotka, A.K., and Lichtenstein, L.M. (1979). Evaluation of the quantity and affinity of human IgG 'blocking' antibodies. *J. Immunol.*, **122**, 965.

Axen, R., Porath, J., and Ernback, S. (1967). Chemical coupling of peptides and proteins to polysaccharides by means of cyanogen halides. *Nature*, **214**, 1302.

Baker, T.S., Abbott, S.R., Daniel, S.G., and Wright, J.F. (1985). Immunometric assays. In: *Alternative Immunoassays* (ed. W.P. Collins). John Wiley & Sons, Chichester, pp. 59–76.

Berkowitz, D.M., and Webert, D.W. (1981). The inactivation of horseradish peroxidase by a polystyrene surface. *J. Imm. Methods*, **47**, 121.

Border, W.A., Ward, H.J., Kamil, E.S., and Cohen, A.H. (1982). *J. Clin. Invest.*, **69**, 451.

Butler, J.E., Peterman, J.H., and Koertge, T.E. (1985). The amplified enzyme-linked immunosorbent assay (a-ELISA). In: *Enzyme-Mediated Immunoassay*. (eds T.T. Ngo and H.M. Lenhoff). Plenum Press, New York, pp. 241–76.

Cantarero, L.A., Butler, J.E., and Osborne, J.W. (1980). The absorptive characteristics of proteins for polystyrene and their significance in solid phase immunoassays. *Anal. Biochem.*, **105**, 375.

Catt, K., and Tregear, G.W. (1967). Solid-phase radioimmunoassay in antibody-coated tubes. *Science (NY)*, **158**, 1570.

Ceska, M., Eriksson, R., and Varga, J.M. (1972). Radioimmunosorbent assay of allergens. *J. All. Clin. Immun.*, **49**, 1.

Challacombe, S.J., Biggerstaff, M., Greenall, C., and Kemeny, D.M. (1986). ELISA detection of human IgG subclass antibodies to *Streptococcus mutans*. *J. Imm. Methods*, **87**, 95.

Chantler, H.M., and Hurrell, J.G.R. (1982). A new enzyme immunoassay system suitable for field use and its application in a snake venom detection kit. *Clin. Chim. Acta*, **121**, 228.

Chen, R., Weng, L., Stizto, N.C., Osorio, B., Hsu, C.J., Rodjers, R., and Litman, D.J. (1984). Ultrasound accelerated immunoassay as exemplified by enzyme-immunoassay of choriogonadotrophin. *Clin. Chem.*, **30**, 1446.

Crosignani, P.G., Nakamura, R.M., Horland, D.N., and Mishell, D.R. (1970). A method of solid-phase radioimmunoassay utilising propylene discs. *J. Clin. Endocrinol. Metab.*, **30**, 153.

Derer, M.M., Miescher, S., Johansson, B., Frost, H., and Gordon, J. (1984). Application of the dot immunobinding assay to allergy diagnosis. *J. All. Clin. Immun.*, **74**, 85.

De Savigny, D., and Voller, A. (1980). The communication of ELISA data from laboratory to clinician. *J. Immunoassay*, **1**, 105.

Engvall, E., and Perlmann, P. (1971). Enzyme-linked immunosorbent assay (ELISA): quantitative assay of IgG. *Immunochemistry*, **8**, 871.

Felgner, P. (1977). Serological diagnosis of extraintestinal amebiasis: A comparison of stick-ELISA and other immunological tests. *Tropenmed. Parasit.*, **28**, 491.

Felgner, P. (1978). A new technique of heterogeneous enzyme-linked immunosorbent assay, Stick-ELISA. 1. Description of the technique. *Zbl. Bakt. Hyg. 1. Abt. Orig. A.*, **240**, 112.

Freytag, J.W., Dickinson, J.C., and Tseng, S.Y. (1984). A high sensitivity affinity-column-mediated immunometric assay, as exemplified by digoxin. *Clin. Chem.*, **30**, 417.

Giegel, J.L. (1985). Radial partition enzyme immunoassay. In: *Enzyme-Mediated Immunoassay* (eds T.T. Ngo, and H.M. Lenhoff). Plenum Press, New York, 343–62.

Giegel, J.L., Brotherton, M.M., and Cronin, P. *et al.* (1982). Radial partition immunoassay. *Clin. Chem.*, **28**, 1894.

Grant, J.A., Goldblum, R.M., Rahr, R., Thueson, D.G., Farnam, J., and Gilaspy, J. (1981). Enzyme-linked immunosorbent assay for immunoglobulin G antibodies against insect venom. *J. All. Clin. Immunol.*, **68**, 112.

Gray, B.M. (1979). Methodology for polysaccharide antigens: protein coupling of polysaccharides for absorption to plastic tubes. *J. Imm. Methods*, **28**, 187.

Guesdon, J-L., and Avrameas, S. (1981). Magnetic solid-phase enzyme immunoassay for the quantitation of antigens and antibodies: Application to human immunoglobulin E. *Methods in Enzymology*, **73**, 471.

Guthrie, G., Kemeny, D.M., and Richards, D. (1987) The development of a micro-radioallergosorbent test (MRAST). *J. Allergy Clin. Immunol.*, **79**, 222 (Abstract).

Hamilton, R.G., and Adkinson, N.F. (1985). Naturally occurring carbohydrate antibodies: Interference in solid-phase immunoassays. *J. Imm. Methods*, **77**, 95.

Hancock, K., and Tsang, U.C.W. (1986). Development of a FAST-ELISA for detecting antibodies to *Schistosoma mansoni*. *J. Imm. Methods*, **93**, 89.

Hendry, R.M., and Herrmann, J.E. (1980). Immobilization of antibodies on nylon for use in enzyme-linked immunoassays. *J. Imm. Methods*, **35**, 285.

Herrmann, J.E., and Collins, M.F. (1976). Quantitation of immunoglobulin absorption to plastics. *J. Imm. Methods*, **10**, 363.

Howell, E.E., Nasser, J., Schray, K.J. (1981). Coated tube enzyme immunoassay: Factors affecting sensitivity and effects of reversible protein binding to polystyrene. *J. Immunoassay*, **2**, 205.

Ishikawa, E., and Kato, K (1978). Ultrasensitive enzyme immunoassay. *Scand. J. Immunol.*, 8 (Suppl 7), 43–55.

Johansson, S.G.O., Foucard, T., Hjelm, H., and Sjokvist, J. (1974). Determination of specific IgG antibody to soluble antigens by absorption to protein A-Sepharose. *Scand. J. Immunol.*, **3**, 881.

Kalimo, K.O.K., Ziola, B.R., Vijanen, M.K., Granfers, K., and Toivanen, P. (1977).

Solid-phase radioimmunoassay of *Herpes simplex* virus IgG and IgM antibodies. *J. Imm. Methods*, **14**, 183.

Kemeny D.M., (1982). The purification of hyaluronidase from bee venom and the assessment of its importance in bee venom allergy. PhD Thesis, London.

Kemeny, D.M., Frankland, A.W., Fahkri, Z.I., and Trull, A.K., (1981). Allergy to castor bean in Sudan: Measurement of serum IgE and specific IgE antibodies. *Clin. Allergy*, **11**, 463.

Kemeny, D.M., and Lessof, M.H. (1987). The immune response to bee venom: II Quantitation of the absolute amounts of IgE and IgG antibody by saturation analysis. *Int. Arch. All. Appl. Immun.* **83**, 113.

Kemeny, D.M., Lessof, M.H., and Trull, A.K. (1980). IgE and IgG antibodies to bee venom as measured by a modification of the RAST method. *Clin. Allergy*, **10**, 413.

Kemeny, D.M., and Richards, D. (1988). Increased speed and sensitivity and reduced sample size of a micro-radioallergosorbent test (MRAST). *J. Immun. Methods*, (in press).

Kemeny, D.M., and West, F.B. (1981). An improved method for washing paper discs with a constant flow washing device. *J. Imm Methods*, **49**, 89.

Klasen, E.A., Rigutti, A., Bos, A., and Bernini, F. (1982). Development of a screening system for detection of somatic mutations. I—Enzyme immunoassay for detection of antibodies against specific haemoglobin determinants. *J. Imm. Methods*, **54**, 241.

Labrousse, H., Guesdon, J.L., Ragimbeau, J., and Avrameas, S. (1982). Miniaturization of β-glactosidase immunoassays using chromogenic and fluorogenic substrates. *J. Imm. Methods*, **48**, 133.

Leaback, D.H., and Creme, S (1980). A new experimental approach to fluorometric enzyme assays employing disposable micro-reaction chambers. *Anal. Biochem.*, **106**, 314.

Leaback, D.H., and Creme, S (1981). Extremely economical micro-ERMA procedures for performaing 'sequential' fluorogenic enzyme assays and fluorogenic enzyme immunoassays on human serum. *Biochem. Soc. Trans.*, **9**, 580.

Lehtonen, O.P., and Viljaken, M.K. (1980). Antigen attachment in ELISA. *J. Imm. Methods*, **34**, 61.

Litman, D.J., Lee, R.H., Jeong, H.J., Tom, H.K., Stiso, S.N., Sizto, N.C., and Ullmann, E.F. (1983). An internally referenced test strip immunoassay for morphine. *Clin. Chem.*, **29**, 1598.

Magnusson, C.G.M., and Masson, P.L. (1982). Particle counting immunoassay of immunoglobulin E antibodies after their elution from allergosorbents by pepsin: an alternative to the radioallergosorbent test. *J. All. Clin. Immun.*, **70**, 326.

Miles, L.E.M., and Hales, C.N. (1968). Labelled antibodies and immunological assay systems. *Nature (Lond.)*, **219**, 186.

Oliver, D.G. *et al.* (1981). Thermal gradients in microtitration plates. Effects on enzyme-linked immunoassay. *J. Imm. Methods*, **42**, 195.

Oreskes, I., and Singer, J.M. (1961). The mechanism of particulate carrier reactions. I. Adsorption of human globulin to polystyrene latex particles. *J. Immunol.*, **86**, 338.

Pachmann, K., and Leibold, W. (1976). Insolubilization of protein antigens on polyacrylamide plastic beads using poly-1-lysine. *J. Imm. Methods*, **12**, 81.

Papadea, C., Check, I.J., and Reimer, C.B. (1985). Monoclonal antibody-based solid-phase immunoenzymometric assays for quantifying human immunoglobulin G and its subclasses in serum. *Clin. Chem.*, **31**, 1940.

Park, H.A (1978). A new receptacle for solid-phase immunoassays. *J. Imm. Methods*, **20**, 349.

Parsons, G.H. (1981). Antibody-coated plastic tubes in radiommunoassay. *Methods in Enzymology*, **73**, 224.

Pecoud, A., Peitrequin, R., Duc, J., Thalberg, K., Schroeder, H., and Frei, P.G. (1986). Application of microtitre plates and fluorescence reading to shorten handling of Phadezym RAST and Phadezym IgE. PRIST. *Clin. Allergy*, **16**, 231.

Pesce, A.J., Ford, D.J., and Gaizutis, M.A. (1978). Quantitative and qualitative aspects of immunoassays. *Scand. J. Immunol.*, **8** (Suppl 7), 1.

Pesce, A.J., Apple, R., Sawtell, N., and Michael, J.G. (1986). Cationic proteins — problems associated with measurement by ELISA. *J. Imm. Methods*, **87**, 21.

Phillips, D.J., Reimer, C.B., Wells, T.W., and Black, C.M. (1980). Quantitative characterization of specificity and potency of conjugated antibody with solid-phase, antigen bead standards. *J. Imm. Methods*, **34**, 315.

Place, J.D., and Schroeder, H.R. (1982). The fixation of anti-HBs Ag on plastic surfaces. *J. Imm. Methods*, **48**, 251.

Reid, M.J., Cheung, N-K.V., and Lewiston, N.J. (1981). Microtitre solid-phase radioimmunoassay for specific immunoglobulin E. *J. All. Clin. Immun.*, **67**, 263.

Reid, M.J., Kwasnicki, J.M., Moss, R.B., and Cheung, N-K.V. (1985). Underestimation of specific immunoglobulin E by microtitre plate enzyme-linked immunosorbent assays. *J. All. Clin. Immun.*, **76**, 172.

Rotman, J.P., and Delwel, H.R. (1983). Cross-linking of *Schistosoma mansoni* antigens and their covalent binding on the surface of polystyrene microtitre plates for use in the ELISA. *J. Imm. Methods*, **57**, 87.

Rotman, J.P., and Scheven, B.A.A (1984). The effect of antigen cross-linking on the sensitivity of the enzyme-linked immunosorbent assay. *J. Imm. Methods*, **70**, 53.

Rubin, R.L., Hardtke, M.A., and Carr, R.I. (1980). The effect of high antigen density on solid-phase radioimmunoassays for antibody regardless of immunoglobulin class. *J. Imm. Methods*, **33**, 287.

Salonen, E.M., and Vaheri, A. (1979). Immobilization of viral and mycoplasmal antigens and of immunoglobulins on polystyrene surface for immunoassays. *J. Imm. Methods*, **30**, 209.

Saltvedt, E., Harboe, M., Følling, I., and Olnes, S. (1974). Interactions between ricinus agglutinin and human IgM and IgG. *Scand. J. Immunol.*, **4**, 287.

Standefer, J.C. (1985). Separation-required (heterogeneous) enzyme immunoassay for haptens and antigens. In: *Enzyme-Mediated Immunoassay* (eds T.T. Ngo and H.M. Lenhoff). Plenum Press, New York, pp. 203–22.

Standefer, J.C., and Saunders, G.C. (1978). Enzyme immunoassay for gentamicin. *Clin. Chem.*, **24**, 1903.

Suter, M. (1982). A modified ELISA techniques for anti-hapten antibodies. *J. Imm. Methods*, **53**, 103.

Tanimori, H., Ishikawa, F., and Kitagawa, T (1983). A sandwich enzyme immunoassay of rabbit immunoglobulin G with an enzyme labelling method and a new solid support. *J. Imm. Methods*, **62**, 123.

Urbanek, R., Kemeny, D.M., and Samuel, D. (1985). Use of enzyme-linked immunosorbent assay for measurement of allergen-specific antibodies. *J. Imm. Methods*, **79**, 123.

Voller, A., Bidwell, D.E., and Bartlett, A. (1979). *The Enzyme-linked Immunosorbent Assay (ELISA)*. Dynatech Europe, UK.

Voller, A., Bidwell, D.E., Huldt, G., and Engvall, E. (1974). A microplate method of enzyme-linked immunosorbent assay and its application to malaria. *Bull. Wld. Hlth. Org.*, **51**, 209.

Walsh, B.J., Wrigley, C.W., and Baldo, B.A. (1984). Simultaneous detection of IgE binding to several allegens using a nicrocellulose 'poly disc'. *J. Imm. Methods*, **66**, 99.

Wide, L., Bennich, H., and Johansson, S.G.O. (1967). Diagnosis of allergy by an *in vitro* test for allergen antibodies. Lancet, **ii**, 1105.

Wide, L., and Porath, J. (1966). Radioimmunoassay of proteins with the use of Sephadex-coupled antibodies. *Biochem. Biophys. Acta*, **130**, 257.

Wreghitt, T.G. and Nagington, J. (1983). Variability in EIA plates used in ELISA tests for detection of antibodies to *Legionella pneumophila* [letter]. *J. Clin. Pathol.*, **36**, 238.

Yalow, R.S., and Berson, S.A. (1960). Immunoassay of endogenous plasma insulin in man. *J. Clin. Invest.*, **39**, 1157.

Zeiss, C.R., Pruzanksy, J.J., Patterson, R., and Roberts, M. (1973). A solid-phase radioimmunoassay for quantitation of human reaginic antibody against ragweed antigen E. *J. Immunol.*, **110**, 414.

ELISA and Other Solid Phase Immunoassays
Edited by D.M. Kemeny and S.J. Challacombe

CHAPTER 3

Quantitative Aspects of Solid Phase Immunoassays

Robert G. Hamilton and **N. Franklin Adkinson, Jr**
University of Texas School of Medicine and Johns Hopkins University School of Medicine

CONTENTS

INTRODUCTION

Since its introduction in the late 1950s by Berson and Yalow (1959) and Ekins (1960), the immunoassay methodology has revolutionized the manner in which analytes are detected and quantitated in the blood of human beings and animals. Immunoassays are generally performed for one of two purposes: either to estimate the amount of a substance (analyte) in an unknown or test specimen or to compare two dissimilar substances in a particular assay system in cross-reactivity or specificity studies. Most immunoassays in clinical and research laboratories are performed for the former purpose, namely measurement of the amount or magnitude of change in the concentration of an analyte in a test specimen.

In this chapter, we examine issues relating to the quantitative aspects of immunoassays in general and microtitre-plate-based ELISAs in particular. The preparation of reference materials (standards or calibrators) will be discussed with an emphasis on their use in homologous and heterologous interpolation schemes for the quantitation of analyte in a test specimen. Semi-quantitative immunoassays that produce a positive/negative result will be contrasted with standardization approaches which generate 'quantitative' results in mass or biological activity per volume units. Finally, a number of immunoassay performance parameters will be examined which influence the accuracy of the final interpolated quantitative measurement.

REFERENCE MATERIALS

Whether an immunoassay is designed to be semi- or fully quantitative, all values derived from the assay are based on a reference material, hereafter called a standard. A standard is a substance that is established by authority as a rule for the measure of quantity, weight, extent, value, or quality of that substance in a test specimen. It is usually chemically identical to the substance to be assayed and it is typically analysed in multiple dilutions. When achievable, well-defined compounds are employed as standards that have the highest obtainable purity and good stability* so that reproducible*, accurate* results can be achieved over time. Three levels of standards have been defined (Table 1) that vary in their degree of documentation and availability (Reimer, 1983).

The international or primary standard is material that has been collected, tested, and aliquoted under the guidance of the World Health Organization (WHO) International Laboratory for Biological Standards. It has been extensively tested with regard to its potency*, stability, and safety using widely accepted (reference) assays for the analyte*. A unit of activity or quantity is assigned to each preparation after extensive collaborative evaluation by multiple well-established reference laboratories. WHO standards are therefore regarded as the most reliable reference preparations available. The major difficulty with internationally prepared standards is the time lag involved in acquisition of the source material and subsequent collection of data required in their documentation.

WHO primary standards are not available in sufficient quantitity for use in routine assays. They are therefore used to cross-standardize secondary standards which are specimens used more routinely due to their wider availability. Secondary standards have not been as extensively tested as the WHO primary standards but they do have a defined potency and purity based on replicate measurements performed by the producer. They are distributed by such institutions as the National Institutes of Health (NIH) in Bethesda, the Centers

* Those terms asterisked in this chapter are explained in 'Notes' on page 82.

TABLE 1 Levels of standards

Standard, reference material, and calibrator	A substance that is established by authority as a rule of quantity, weight, extent, value, or quality for an analyte in a test specimen
Primary standard:	Material collected, tested, and aliquoted under guidance of an internationally recognized body (e.g. WHO); extensively tested with regard to potency, stability of analyte and overall safety; regarded as most reliable reference preparation
Secondary standard:	Cross-standardized to a primary standard; potency is based on replicate measurements by producer (NIH, CDC, NIMR).* Purity assessed by cross-reactivity or specificity studies. More plentiful than primary standards
Tertiary or working standard:	Typically a pool of specimens, initially selected from routine screening immunoassays, based on their high level of analyte. The serum pool is cross-standardized by analyzing 5–10 dilutions of both the working and primary or secondary standards in 5 to 10 independent immunoassays. Re-standardization is needed every 3–6 months
Characterized in-house antibody standards:	Referred to in text for preparation of antigen-specific antibody standards which are not available from international or national sources Characterized in-house antibody standards can be prepared by one of three methods: 1. Scatchard analyses for pure antigen systems 2. Elution techniques for complex antigen systems 3. Depletion analysis when specific antibody is a large percentage of the total antibody content

* NIH = National Institute of Health, Bethesda, MD, USA; CDC = Center for Disease Control, Atlanta, GA, USA; NIMR = National Institute for Medical Research, Mill Hill, UK

for Disease Control (CDC) in Atlanta, and the Division of Biological Standards at the National Institute of Medical Research (NIMR) in London. Several private groups, such as the College of American Pathologists, also provide standard materials in their quality control programme. These preparations are widely used and they provide an invaluable tool for calibration because there is generally extensive data on their quality and potency that allows the establishment of normal and abnormal ranges.

The third level of reference preparations are those prepared in-house called 'working standards'. These preparations are selected from screening immunoassays as having a high quantity of the test analyte (drug, hormone, specific antibody) and thus they are acquired without any prior reliable potency estimates. In many respects, tertiary standards are the most important because they constitute the basis for the accuracy of the routine assay. Each investigator is responsible for defining the potency of his working standards. This is generally done by analyzing 5–10 dilutions of the working standard and a primary or secondary standard in 5–10 independently performed immunoassays and calculating a mean potency estimate with a confidence interval or precision profile error envelope (Ekins, 1981). Potency of the working or tertiary standards should be re-evaluated at least every 3–6 months by direct comparison with a primary or secondary standard and when new lots of assays reagents are employed (e.g. conjugate, solid phase antigen or antibody). Small differences in reagent quality (e.g. antibody specificity) can alter the apparent potency of a working standard.

PREPARATION OF CHARACTERIZED STANDARDS

There are many cases where primary and secondary standards are not available for a given analyte from international or national sources. This is especially true of antibody standards which are particularly important to the ELISA laboratory for setting positive thresholds or creating standard curves from which quantitative measurements of human and animal antibody may be performed. Thus, the relatively simple process of cross-standardizing an in-house standard for routine use becomes an impossible task. The investigator must decide to prepare his own characterized in-house standard (Table 1).

The time and expense involved in the characterization of an in-house standard depends on the availability, degree of purity required, and the intrinsic heterogeneity of the analyte of interest. Drugs and hormones are often available in pure form from chemical houses or foundations (National Pituitary Foundation, Baltimore, MD) at a modest cost. For these, it is a matter of adding weighed amounts of analyte to a protein matrix to prepare a standard. Analytes that are inherently heterogeneous in spite of a similar molecular weight and/or biological activity are difficult to obtain from commercial sources. Heterogeneous analytes occur naturally due to: (a) genetic variation within a given species (isohormones, isoenzymes); (b) precursor and metabolic forms (e.g. proinsulin, insulin); (c) breakdown of homogeneous analytes in biological specimens due to enzymatic degradation, handling, and storage; and (d) intrinsic heterogeneity as is the case with glycoproteins and immunoglobulins. Preparation of standards from the first three groups is extensively discussed by Bangham (1983). The latter group of heterogeneous analytes includes antigen-specific human and animal antibodies that are of special interest to laboratories that employ microtitre-plate-based ELISA techniques. For this reason, preparation of antigen-specific immunoglobulin standards will be the major focus of discussion below.

Three general methods which have been reported for the preparation of antigen-specific antibody standards will be discussed: Scatchard/saturation analysis; antibody elution techniques; and depletion analysis. Each method

TABLE 2 Antibody content by saturation and saturation analysis

(1) Assumptions and given data

The data used in this example have been extracted from Figure 1. It is assumed at saturation that 2 moles of Ag are bound to each mole of antibody

Given: MW of IgG antibody = 150 000 grams/mole
MW of insulin = 5734 grams/mole

(2) Antibody content computation from saturation data (Figure 1, right panel)

1.6 ng/ml of total insulin bound was bound at the saturation plateau

$(1.6 \times 10^{-6}$ grams/litre$)$ / 5734 grams/mole $= 2.7 \times 10^{-10}$ mole/litre = total insulin bound at saturation

2.7×10^{-10} mole/litre / 2 = 1.35×10^{-10} mole/litre of antibody present

(3) Antibody content computation from Scatchard data (Figure 1, left panel)

Scatchard analysis equation: $B / F = K [Ab_o - B] = (-K \times B) + (K \times Ab_o)$

B/F = ratio of bound to free antigen at equilibrium
Ab_o = total antibody binding sites
K = equilibrium or association constant
B = bound antigen (or twice antibody using an $Ab\text{-}Ag_2$ ratio) at equilibrium

Using the Scatchard plot (B/F v. bound antigen)
Slope = $-K$ *(the negative of the affinity constant in litres*/mole)
Y intercept = $K \times Ab_o$ (affinity constant × total antibody binding sites)

From Figure 1 (left panel: Y intercept = 1.1 and $K = 4.1 \times 10^9$ litre/mole)

$B/F = K \times Ab_o$ or $1.1 = 4.1 \times 10^9$ litre/mole $\times Ab_o$ (in mole/litre)

$(1.1 / 4.1 \times 10^9$ litre/mole$) / 2 = 1.35 \times 10^{-10}$ mole/litre of antibody

(4) Conversion of molar concentration of antibody into weight/volume units

$(1.35 \times 10^{-10}$ mole/litre IgG$) \times 150\,000$ grams/mole = 20 ng/ml of IgG specific for antigen in the final reaction mixture

If the final reaction mixture volume is 0.3 ml and 0.1 ml of antibody reagent was used, then there is 20 ng/ml × 3 or 60 ng of specific antibody per ml in the starting antibody solution

will be discussed within the context of its appropriate application to the standardization of specific antibody standards when the antigen of interest is a single purified protein or a complex protein mixture.

Scatchard or saturation analysis

Saturation analysis was initially reported by Scatchard (1949) for analysis of the attraction of proteins for small molecules. More recently, it has been successfully applied to the quantitation of the amount of antibody in serum pools that is specific for a purified antigen using the radioimmunoprecipitation (RIP) assay. Two exemplary human antibody standards that have been prepared using this method include human IgG anti-bovine insulin in the serum of diabetics (Hamilton and Adkinson, 1980) and human IgG anti-phospholipase A_2 (Adkinson, Sobotka, and Lichtenstein, 1979) in the serum of Hymenoptera-venom-allergic patients where phospholipase A_2 is the major allergen in honey bee venom. Saturation analysis, in general, relates the amount of bound labelled antigen at saturation to the number of antibody binding sites and ultimately to the mass of antibody (Table 2). A more detailed review of the calculations and assumptions involved in saturation analysis is presented elsewhere (Thorell and Larson, 1978).

The RIP assay is a liquid phase non-competitive double antibody immunoassay. A fixed concentration of serum is evaluated with increasing concentrations of radioiodinated antigen of a known specific activity (disintegrations per minute (dpm)/ng, until saturation of the antibody binding sites is achieved. A second antibody is then added to precipitate the antibody class (e.g. IgG, IgM) being studied. Bound and free ^{125}I antigen is determined and analysed according to the method of Scatchard (1949). In addition, the maximum ^{125}I antigen bound by a given dilution of antiserum are determined and used to calculate antibody content directly. Both methods require assumptions of the achievement of equilibrium and an antigen–antibody ratio of 2 in extreme antigen excess. A schematic representation of the two forms of data presentation (Scatchard and saturation plots) is presented in Figure 1. Both lead to the same estimates for the antibody affinity constant and amount of antibody binding sites. More often the Scatchard plots are not linear as indicated in Figure 2 which indicates a heterogeneous population of antibodies with differing affinity constants.

In addition to the assumption of equilibrium and an antigen–antibody ratio of 2, the antigen must be homogeneously labelled or differential labelling will lead to erroneous estimates of antibody content. While preparation of homogeneously labelled antigen is readily accomplished with single pure antigens, differential radioiodination typically occurs with mixtures of protein antigens. In Figure 3, the differential radioiodination is depicted as varying numbers of white circles within each antigen symbol. This condition leads to dilution curves which are non-parallel and saturation plateaux at different heights (Figure 4). Both conditions will lead to variable and erroneous antibody estimates by saturation analysis. The problem of differential labelling of the antigen experienced in the RIP can be overcome by employing a solid phase antigen elution technique to estimate antibody which eliminates the need for labelling antigen.

Solid phase elution techniques

Solid phase elution techniques have been reported for quantitation of antigen-

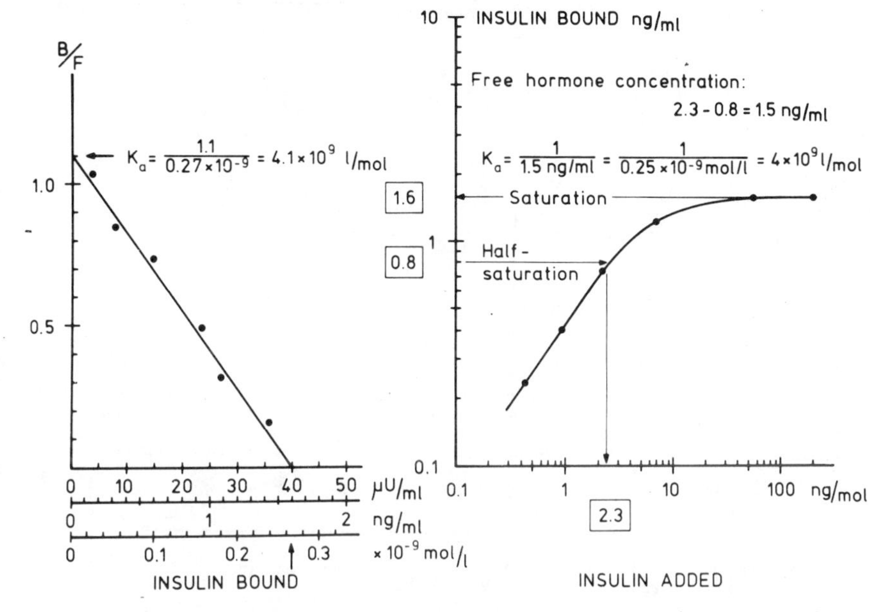

FIGURE 1 Schematic of the Scatchard plot (left) and saturation plot (right) for determination of the relative antibody affinity and quantity of specific antibody (number of antigen-binding sites). The slope of the Scatchard plot represents the affinity constant of the antibody population. Inverse of the concentration of unbound antigen at half-saturation of antibody in the plot on the right permits computation of the affinity constant. See Table 2 for computation of the number of antibody receptors. (Reproduced with permission from Thorell and Larson. 1978)

specific IgE and IgG antibodies specific for a complex mixture of proteins (Schellenberg and Adkinson, 1975; Hamilton and Adkinson, 1980, 1981). A schematic diagram for the IgG protein A elution technique is presented in Figure 5. In these methods antigen insolubilized on agarose is used in sufficient quantitity to remove >90 per cent of specific antibody from an aliquot of human serum. From 30 to 60 per cent of the antibody is then eluted from the washed antibody–antigen sorbent complex by means of a pH change. A total immunoglobulin content of the eluate is determined by a sensitive solid phase RIA. The partially eluted antigen–sorbent together with a sham control is then washed and incubated further with the radiolabelled detection protein in order to calculate the elution efficiency. With these data, it is then possible to calculate specific immunoglobulin content in weight per volume units.

In complex antigen mixtures, the elution methods offer an opportunity to quantitate specific antibody without the need for radioiodination of antigenic mixtures. The method is labour intensive, requires a very sensitive total IgG or IgE assay, and is subject to error at multiple stages in the procedure. Replicate

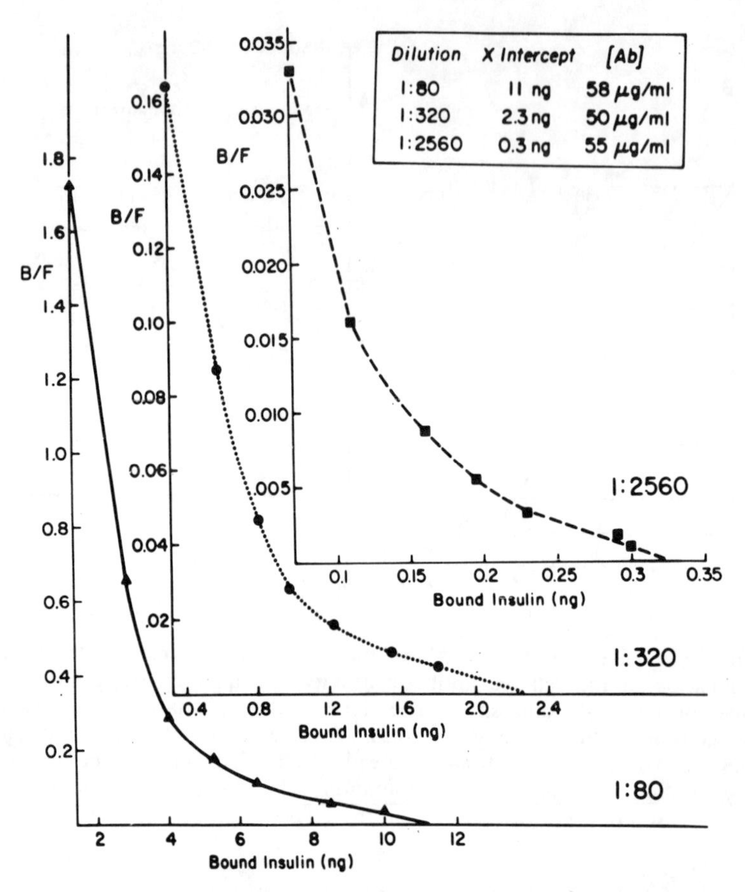

FIGURE 2 Scatchard analysis of IgG anti-porcine insulin antibody levels in three dilutions of a potent serum from an insulin-resistant diabetic patient. Bound/free ratios are plotted on the ordinate as a function of antibody bound on the abscissa. In this assay, 0.1 ml of ^{125}I-insulin was mixed with 0.1 ml of serum diluted at least to 1:40. Following 4 hr at 25 °C, 0.1 ml of an optimal dilution of goat anti-human IgG was added and the incubation continued overnight. The precipitate formed was washed three times with cold borate buffered saline and pelleted radioactivity was counted. (Reproduced with permission from Hamilton and Adkinson, 1980)

measurements are therefore important to provide reproducible and valid measurements.

Depletion method

There are select cases where one needs to prepare an absolute standard containing a known amount of antigen-specific antibody and the specific antibody is a substantial percentage of the total (>10 per cent). The depletion

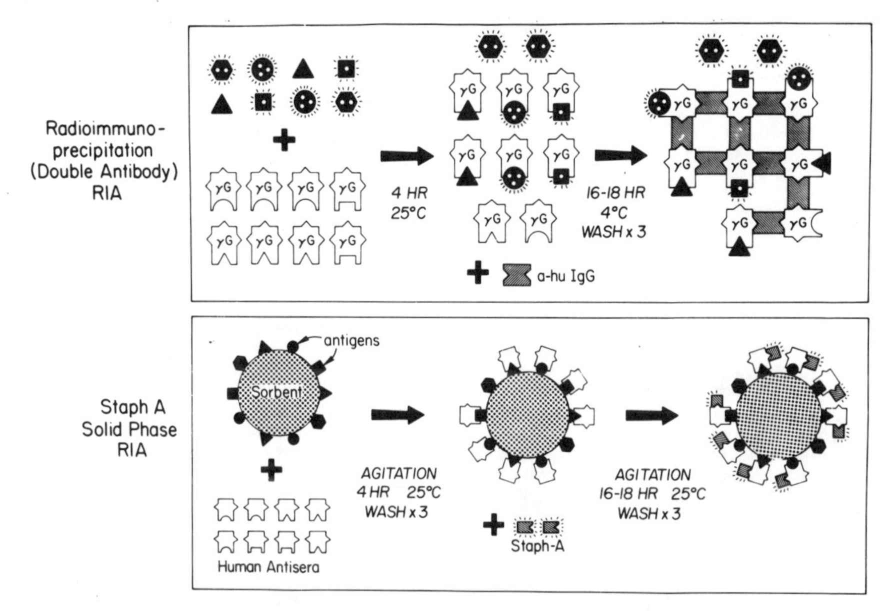

FIGURE 3 Schematic diagram illustrating the principles of quantitation of antigen-specific IgG in complex antigen system using the radioimmunoprecipitation assay. Some potential problems of the RIP are illustrated including (1) differential radio-idionation of protein antigens [four closed symbols with 0 to 3 radioactive atoms (internal white circles)]; and (2) antigen-limiting conditions leading to underdetection of certain antibody specificities. Solid phase elution methods avoid radiodination of antigen mixtures and operate in large antigen excess to avoid these problems

method, which employs the same solid phase antigen preparation described for the elution methods above, may be a simpler alternative approach. Depletion analysis as originally described by Gleich, Jacob, and Yunginger (1977) involved incubation of dilutions of a serum containing ragweed-specific IgE antibody with solid phase cellulose coupled with ragweed or an irrelevant protein as a sham control. After this incubation, the beads were sedimented with centrifugation and the amount of total IgE was determined in both the antigen absorbed and sham control supernatants using a total IgE radio-immunoassay. The difference in the IgE content of the absorbed and un-absorbed (sham control) supernatants was the absolute amount of IgE anti-ragweed in the serum.

The non-specific absorption as assessed by detection of IgE bound to the sham control solid phase is generally negligible due to the low amount of total IgE in human serum. This method requires a sensitive assay for quantitation of total human IgE and a high ratio of specific to total antibody. While this method has been successfully employed to determine the amount of specific IgE in a serum (Gleich, Jacob, and Yunginger, 1977; Hamilton, Rendell and Adkinson, 1980), it has not generally been useful for quantitation of specific

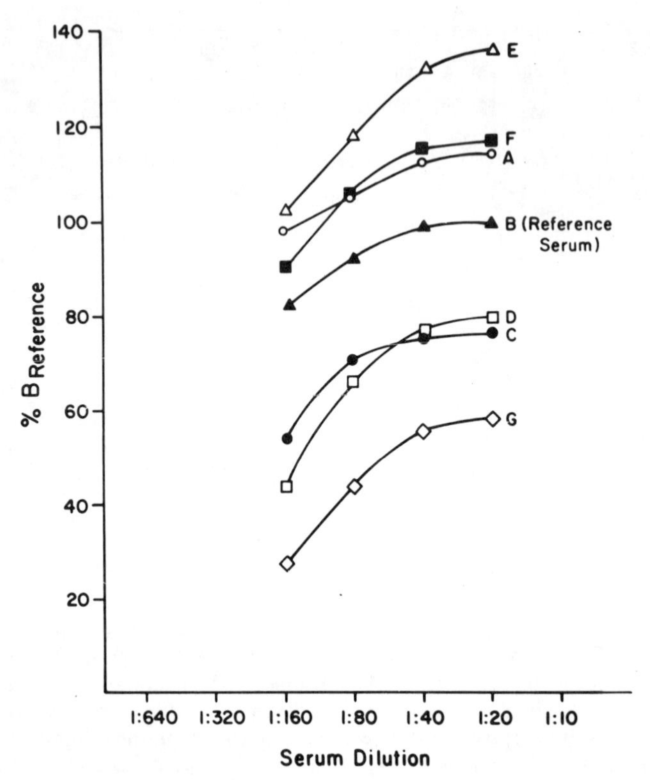

FIGURE 4 Serum dilution curves for IgG anti-yellow-jacket venom in six sera and a reference serum (closed triangles) by the RIP assay. Per cent bound reference is per cent of net counts per minute bound with 1:20 of reference serum (B) taken as 100 per cent. (25 ng ^{125}I-yellow-jacket venom antigen). NSB (normal human serum, negative control) = 161 cpm; 100 per cent B_{ref} = 1817 cpm. (Reproduced with permission from Hamilton and Adkinson, 1981)

IgG in human serum. The reason stems from the fact that specific IgG immune responses tend to account for a small percentage of the total IgG in serum. A small decrease in the total IgG resulting from extraction of small amounts of specific IgG can be masked by inherent statistical variation and in some cases high non-specific binding in total immunoglobulin assays.

PARALLELISM* OF STANDARD AND TEST SPECIMEN DILUTION CURVES

The basic assumption underlying the use of a standard to confer a quantitative quality on an immunoassay is that analytes in the standard and test specimens exhibit equivalent ability to displace the labelled ligand from solid phase

* See 'Notes' on page 82.

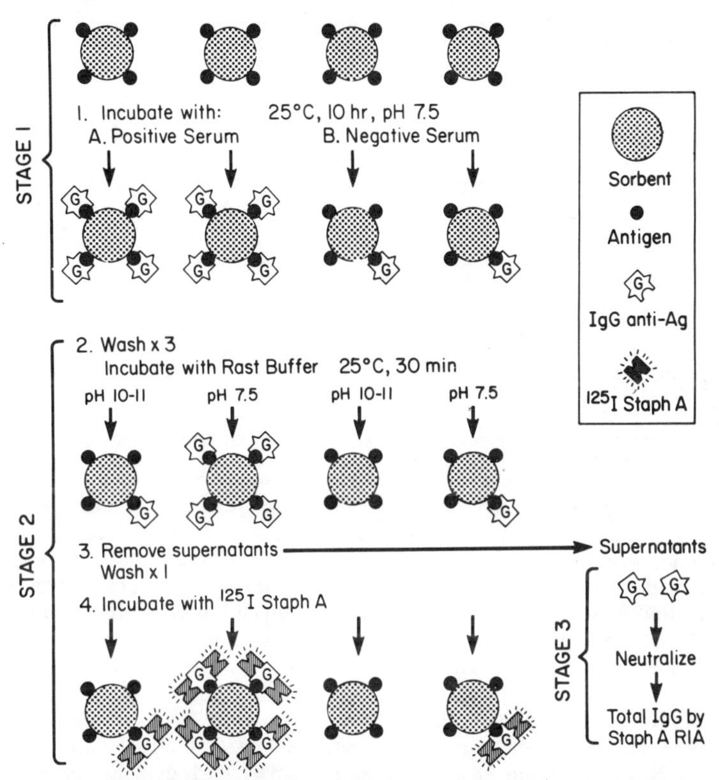

FIGURE 5 Schematic diagram of the IgG–Staph A solid phase elution technique for the quantitation of antigen-specific IgG in human sera. (Reproduced with permission from Hamilton and Adkinson, 1980)

receptor (competitive format) or bind to solid receptor (non-competitive format). Ideally, the analytes of interest in the standard and test samples should be identical in their molecular configuration. Additionally, the diluent (matrix) which is selected for the reference preparation should closely resemble the sample matrix. This can be difficult to achieve for all test samples because the constituents of serum (proteins, lipids, ions) can vary widely between normal individuals. These variations in the matrix give rise to non-specific effects that alter the kinetics of the antibody–antigen reaction and can result in artefacts (non-specific binding, differential binding plateaux; Figure 4). In some cases, standard material is added to a protein base (e.g. extracted or stripped serum), keeping the base or protein matrix constant over the range of the analyte concentrations. Expectedly, serum from patients with pathology will have different matrices depending on whether they have hypo- or hyper-states. However, it is not reasonable to custom tailor a reference dose curve to each test specimen's protein matrix. Thus, serum from a patient who has hypergammaglobulinaemia (elevated immunoglobulin level) which is being

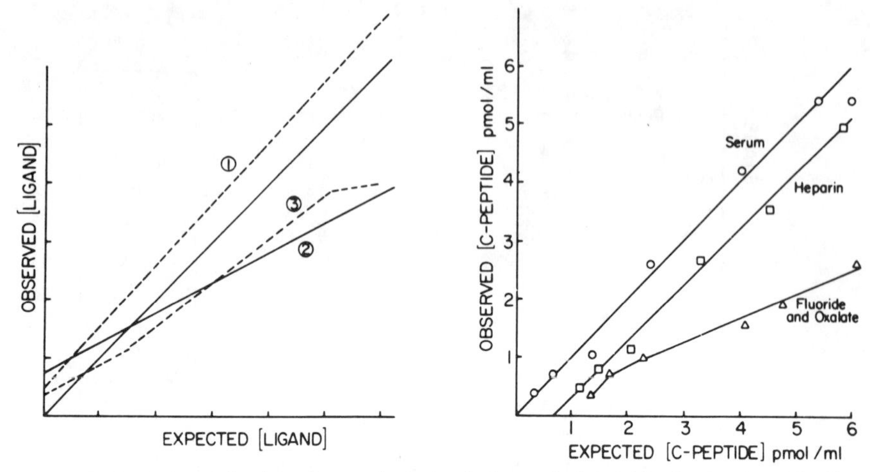

FIGURE 6 Left panel displays theoretical parallelism relationships obtained by plotting the expected analyte (ligand) concentration versus the observed analyte concentration interpolated from the standard dose–response curve. Curve 1 (slope = 1) is parallel to the standard observed versus expected curve that has an intercept = 0 and a slope = 1. Curves 2 and 3 are non-parallel which leads to invalid interpolated results in the test specimen. See text for detailed explantion. Right panel depicts parallelism results from a C-peptide immunoassay. Known amounts of C-peptide were added to serum, heparinized plasma, and fluoride and oxalate plasma. Serum and heparinized plasma matrices produced parallel dilution curves (slopes = 1). Recovery of C-peptide from serum was slightly better than from plasma. Effect of fluoride and oxalate on the C-peptide is quite pronounced and non-parallelism is clearly evident. (Reproduced with permission from Perlstein, Chan and Bill, 1980)

analysed for specific antibody may produce a higher non-specific binding level in the immunoassay than a normal patient's serum. The higher NSB which is not controlled for by the negative serum control may lead to a false-positive result.

Many of these undesirable assay properties result from differences in the composition of the standard and test sample. A parallelism study is a useful tool for identifying and monitoring any changes in the assay performance due to reagents. Ideally, the standard and test specimen dilution curves should be parallel. The practical importance of parallelism in the clinical assay is that if any test sample dilution is analysed, it will generate the same final dose estimate after interpolation from the standard curve following correction for specimen dilution.

When the test sample and reference specimen dose–response curves do not dilute out in parallel, several trends may be observed. Figure 6 (left panel) displays theoretical expected versus observed analyte (ligand) concentration relationships obtained following interpolation from a standard curve. The standard interpolated from its own dose–response curve produces an observed versus expected curve with an intercept of 0, slope of 1. To show parallelism, the slope of the observed versus expected concentration of analyte must be close to 1.0 as is displayed in Curve 1. Curves 2 and 3 are not parallel and assays producing these relationships between standard and test specimen binding will produce variable interpolated results that vary as a function of the sample dilution analysed. Curve 2 can result from the presence of endogenous binders (human antibodies) in the test specimen that cause an apparent increase in the binding site concentration. These effects are apparent over a large portion of the test sample concentration and may or may not be corrected by non-specific binding analysis performed on the test sample.

Curve 3 demonstrates non-parallelism primarily at the extreme low and high concentrations of test sample. At low test sample concentrations, differences in the ionic strength (molarity) between the test sample and standard may alter the extent of the binding reaction. This is usually more pronounced at the low analyte concentrations but it can also occur at higher doses. A flattening of the observed versus expected curve at high doses typically indicates that the undiluted test sample falls outside the useful limits of the standard curve and the computed values will not be valid. A parallelism study demonstrating the effect of anticoagulants on the stability of the analyte in the standard is presented in Figure 6 (right panel). While parallelism is a necessary peformance parameter for a valid assay, it is not sufficient in itself to prove the validity of an assay or to show identity between an analyte in two specimens.

APPLICATION OF STANDARDS

Semi-quantitative screening immunoassays

Screening immunoassays produce semi-quantitative data that can provide valuable results for investigators in both research and clinical laboratories. Questions such as is an autoantibody present, is a woman pregnant, or does the individual have a hepatitis B infection require determination of the presence or absence of autoantibody, beta human chorionic gonadotropin, or viral antigen in blood, respectively. Semi-quantitative assays also provide a useful selection criterion for hybridoma clones and they may serve as an initial assay for identifying sera that can be used in preparing in-house (tertiary) standards.

Screening immunoassays involve the analysis of unknown specimens at several dilutions and direct comparison of resultant assay response levels (CPM bound, optical density (OD), fluorescence signal units (FSU)) with a pre-selected positive threshold level established by positive and negative control specimens analysed in the same assay. In the case of the ELISA, most microtitre plate readers can provide a matrix format that produces a '+' result for results above an investigator-defined response (optical density). In addition to the '+ or −' print format, the magnitude of the response variable (net cpm, net OD, net FSU) can be reported as a ratio or number of standard deviations above the negative control binding levels (Homburger and Jacob, 1982; Hamilton and Adkinson, 1986a). This latter approach, while useful, can be misleading if the height of the net response is interpreted as being equivalent to the dose. This is in fact not the case because the response variable rarely changes in direct proportion to the dose especially as one approaches either the high or low extremes of a sigmoidal dose–response curve.

Quantitative immunoassays

While the semi-quantitative assay can produce useful results for answering some clinical questions, the most desired end result of an immunoassay is a quantitative measurement which indicates how much analyte is present. Quantitative results can be either relative, as in the comparison of amounts of an analyte in several test specimens by dilution analysis, or absolute in weight per volume units using a pre-calibrated standard.

Relative quantitative measurements are best exemplified by the endpoint or 50 per cent maximum binding approaches using dilutional analysis. Endpoint titration involves the analysis of replicate serial dilutions of test samples in an immunoassay and comparison of the reciprocal of the dilution of each that converts from a positive to negative result. Alternatively, the reciprocal of the dilution at 50 per cent maximum binding can be compared. The former method requires clear determination of the dilution that converts from positive to negative while the latter approach requires clear determination of the maximum binding level as indicated by a saturation plateau. Many dilutions of the test sample must be analysed using either criterion and there must be good parallelism between the dilution curves. In dilutional analysis, positive and negative controls provide a definition of the positive threshold level.

Comparison of the titres (reciprocal dilution at endpoint or 50 per cent maximum binding) have been particularly useful in hybridoma laboratories where the potency of monoclonal antibody in ascites or culture medium must be monitored on a routine basis and compared from lot to lot (Hamilton and Rodkey, 1987). Figure 7 displays such a comparison of six monoclonal antibodies specific for human IgG and its subclasses in ascites using dilutional analysis. If the dilutions of the ascites at 50 per cent maximum OD are compared the HP6050 (crosses) displays a potency of 1: 100 000 while HP6017 (closed boxes) has a potency of 1: 400 000 or four fold greater. The dilution curves are parallel which is a requirement in dilutional analysis.

Absolute quantitation in weight or mass per volume or biological units per volume requires a pre-calibrated standard (see above). Two approaches have been employed in translating a response into a dose estimate: homologous interpolation and heterologous interpolation.

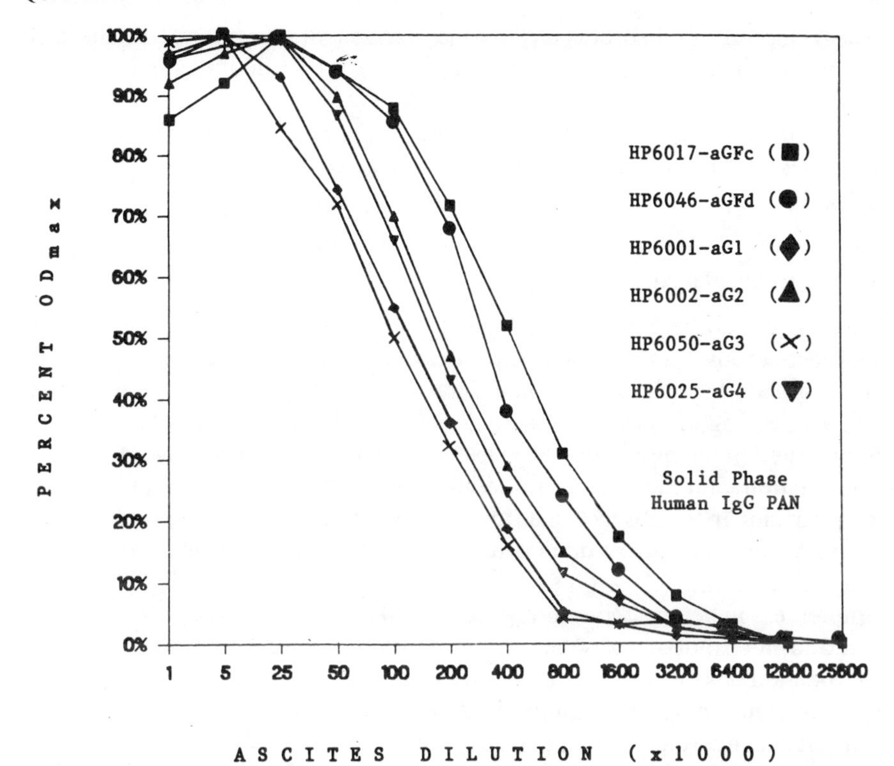

FIGURE 7 Relative potency estimates of monoclonal anti-human IgG antibodies in ascites by ELISA dilutional analysis. Human IgG myeloma subclass 1, 2, 3, 4, or PAN was coated in microtitre wells (10 µg-ml, 2 hr, 37 °C). Following blocking with BSA and washing, multiple dilutions of ascites (1:1,000 to 1:25 600 000) were added in duplicate and allowed to incubate (2 hr, 37 °C). Following a second wash, horseradish-peroxidase-conjugated goat anti-mouse IgG (pre-absorbed against human IgG) was added (2 hr, 37 °C) followed by washing and addition of *o*-phenylenediamine and, 15 min later, 2M H_2SO_4. The percentage of maximum optical density (plateau in monoclonal antibody excess) is plotted against the assayed ascites dilution. Dilution curves were parallel and the 50 per cent OD_{max} dilutions were compared for relative potency determination

Homologous interpolation

Homologous interpolation involves the use of a standard containing the homologous or identical analyte being measured in the test sample. Analysis of multiple dilutions of the homologous standard in replicates allows construction of a dose–response curve. Mathematical algorithms and data transformation are then used to fit a curve to the standard data points. The magnitude of the response in the test sample can then be interpolated from the standard dose–response curve producing a concentration of the analyte in the test sample. When the test sample is analysed in two or three dilutions and the assay system demonstrates parallelism, the interpolated dose from each dilution, corrected for its respective dilution, will give the same final concentration of analyte within reasonable statistical error (e.g. <10 per cent inter-dilutional CV).

Heterologous interpolation

There are cases where it is not convenient or too costly to include a homologous reference serum in an immunoassay for each analyte. It may be more efficient to interpolate test sample response data from a reference curve constructed with a heterologous serum that is similar but not identical to the other analytes being tested in the immunoassay. This process has been termed 'heterologous interpolation' and it can generate weight per volume estimates of the analyte concentration in the absence of an absolute standard. A heterologous standard contains a known amount of analyte similar to the one being measured in the assay (e.g. human IgG antibody of several different specificities). The major requirement in heterologous interpolation is that the dilution curves of the test sera and heterologous standard serum must be parallel to ensure that the interpolated dose is the same no matter what test sample dilution is analysed. This procedure produces results which can be reported in weight per volume unit equivalents (e.g. μg/ml equiv.).

Data from a solid phase radioallergosorbent test (RAST), which measures IgE specific for Hymenoptera venoms (e.g. yellow jacket (YJ) and honeybee (HB) venoms), is presented in Table 3 to demonstrate results obtained from an immunoassay using heterologous interpolation (Hamilton and Adkinson, 1986a). The RAST is configured using a YJ or HB venom protein on a solid phase (allergosorbent) which binds specific antibody from human serum. Radiolabelled anti-human IgE(Fc) is then added to detect bound IgE such that cpm bound are proportional to concentration of the venom-specific IgE in serum. IgE anti-HB venoms are interpolated from a dose–response curve generated with a 'heterologous' serum containing a known amount of IgE anti-YJ venom antibody. Similar measurements can be successfully performed using the ELISA format with antigens coated on microtitre plate wells and enzyme-conjugated anti-human IgE (Fc) as the detection antibody (Kemeny *et al.*, 1985).

The heterologous interpolation method of solid phase immunoassay data processing in Table 3 involves initial subtraction of the homologous venom sorbent non-specific binding level from total test serum binding and interpolations of the test serum's 'net binding' from an anti-yellow-jacket reference curve. The standard curve at the top of Table 3 was constructed using nine serial dilutions of the YJ venom standard serum analysed in each assay.

Multiple dilutions of the positive and test sera (commonly neat and 1: 3) were then analysed on the different venom sorbents. Net binding results interpolated from the reference curve produced an average dose which was then corrected for dilution. The term heterologous interpolation is employed because HB venom RAST binding levels were being interpolated from an IgE anti-YJ venom standard cuvre. The mean and standard deviation of the dilution corrected doses for the two or three dilutions of serum were computed and used to calculate the inter-dilutional percentage CV ($SD_{dose}/mean_{dose} \times 100$) which is an indicator of parallelism. An inter-dilutional percentage of CV > 25 per cent indicates non-parallelism (poor agreement between interpolated doses of different dilutions of the test serum) which in the RAST commonly stems from specific IgG antibody interference.

Thus far, heterologous interpolation has been discussed within the context of antigen-specific IgG and IgE measurements. More recently, monoclonal antibody reagents have become available which permit the dissection of the antigen-specific IgG immune response into its four subclasses (Reimer *et al.*, 1984). While semi-quantitative detection of antigen-specific IgG subclass antibody has been possible with these monoclonal antibodies, it has been difficult to quantitate the absolute amount of each IgG subclass that comprises the total antigen-specific IgG immune response (Hamilton and Adkinson, 1986b). One approach to the standardization of these immunoassays would be the preparation of an antigen-specific IgG standard of a single subclass using affinity chromatography. Once all the IgG from a given subclass has been separated from the other three IgG subclasses in a serum pool, the amount of antigen-specific IgG can be quantitated in absolute terms using one of the methods discussed above. This approach is unfortunately expensive because it requires large amounts of antisera specific to each of the four human IgG subclasses to construct human IgG subclass affinity columns. Thus, this approach to the preparation of absolute IgG subclass standards may be limited to few laboratories that have available sources of monoclonal anti-human IgG antibody. For those that do not possess large amounts of these antisera, quantitation of antigen-specific IgG subclass levels in human serum has been addressed in part by several interesting modifications of the heterologous interpolation schema discusssed above. Each will now be discussed within the context of the importance of parallelism.

Quantitation of human IgG subclass antibodies

Kemeny *et al.* (1987) have developed microtitre-plate-based ELISA assays for several allergen systems, including bovine alpha casein and *Streptococcus mutans* in which IgG_2 antibody responses are significant. First, the binding and disassociation of anti-IgG subclass antibodies from insolubilized IgG was shown to be similar for all IgG subclass-specific monoclonal antibodies tested

TABLE 3 RAST data processing. Heterologous interpolation method

Dose (ng/ml)	Net cpm	% Bmax	SD	SEM	N	CV
0.253	321	0.889	0.021	0.021	3	0.024
0.505	420	1.149	0.035	0.020	3	0.030
1.011	572	1.604	0.015	0.000	3	0.009
2.002	917	2.570	0.115	0.067	3	0.045
4.004	1321	3.703	0.305	0.176	3	0.082
8.088	2168	6.070	0.014	0.000	3	0.002
16.750	3243	9.094	0.294	0.170	3	0.032
32.350	5109	14.326	0.253	0.146	3	0.018
64.700	7295	20.455	1.010	0.583	3	0.049

TOTALS	84510	Bm/T	(%)	42.387	IgE anti-YJV reference curve
Bmax	37222	N /T	(%)	0.188	
NSB	159	N /Bm	(%)	0.444	
T-N	84361	Bm-N		35662	

YELLOW JACKET

Group		Patient ID	Net cpm	N	Precision CV (cpm)	t′	df	Average dose	Precision CV (dose estimate)	Dilution of serum	Corr. dose (undiluted serum)	Inter-dil. CV
1	1	Pos. contr. (YJV)	4352	2	0.023	45.983	2	24.80	0.035	16,000	396.72	
1	2		3002	2	0.129	10.214	1	14.29	0.260	32,000	460.47	
1	3		2100	2	0.020	30.769	4	7.69	0.037	64,000	492.21	
												0.108
2	1		310	2	0.021	5.063	4	0.25	0.000	1,000	0.25	
2	2	A-001 (YJ)	271	2	0.030	4.408	4	0.25	0.000	3,000	0.76	
											0.25*	0.000
3	1		1732	2	0.037	21.998	3	5.63	0.058	1,000	5.63	
3	2	A-002 (YJ)	1058	2	0.063	12.949	3	2.56	0.132	3,000	7.65	
											6.65*	0.218
4	1		760	2	0.064	10.315	4	1.47	0.116	1,000	1.47	
4	2	A-003 (YJ)	483	2	0.004	7.937	4	0.69	0.012	3,000	2.07	
											1.77*	0.241

												1.16*	0.123
HONEYBEE	6	1	NSB-HB	0	3	0.115	00.000	4	0.25	0.000	1,000	0.25	
												0.25*	0.000
	7	1	Pos. contr. (HB)	2941	2	0.015	77.013	1	13.72	0.033	4,000	54.86	
	7	2		1671	2	0.086	14.064	1	5.73	0.137	8,000	42.95	
												48.91*	0.172
	8	1		9855	2	0.005	235.827	1	65.70	0.000	1,000	64.70	
	8	2	A-001 (HB)	7211	2	0.029	46.083	1	63.78	0.000	3,000	191.33	
												191.33*	0.000
	9	1		110	2	0.065	4.524	2	0.25	0.000	1,000	0.25	
	9	2	A-002 (HB)	32	2	0.017	1.776	2	0.25	0.000	3,000	0.76	
												0.25	0.000
	10	1		27	2	0.221*	0.560	1	0.25	0.000	1,000	0.25	
	10	2	A-003 (HB)	38	2	0.031	2.020	2	0.25	0.000	3,000	0.76	
												0.25*	0.000

Group 6 used as blank for groups 7 to 10 261.00 +/− 30.116

* CV 15–25%
** CV > 25%
*** CV > 50%

Reference curve

Totals: total cpm of ^{125}I Ra-a-human IgE added to each tube
Bmax: immunoreactive cpm of ^{125}I added per tube
NSB: non-specific binding to YJ sorbent
Bm/T (%): percent of total counts added that are immunoreactive
N/T (%): percent of total counts added that are non-specifically bound
Dose: ng/ml of IgE anti-YJ antibody in reference serum
net cpm: total cpm minus NSB
%Bmax: net cpm/B_{max} × 100
SD: standard deviation of %B_{max} for the response replicates
SEM: standard error of the mean of %B_{max} for the response replicates
N: number of replicates
CV: CV of the response replicates: SD/mean

Control and test samples: positive specific IgE control serum and patient test serum

Net cpm: total cpm bound minus non-specific bound cpm for that sorbent
N: number of replicate tubes
Precision (cpm) CV: coefficient of variation of the response variable (cpm) bound which is an estimate of intra-assay precision
t′: t prime statistic from the Student's t test
df: degree of freedom (rounded to the nearest number)
Average dose: dose interpolated from reference curve using net cpm
Precision (CV) dose estimate: CV of interpolated dose (ng/ml), estimate of intra-assay precision
Dilution of serum: correction factor for serum dilution analyzed
Corr. dose (undiluted serum): interpolated dose corrected for serum dilution
Inter-dil. CV: inter-dilutional coefficient of variation.

as well as the monoclonal anti-PAN IgG which alone binds all four IgG subclasses. Furthermore, the dilution curves for all single IgG subclasses and the PAN IgG monoclonals were parallel. The subclass antibody results were thus expressed in arbitrary units by interpolation from a reference curve for total IgG antibody constructed using solid phase antigen and labelled anti-IgG PAN monoclonal antibody. The indirect validation of this approach stems from a comparison of the summation of the four subclass values with the total antigen-specific level obtained using anti-IgG PAN conjugate. Where such comparisons could be conducted, correlation coefficients ranged from 0.83 to 0.98. Parallelism is a mandatory requirement for the use of this method.

A second heterologous interpolation approach used in the quantitation of antigen-specific human IgG subclass antibody levels has been termed equipotency (Djurup *et al.*, 1985). This method involves designing each subclass immunoassay so that equivalent concentrations of subclass antibodies produce the same response signal (cpm bound, OD, FSU). IgG myeloma of the four subclasses was individually coupled to nylon balls. The insolubilized subclass proteins were then used to determine dilutions of monoclonal anti-subclass antisera that gave a comparable readout of approximately 14 per cent specific binding in a radioimmunoassay using solid phase myeloma and labelled anti-IgG subclass monoclonal antibodies. With any method for quantitating IgG subclasses, the ratio of the subclass content should be independent of the serum dilution tested. Unfortunately for the equipotency approach, there was a trend towards decreasing ratios of specific IgG subclass 4 to IgG subclass 1 with increasing serum dilution. This indicated that these sera did not give fully parallel results over the range of antibody tested. The IgG_2 and IgG_3 subclass assays have not been tested yet using this method. Despite the failure to achieve parallelism which is required for comparison of the amounts of the different IgG subclass antibodies, the approach remains viable and worthy of further efforts.

A third approach has involved the construction of an IgG subclass standard curve by insolubilizing known concentrations of the appropriate myeloma proteins on a solid phase (Wahlgren *et al.*, 1983). The assay is performed by incubating dilutions of patients' sera on solid phase antigen and detecting bound IgG subclass 1, 2, 3, or 4 antibodies with labelled monoclonal IgG subclass antisera. The net binding results are then interpolated from a calibrated curve prepared by adding the same labelled antibody to varying amounts of insolubilized myeloma protein. The assumption is made that labelled monoclonal human IgG subclass antisera will bind to physically absorbed or covalently coupled myeloma protein with the same kinetic course and affinity as they do to IgG subclass antibodies bound to insolubilized antigen. There is considerable evidence now, as discussed in previous chapters by Kemeny, to suggest that this critical assumption is not achieved, at least for insolubilized antigens. A modification of this method is to capture human IgG myeloma of defined amounts with a monoclonal IgG subclass antibody instead of using absorbed or covalently coupled IgG myeloma proteins (Zollinger and Boleto, 1981). This approach appears to display the IgG antibody in a stereochemically similar manner to that in which an antibody bound to antigen would be viewed by a labelled monoclonal IgG subclass antiserum. This approach would remain feasible even if each subclass diluation curve was not parallel to every other subclass assay, provided that within a given subclass assay, the standard curve and test samples diluted in parallel.

FACTORS THAT INFLUENCE ACCURACY

Thus far, a pre-calibrated reliable standard and the doctrine of parallelism have been discussed as essential to producing an accurate concentration estimate of the analyte in a test specimen using an immunoassay. There are other interrelated factors that have a profound influence on the accuracy of the quantitative measurement. These include the immunoassay performance parameters: precision (intra-assay and inter-assay) and non-specific binding; incubation-time-dependent drift and the strategy of data transformation; curve fitting and interpolation of data from a standard curve.

There is often confusion over the word 'accuracy'* which may be defined as the production of an anlyte measurement that is essentially 'devoid of error' (Ekins, 1981). 'Bias'* and 'precision' are two terms which in combination more clearly define accuracy. These terms reflect the fact that errors in immunoassay measurements are generally of two kinds: systematic errors which are manifested in the form of bias and random errors which are reflected in poor reproducibility or 'imprecision'. Both bias and imprecision are statistical parameters that require analysis on a large number of replicate measurements for their evaluation. The major goal therefore in designing a quantitative assay is to minimize both systematic and random errors.

Systematic error leading to bias is often difficult to perceive since the measurement of bias presupposes that the 'true' value of an analyte is known in a sample. The standard as discussed above is that specimen in which the 'true' value of an analyte is known or defined. By analysing the standard in each assay, the systematic errors leading to bias can be minimized. The assessment and minimization of random errors of measurement which lead to imprecision and the diminished accuracy of a quantitative measurement will now be discussed.

The precision with which any individual specimen is measured within a given assay is governed by the physical charateristics of the specimen (e.g. turbidity

* See 'Notes' on page 82.

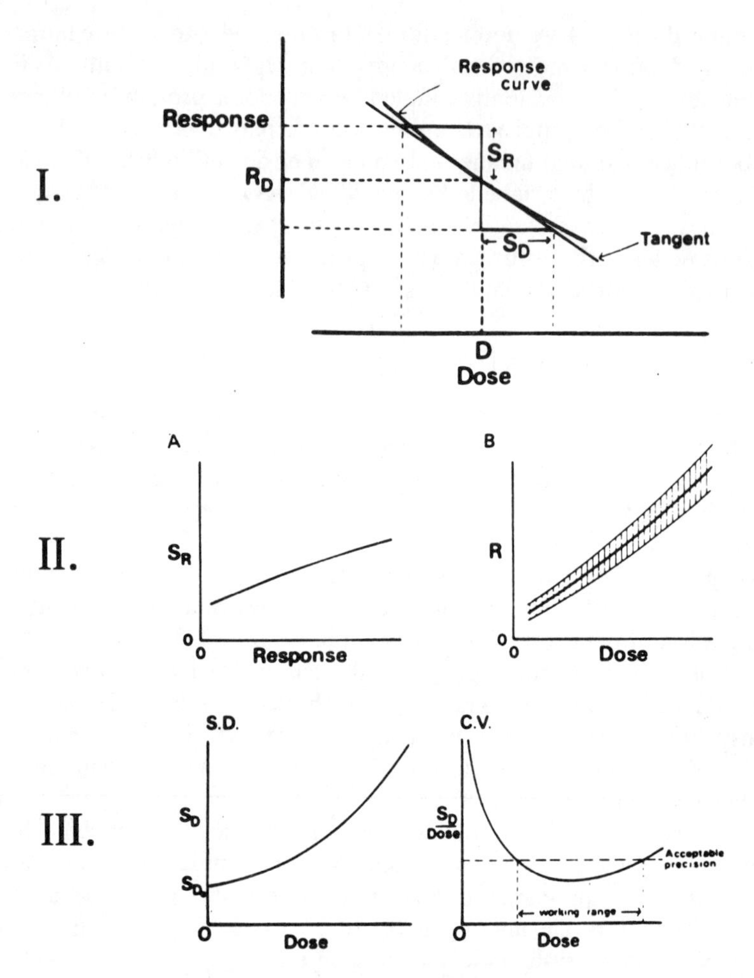

FIGURE 8 Immunoassay precision Panel I. Relationship between the random error of measurement of the response variable (S_R) and consequent random error of measurement of dose variable (S_D). S_D is approximately equal to S_R/slope of the response curve. R_D is the mean value of the observed or measured response and D is the corresponding value of the dose. Slight variations in the slope of the standard (dose–response) curve will result in error in the interpolated dose

Panel IIA: Schematic of a response–error relationship (RER) which is a curve relating S_R (error in the response) to the value of the response variable. In IIB, the error envelope surrounds the dose–response curve. The form and width of the error envelope is determined by the RER shown in IIA

The precision profile is plotted using two sets of parameters in Panel III. IIIA: The standard deviation of the interpolated dose versus dose; IIIB: The coefficient of variation of the dose estimate versus dose. The standard deviation at zero dose $(S_D)_O$ establishes the sensitivity of the assay. The working range of the immunoassay may be defined as the range over which assay results are more precise than a pre-defined accepted level (IIIB). (Extracted from Ekins, 1981) or viscosity), quality of the pipettes and other instruments used, and the skill of the immunoassayist. The error incurred in the measurement of a response (error in OD measurements due to instability of the light source or in counting of radioactivity due to decay probability) depends on the instrumentation. The principal objective of immunoassay design is to monitor all these random errors and minimize their effect on the quality of the final assay result.

Figure 8 (Panel I) schematically portrays a segment of a standard dose–response curve and the relationship between the random error of a measurement response variable (S_R) (e.g. OD) and the resulting random error of measurement of the dose (S_D) at a single dose concentration. The magnitude of the random error, as measured by the standard deviation of the dose estimate, varies as a function of the slope of the response curve and thus it may not be constant over the working range of the assay. The changing random error or non-uniformity of variance in the response over working range of the dose–response curve is shown by a response–error relationship (RER) (Figure 8, Panel II) and it is frequently referred to as heteroscedasticity. It is thus important to monitor the profile of the random error in the interpolated dose over the desired working range of the assay which is done using a 'precision profile'*. The precision profile is a plot of the standard deviation or CV of the interpolated dose as a function of dose (Figure 8, Panel III). By selecting a degree of random error (SD_D/dose or CV threshold), decisions can be made about optimizing the immunoassay. A schematic of expected intra-assay, inter-assay, and inter-laboratory precision profiles for a particular assay is presented in figure 9 (Panel II). The difference between the profiles reflects the added sources of error that are introduced into an assay when it is re-run in the same or different laboratories. The broadening of the acceptable working range of an assay can be made by improving intra-assay precision through analysis of duplicates or quadruplicates instead of singlicates (Figure 9, Panel I).

The non-specific binding (NSB) of an analyte to a solid phase varies as a function of the protein insolubilized (Hamilton and Adkinson, 1986). Homologous blanks run with negative sera for each solid phase antigen ensure the subtraction of the appropriate NSB. If this is not done, non-parallelism and inaccuracies in interpolated dose measurements will occur which lead in some cases to false-positive or -negative results.

If the time of serum incubation with solid phase antigen or antibody is too short ($\leq$ 1 hr), a trend of decreasing interpolated results can occur when the same dilution of a specimen is pipetted from the top and bottom of a microtitre plate (Figure 10). This phenomenon is referred to as time-dependent drift and it diminishes the quantitative nature of a plate-based immunoassay. When such a trend is observed, it can be minimized by lengthening the time of incubation of the serum with solid phase ligand.

* See 'Notes' on page 82.

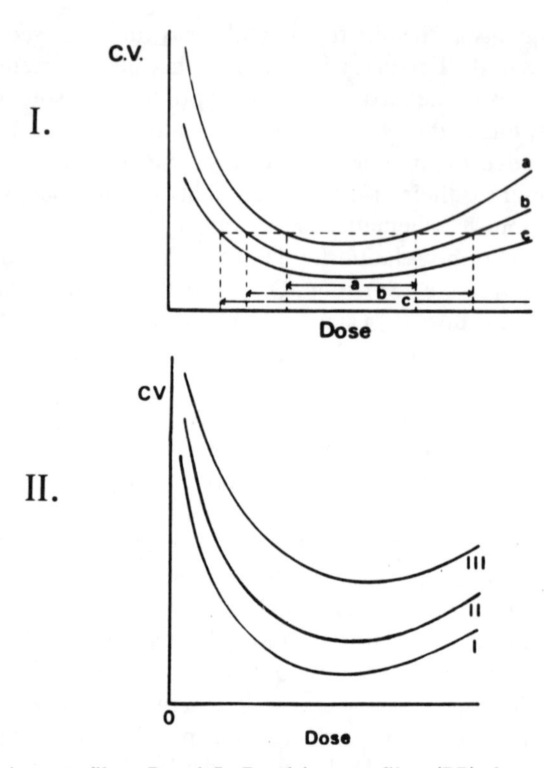

FIGURE 9 Precision profiles. Panel I. Precision profiles (PP) for an assay where replicate measurements of samples were performed in singlicate (a), duplicate (b), and quadruplicate (c). The mean precision (standard deviation of mean dose estimate/dose estimate or CV_{dose} on Y axis versus dose on X axis). The precision improves with the use of replicate assay tubes and the consequence is an extension of the assay working range

Panel II. Precision profiles of the variation of dose (Y axis: CV_{dose}) versus dose (X axis) for replicate measurements performed within one assay (i:intra-assay PP), between assays (ii: inter-assay PP), and between laboratory measurements (iii: inter-laboratory PP). The separation between the profiles reflects the added sources of error which are introduced when an assay is re-run in the same or in different laboratories. (Extracted from Ekins, 1981)

Finally, the data transformation, curve fitting, and interpolation methods all influence the response–error relationship and error envelope of an immunoassay. Identical data presented in terms of three different response variables and transformations will produce different error envelopes depending upon the combination of the algorithm used in processing and interpolation of test data from a dose–response curve. With the ready availability of personal computers, any of a wide selection of four and five parameter logistic, spline, quadratic, and point-to-point fitting routines for standard curve preparation can be readily employed (Hamilton, 1985; Cernosek, 1985). Each has its characteristic error envelopes (Ekins, 1981).

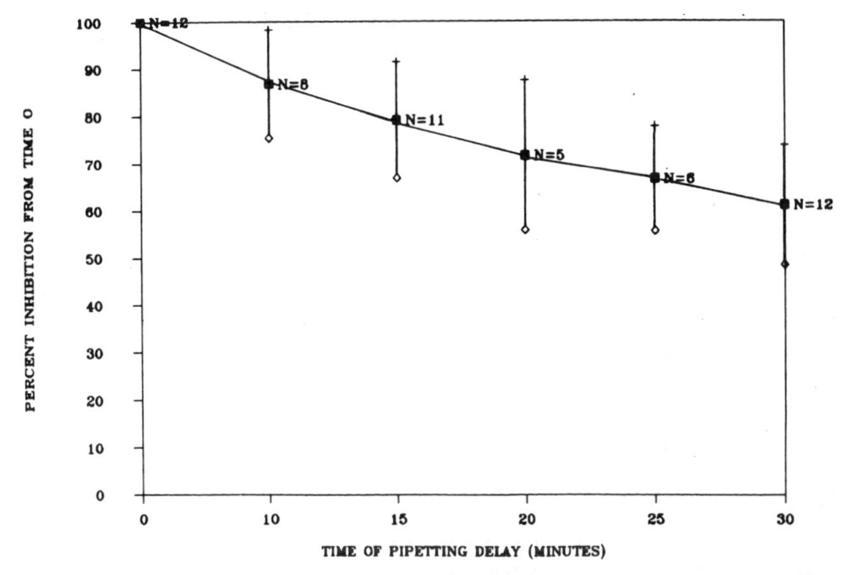

FIGURE 10 Time-dependent drift analysis. Multiple sera were pipetted into a microtiter-plate-based fluorescent allergosorbent test (FAST) immediately following the reference curve (time 0) and then 10, 15, 20, 25, and 30 minutes following the initial pipetting. The mean ± 1 SD per cent reduction (inhibition) in the binding level as compared to time 0 are plotted as a function of time following the first pipetting. N = the number of measurements used to construct each data point. A 20-min delay reduces the binding levels detected in the FAST by 25 per cent. A careful worker will take 20 min to pipette a 96-well microtitre plate

SUMMARY

The standard is that component which confers a quantitative nature upon the immunoassay. Several approaches have been outlined in this chapter for the preparation of characterized standards with emphasis placed on reference preparations containing known amounts of specific antibody. The major assay requirement for the successful use of a standard is parallelism which ensures that the final interpolated test sample result does not vary as a function of the serum dilution analysed. In selected cases, it is more expedient to employ a 'heterologous' standard which is similar but not identical to the test analyte. Multiple forms of the heterologous interpolation scheme are exemplified using the antigen-specific IgG subclass solid phase immunoassays as a model.

Finally, several performance parameters that influence the quality of the interpolated immunoassay results have been discussed. Once optimized and standardized, a solid phase immunoassay can provide unbiased and precise measurements for analytes of interest throughout the many fields of biology and medicine.

NOTES

Accuracy Closeness to the 'true' or real value. It means to yield analyte measurements that are essentially 'devoid of error'. The terms precision and bias may be substituted for accuracy.

Analyte A substance detected or quantitated by an assay system.

Bias The deviation of the expected value of a statistical estimate from the quantity it estimates. It is systematic error introduced into sampling or testing by selecting or encouraging one outcome or answer over others.

Non-competitive solid phase immunoassays Property of the dose–response curves generated by test and standard analytes binding in equivalent manner from insolubilized receptor and producing curves that extend in the same direction, do not meet, and are everywhere equidistant.

Parallelism Competitive solid phase immunoassays Property of the dose–response curves generated by test and standard analytes displacing labelled ligand in equivalent manner from receptor binding sites and producing curves that are extending in the same direction, not meeting and everywhere equidistant.

Potency The amount of an analyte in the sample. It is normally determined by direct comparison with a calibrated standard. It is a synonym for titre which is a measurement of antibody concentration that will bind 50 per cent of added labelled ligand or solid phase ligand.

Precision or reproducibility Degree of agreement of repeated measurements of a quantity within (intra) or between (inter) assays. It is usually expressed as a coefficient of variation (mean / standard deviation × 100) for repeated measurements of the same specimen. The term precision is a synonym.

Precision profile Plot of the precision envelope (mean ±1 SD of the coefficient of variation (CV) of multiple replicates) or the mean precision (mean CV) for multiple dilutions of dose throughout the working range of the immunoassay.

Standard A substance set up and established by authority as a rule for the measure of quantity, weight, extent, value or quality; usually chemically identical to the substance to be assayed; typically added in multiple dilutions to reference tubes in an assay to serve as a criterion for quantitation of contents in samples.

Stability Maintenance of integrity of the analyte over long-term storage.

REFERENCES

Adkinson, N.F. Sobotka, A.K., and Lichtenstein, L.M. (1979). Evaluation of the quantity and affinity of human IgG 'blocking antibodies'. *J. Immunol*, **192**, 965.

Bangham, D.R. (1983). Reference materials and standardization. In: *Principles of Competitive Protein-Binding Assays* (eds W.D. Odell and P. Franchimont). John Wiley & Sons, NY, p. 85.

Berson, S.A., and Yalow, R.S. (1959). Quantitative aspects of the reaction between insulin and insulin binding antibody. *J. Clin. Invest.*, **38**, 1996.

Cernosek, S.F. (1985). Data reduction in radioimmunoassay: Computerized data reduction. *J. Clin. Immunoassay*, **8**, 203.

Djurup, R., Malling, H.J., Sondergaard, I.L., and Weeke, B. (1985). The IgE and IgG

subclass antibody response in patients allergic to yellow jacket venom undergoing different regimens of venom immunotherapy. *J. Allergy Clin. Immunol.*, **76**, 46.
Ekins, R.P. (1960). The estimation of thyroxine in human plasma by an electrophoretic technique. *Clin. Chim. Acta*, **5**, 463.
Ekins, R.P. (1981). The precision profile: its use in immunoassay assessment and design. *J. Clin. Immunoassay* (formally *Ligand Quarterly*) **4**(2), 33.
Gleich, G.J., Jacob, G.L., and Yunginger, J.W. (1977). Measurement of absolute levels of IgE antibodies in patients with ragweed hay fever. *J. Allergy Clin. Immunol.*, **69**, 245.
Hamilton, R.G. (1985). Application of personal computer integrated software to the immunoassay laboratory. *J. Clin. Immunoassay*, **8**, 220.
Hamilton, R.G., and Adkinson, N.F. Jr. (1980). Quantitation of antigen-specific IgG in human serum. I. Standardization by a Staph A solid phase radioummunoassay elution technique. *J. Immunol.*, **124**, 2966.
Hamilton, R.G., and Adkinson, N.F. Jr. (1981). Quantitation of antigen-specific IgG in human serum. II. Comparison of radioimmunoprecipitation and solid phase RIA techniques for the measurement of IgG specific for a complex antigen mixture (yellow jacket venom). *J. All. Clin. Immunol.*, **67**, 14.
Hamilton, R.G. and Adkinson, N.F. Jr. (1986a). Laboratory methods in the quantitation of hymenoptera-venom-specific IgE and IgG human antibodies. *Folia Allergol. Immunol. Clin.*, **33**, 31.
Hamilton, R.G., and Adkinson, N.F., Jr. (1986b). Assessment and management of human allergic disease. In: *Clinics in Laboratory Medicine*, (ed. R. Nakamura). W.B. Saunders Company, Philadelphia, Vol 5, pp. 1–22.
Hamilton, R.G., Rendell, M., and Adkinson, N.F., Jr. (1980). Serological analysis of human IgG and IgE anti-insulin antibodies using solid phase radioummunoassays. *J. Lab. Clin. Med.*, **96**, 1023.
Hamilton, R.G., and Rodkey, S.L. (1987). Quality control of murine monoclonal antibodies using isoelectric focusing affinity immunoblot analysis. *Hybridoma*, **6**, 205
Homburger, H.A., and Jacob, G.L. (1982). Analytic accuracy of specific immunoglobulin E antibody results determined by a blind proficiency survey. *J. All. Clin. Immunol.*, **70**, 474.
Kemeny, D.M., Urbanek, R., Samuel, D., and Richards, D. (1985). Increased sensitivity and specificity of a sandwich ELISA for measurement of IgE antibodies. *J. Imm. Methods* **78**, 212.
Kemeny, D.M., Urbanek R., Richards, D. and Greenall, C. (1987). The subclass of IgG in human immune responses: I. Development of a quantitative enzyme-linked immunoabsorbent assay (ELISA). *J. Imm. Methods,* **96**, 47.
Perlstein, M.T., Chan, D.W., and Bill, M.J. (1980). Parallelism: a useful tool for troubleshooting. *J. Clin. Immunoassay*, **3**, 34.
Reimer, C.B. (1983). In: *Diagnostic Immunology: Technology Assessment and Quality Assurance* (eds. J.H. Rippey and R.M. Nakamura). College of American Pathologists, Skokie, IL., p. 130.
Reimer, C.B., Phillips, D.J., Aloisio, C.H., Moore, D.D., Galland, G.G., Wells, T.W., Black, C.M., and McDougal, J.S. (1984). Evaluation of thirty-one monoclonal antibodies to human IgG epitopes. *Hybridoma*, **3**, 263.
Scatchard, G. (1949). The attraction of proteins for small molecules and ions. *Ann. NY Acad. Sci.*, **51**, 660.
Schellenberg, R.R., and Adkinson, N.F. Jr. (1975). Measurement of absolute amounts of antigen-specific IgE by a radioallergosorbent test (RAST) elution technique. *J. Immunol.*, **115**, 1577.

Thorell, J.I., and Larson, S.M. (1978). Data presentation and quality control. In: Radioimmunoassay and Related Techniques (eds. J.I. Thorell and S.M. Larson). C.V. Mosby Company, St Louis, pp. 29 and 75.

Wahlgren, M., Berzins, K., Perlmann, P., and Persson, M. (1983). Characterization of the humoral response in *Plasmodium falciparum* malaria. II. IgG subclass levels of anti-P. *Falciparum* antibodies in different sera. *Clin. Exp. Immunol.*, **54**, 135.

Zollinger, W., D., and Boleto, J., W. (1981). A general approach to the standardization of solid phase radioimmunoassays for quantitation of class specific antibody. *J. Imm. Methods*, **46**, 129–140.

ELISA and Other Solid Phase Immunoassays
Edited by D.M. Kemeny and S.J. Challacombe

CHAPTER 4
Amplification by Second Enzymes

Axel Johannsson and **David L. Bates**
IQ (Bio) Limited, Cambridge

CONTENTS

INTRODUCTION

The specificity of immunological reactions has long been applied to the detection and measurement of antibodies resulting from infection and to the direct detection of bacterial and viral antigens. Simple methods of limited sensitivity employing visual precipitation or agglutination reactions have usually been adequate for this purpose.

The indirect technique of radioimmunoassay, pioneered in the late 1950s, combines the specificity of antibodies with the sensitivity with which radioisotopes can be detected. Thus assays for a large number of analytes have been devised, particularly in the fields of clinical chemistry, serology, and microbiology. The 1970s have seen the use of immunoassays extended to most fields of molecular analysis with radioisotopes being replaced by alternative labels in order to achieve greater speed, sensitivity and convenience. The use of enzyme

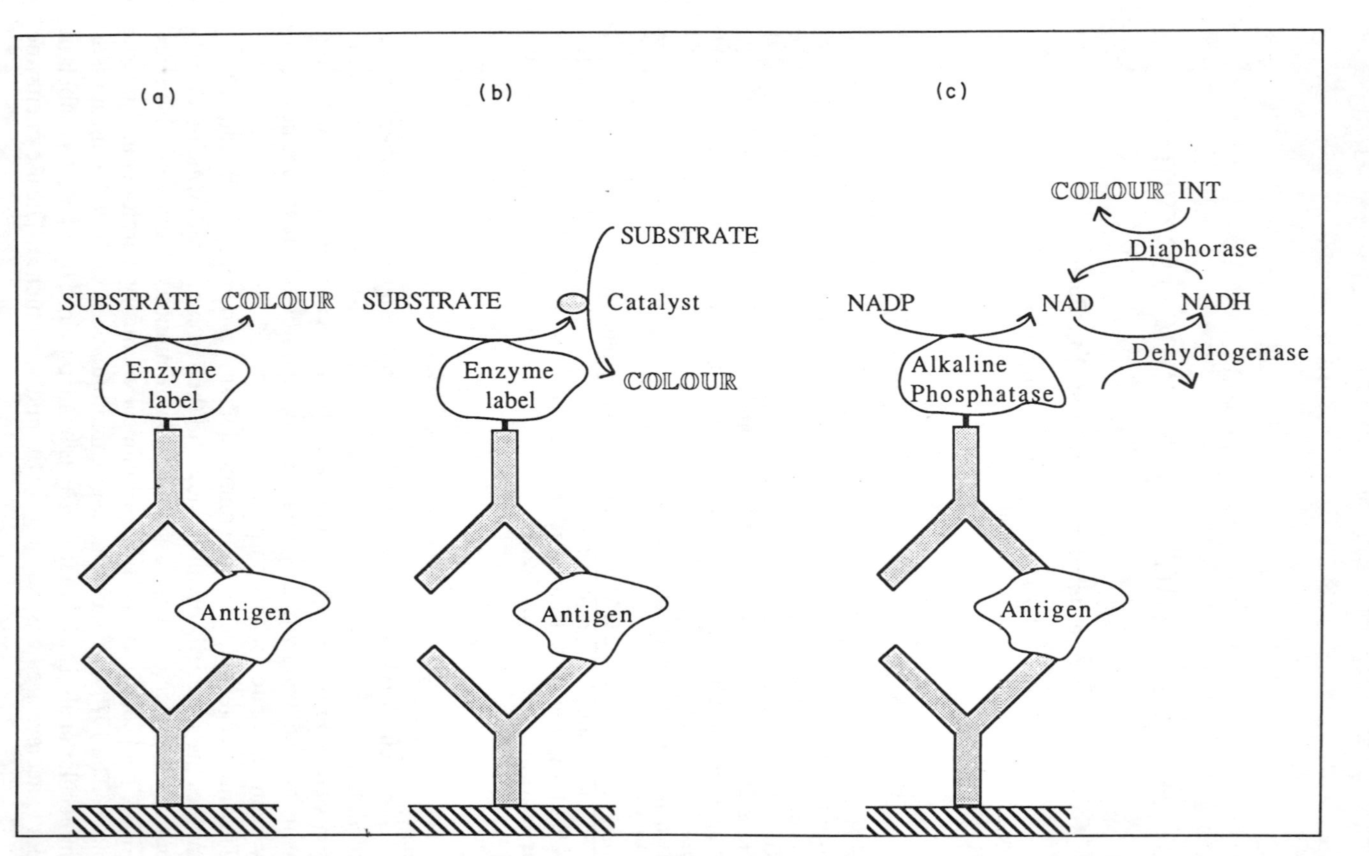

FIGURE 1 Sandwich enzyme immunoassays illustrating different detection methods. (a) Conventional direct colour formation by the enzyme label. (b) Principle of amplified enzyme immunoassay. (c) Enzyme amplification by an NAD-activated redox

labels has eliminated the potential hazards of ionizing radiation and has led to reagents with greater shelf-life. Furthermore, significant increases in assay speed and sensitivity have recently been achieved in enzyme immunoassays with amplification by 'second' enzymes (Stanley, Johannsson and Self, 1985).

The detectability of radioisotopes is in practice limited to about 10 disintegrations per minute (dpm) which for ^{125}I corresponds to about 10^7 molecules. The introduction of enzyme labels has been widely expected to increase further the sensitivity of immunoassays since enzymes are catalysts and can therefore, in theory, be measured with infinite sensitivity; indeed there are reports, although not directly applicable to immunoassays, of the detection of single enzyme molecules (Rotman, 1961). The limited assay speed and sensitivity of conventional enzyme immunoassays has frequently been due to the difficulty in detecting, in a given time, the limited number of coloured product molecules generated by the enzyme label.

A simple principle for increasing the 'detectability' of labels for immunoassay has been proposed by Self (1982): the label is a catalyst which gives rise not to a directly measurable product, but a second catalyst each molecule of which in turn gives rise to many molecules which are detected. Thus by coupling two catalytic processes in series it is possible to achieve amplification of colour formation by several orders of magnitude compared with the one-step colour formation of conventional enzyme immunoassays (Self, 1985; Stanley, Johannsson, and Self, 1985; Johannsson *et al.*, 1986). Figure 1 shows a sandwich immunoassay illustrating: (a) conventional one-step colour formation; (b) the principle of enzyme amplification; (c) the enzyme amplification system described in detail in this chapter.

The signal from bound enzyme can also be increased by multiplying the number of enzyme labels immunochemically, prior to the colour formation as described by Butler, Peterman, and Koertge (1985). See also this volume Chapter 5. It should be emphasized that amplification of the signal from enzyme labels only achieves improved assay sensitivity when other assay conditions are not limiting. For example, the sensitivity of competitive (and inhibition) immunoassays is usually governed by the affinity of the antibodies employed and not by the specific activity of the label (Jackson and Ekins, 1986). However, improving the detectability of the label can often lead to increased assay sensitivity if the immunoreactions are far from equilibrium. Other factors which may improve assay performance include antibody affinity and purity (monoclonal antibodies) and the reduction of non-specific binding by careful optimization of incubation and washing buffers and the use of high quality enzyme–antibody conjugates (Ishikawa *et al.* 1983).

Labels of improved detectability will have the greatest impact in immunometric assays (when labelled antibody is an excess reagent) which use labelled antibodies of high affinity and/or with low non-specific binding, and in very rapid assays.

In this chapter we will outline different forms of enzyme amplification before concentrating on the theoretical aspects of substrate and cofactor cycles as amplification systems. Protocols for the preparation of amplification reagents based on an NAD redox cycle for use with alkaline phosphatase labels will be presented followed by a detailed characterization of this detection system in isolation and applied to an 'ultrasensitive' assay for TSH.

ENZYME AMPLIFICATION

There are a number of textbook examples of enzymes that can be used to generate molecules which catalytically activate another reaction giving rise to a detectable product. The 'second catalyst' may be another enzyme, a modulator of enzyme activity, or a substrate or cofactor taking part in a cyclic sequence of reactions.

The cascade amplification of blood clotting is an example of sequential activation of enzymes ultimately leading to a response (fibrin) that is many orders of magnitude greater than the initial triggering events (activation of Factor VII). Decompartmentalization of enzymes, e.g. their release from liposomes, has been applied successfully to a number of immunoassays. Other examples of amplification involving enzymic activation of a second enzyme include:

(1) formation of chymotrypsin by pepsin (proteolytic activity);
(2) activation of phosphorylase kinase by protein kinase (kinase activity).

Modulators of enzyme activity, e.g. allosteric effectors, are not generally suitable for two-stage amplification for immunoassays because the concentration of effector required for modulation of enzyme activity is relatively high — thus a significant change in its concentration is unlikely to be achieved by the low concentration of enzyme label in an assay.

The product of the reaction catalysed by an enzyme label is in effect a catalyst when it takes part in an irreversible cyclic sequence of reactions where each turn of the cycle produces a stoichiometric amount of second product. The degree of amplification can be defined as the ratio of the amount of the second product to the amount of product generated by the enzyme label. In practice, this definition has little value and direct comparisons will be made later of conventional and amplified detection systems.

A number of potentially useful cyclic reactions are known, some of which are depicted in Figure 2. The fructose-6-phosphate/fructose-bis-phosphate (F6P/FBP) 'futile cycle' (a) plays an important regulatory role in glycolysis. One molecule of glucose feeding into this cycle can give rise to a large number of adenosine diphosphate and phosphate molecules. An enzyme-amplified immunoassays based on this cycle requires an enzyme label capable of generating

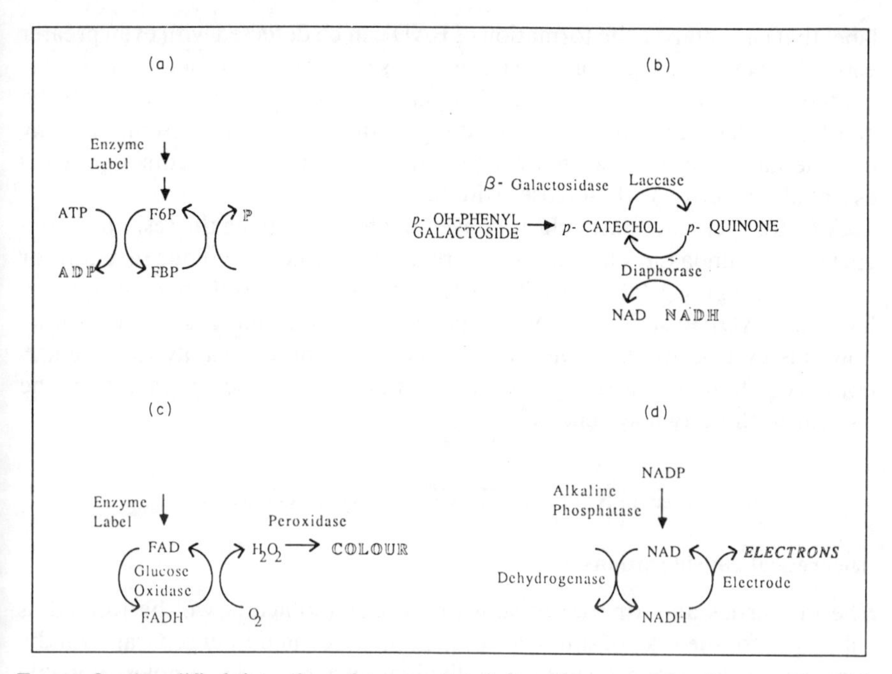

FIGURE 2 Amplified detection of enzyme labels by substrate and cofactor cycles. (a) Enzyme label: e.g. hexokinase; amplification system: phosphoglucose isomerase, phosphofructokinase, fructose bisphosphatase; final product: e.g. inorganic phosphate which can be stoichiometrically converted to a coloured compound. (b) Enzyme label: beta-galactosidase; amplification system: laccase and diaphorase; final product: NAD which can be determined by spectrophotometric measurement of NADH. (c) Enzyme label: an enzyme capable of generating FAD; amplification system: glucose oxidase; final product: hydrogen peroxide which can be stoichiometrically converted to a coloured product via a peroxidase-catalysed reaction. (d) Enzyme label: alkaline phosphatase; amplification system: a dehydrogenase and electrode with suitable voltage applied and electron transfer catalyst; final product: an electric current

(directly or indirectly) F6P (of FBP) and a method for measuring either the build-up of phosphate or adenosine diphosphate (Harper and Orengo, 1981).

The second amplification cycle (b) utlizes the enzymes laccase and diaphorase which interconvert *p*-catechol and *p*-quinone with oxidation of NADH to NAD for each turn of the cycle. The decrease in NADH concentration can be monitored spectrophotometrically at 340 nm. The *p*-catechol is formed from *p*-OH-phenyl galactoside by the action of β-galactosidase, a commonly used enzyme label.

The third cycle (c) is slightly different in that the product of the reaction catalysed by the label takes part in a redox cycle mediated by a single enzyme, glucose oxidase. A variety of sensitive colorimetric means exist to measure the hydrogen peroxide formed. This system can detect low concentrations of FAD and has been used in many immunoassay applications. However, an enzyme label that can catalyse the formation of FAD can be detected with even greater ease. For example, FAD-phosphate, which is inactive as a cofactor for glucose oxidase, may be dephosphorylated by alkaline phosphatase to activate the glucose oxidase and thus lead to catalytic colour formation. Similarly, any enzyme label can be used that can generate FAD from a precursor which is essentially inactive with glucose oxidase.

Cofactor cycles are particularly convenient for amplifying the response from enzyme immunoassays because they are easily coupled to colour reactions or electrodes. Figure 1 (c) shows the colorimetric system based on dephosphorylation of NADP to activate an NAD-specific redox cycle and Figure 2 (d) shows how this cycle can be coupled to an electrode where a catalytic current is obtained. There is an analogy between the rate of colour development and the current in these two systems.

SUBSTRATE AND COFACTOR CYCLES

Theoretical considerations

The properties and kinetics of substrate cycles (cofactors will be treated as substrates for the sake of simplicity) can be analysed in purely general terms by considering the properties of the individual enzymes. The simplest possible scheme for enzyme amplification by substrate cycling is presented in Figure 3.

In this system, the primary label e_o converts a substrate S_o into a product P which is not detected directly, but which can exist in two forms, P_a and P_b. These forms are interconverted by the enzymes of the amplifier cycle e_1 and e_2 which couple the conversion of two further substrates to products P_1 and P_2, one of which must be amenable to a simple detection process such as spectrophotometry. Depending on the nature of the enzymes chosen and the

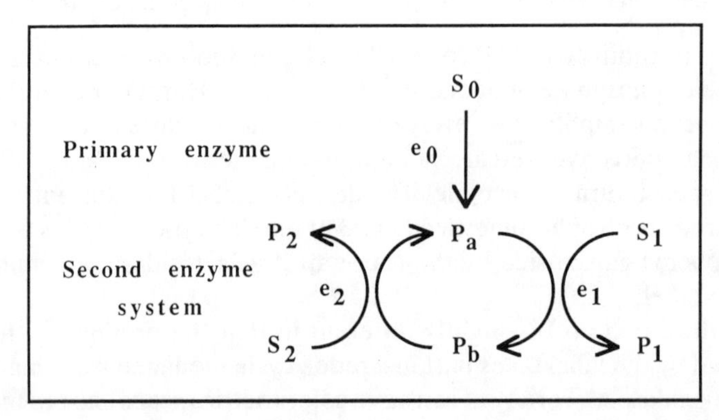

FIGURE 3 An amplification system based on substrate or cofactor cycling

compatability of their associated cofactors, substrates, and optimum working conditions, it may be possible to allow all three enzymes to work simultaneously in a single step or it may be preferable to separate the primary reaction and secondary cycling system into two sequential steps (see the section on Amplified ELISA below).

To achieve maximum response it is necessary for all three substrates S_o, S_1, and S_2 to be provided in high concentration ($> K_m$) to enable the three enzymes to work efficiently. Since e_o is the rate-limiting enzyme, only a small fraction of S_o will be converted into P_a which places strict limitations on the purity of the substrates and the specificity of the enzymes. In particular the enzymes e_1 and e_2 must be able to cycle a small concentration of P in the presence of a much higher concentration of S_o. This means that S_o must not be a substrate for either e_1 or e_2 nor must it compete with P_a or P_b. Clearly, S_o, e_1 and e_2 must not be contaminated with P_a or P_b, and the non-enzymic reaction between S_1 and S_2 must not occur at any significant rate. In practice these considerations severely limit the range of enzyme systems suitable for use as substrate cycle amplifiers.

Kinetics

The kinetics of the above substrate cycle can be derived from a few simple assumptions and it can readily be shown that the amplifier cycle behaves as a simple catalyst which responds linearly to the product of the primary enzyme, which is produced at a constant rate. The following assumptions have been made:

(1) Only initial rates are measured.
(2) Both S_1 and S_2 are present at high concentrations, which are effectively constant.
(3) As a result of (2), the enzymes e_1 and e_2 behave as simple Michaelean enzymes with single substrates.
(4) A steady state is rapidly establised with a constant concentration of cycling intermediates.
(5) The concentrations of both P_a and P_b are low and much less then the apparent K_ms of e_1 and e_2 respectively.

It follows from these assumptions that the velocity equations for both e_1 and e_2 simplify to the following form:

$$v_e = k.e.[S]$$

where $k = k_{cat}/K_m$, $k.e. = V_{max}/K_m$, and K_m is the apparent Michaelis constant for the cycling intermediate in the presence of a fixed concentration of the second substrate under the prevailing conditions.

Since the velocity of each half of the cycle must be equal to the overall cycle velocity (v_c), it follows that:

$$v_1 = k_1.e_1.[P_a] = v_2 = k_2.e_2.[P_b] = v_c \qquad [1]$$

The individual concentrations of P_a and P_b are unknown, but they are simply related to the concentration of P, the product of the primary enzyme:

$$[P_a] + [P_b] = [P]$$

Substituting in equation 1:

$$k_1.e_1.[P_a] = k_2.e_2.([P] - [P_a])$$

Rearranging:

$$[P_a] = \frac{k_2.e_2}{k_1.e_1 + k_2.e_2} \cdot [P]$$

Substituting in equation 1:

$$v_c = \frac{k_1.e_1.k_2.e_2}{k_1.e_1 + k_2.e_2}.[P] \qquad [2]$$

Since $k.e = V_{max}/K_m$, this equation can be rewritten as:

$$v_c = \frac{V_{max1} \cdot V_{max2}}{V_{max1} \cdot K_{m2} + V_{max2} \cdot K_{m1}}.[P] \qquad [3]$$

It may also be shown (to be published elsewhere) that at higher substrate concentrations (P_a and P_b) where assumption (5) no longer applies the cycle velocity is no longer linear with respect to $[P]$, but saturates in a hyperbolic manner analogous to the Michaelis equation. Under these conditions, the cycle velocity is given by:

$$v_c = \frac{V_{max1} \cdot V_{max2} \cdot (K_{m1} + K_{m2})}{([P] + K_{m1} + K_{m2}) \cdot (V_{max1} \cdot K_{m2} + V_{max2} \cdot K_{m1})} \qquad [4]$$

Likewise at very high enzyme concentrations, where the total enzyme concentration approaches K_m, the velocity no longer increases in direct proportion to the enzyme concentration (equation 1) but saturates as the free substrate concentration becomes limiting.

Under the simplified conditions of equation 2, however, the kinetics of the amplifier reduce to a linear relationship between rate and the concentration of the cycling intermediate P:

$$v_c = k_c.e.[P] \qquad [5]$$

It may be shown that for a given total amount of amplifier enzymes ($e_1 + e_2 = e$), the second-order rate constant k_c takes on a maximum value of:

$$k_c = \frac{k_1.k_2}{(\sqrt{k_1} + \sqrt{k_2})^2}$$

when the enzymes are mixed in the ratio:

$$\frac{e_1}{e_2} = \frac{\sqrt{k_2}}{k_1}$$

The essential properties of the amplification cycle are encapsulated in equation 5. This shows that the velocity measured is proportional to the concentration of amplifier enzymes and, at low concentrations of P, the cycle rate is proportional to $[P]$. Although the nature of the signal produced by the enzyme-amplified system (absorbance, fluorescence, etc.) could vary according to the molecule being detected (either P_1 or P_2), the overall kinetics depend on whether the two phases of the process are coupled directly in a one-step system or sequentially in two steps. In the two-step configuration, the kinetics are linear since $[P]$ is constant in the second phase. The rate measured in the second phase reflects both the amount of the primary enzyme (the label) and the length of the first incubation when S_o is converted to P. The total reagent blank will be influenced by the relative contribution of each phase, but the maximum signal will be obtained when each phase uses half the available time.

In a one-step process, where both components in the amplified system work simultaneously, the cycling system is presented with a concentration of P that increases linearly with time. Hence the rate of signal development also increases and the signal kinetics follow a quadratic curve.

This type of kinetics is characteristic of an amplification system where the product of one reaction affects the rate of a subsequent step and, for any complex scheme with two or more phases linked in series, each phase will contribute a time-dependent term to the kinetics.

Preparation of substrate and amplifier reagents

Amplification reagents for use with alkaline-phosphatase-based enzyme immunoassays are commercially available, alternatively suitable reagents can be prepared with enzymes and substrates from a number of commercial sources. The methods that follow can be used to prepare two-step reagents in liquid formulations that will remain stable at 2–8 °C for several hours or much longer at −20 °C. The two-part formulation represents the most convenient deployment of the reagents which are referred to as substrate (NADP in 'substrate diluent') and amplifier (dehydrogenase and diaphorase enzymes in 'amplifier diluent'). Suitable diluent formulations are:

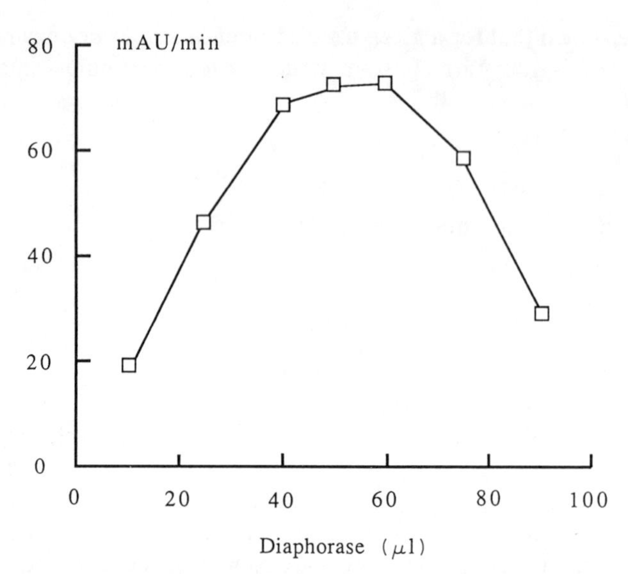

FIGURE 4 Optimization of enzyme ratio. The optimum eznyme ratio in the amplifier was determined by preparing various mixtures of diaphorase (10–90 μl) and alcohol dehydrogenase (90–10 μl) in a total of 100 μl. The mixtures were diluted to 1 ml and assayed with 100 nM NAD as described in the text. The rate of increase in absorbance at 492 nm is plotted against the diaphorase volume

Substrate diluent: 50 mM diethanolamine pH 9.5
1mM $MgCl_2$
4 per cent (v/v) ethanol

Amplifier diluent: 20 mM sodium phosphate pH 7.2
1mM INT-violet

Sodium azide (0.1 per cent) may be added as a preservative to these solutions which are stable for several weeks stored in the dark at 2–8 °C.

Alcohol dehydrogenase (EC 1.1.1.1) from yeast (70 mg) is dissolved in 7 ml 20 mM sodium phosphate buffer, pH 7.2, and dialysed extensively against the same buffer at 4 °C. The stability of the enzyme can be improved by the addition of inert proteins before dialysis; after dialysis it is advisable to filter or centrifuge the enzyme to remove any insoluble material. Diaphorase (NADH: dye oxidoreductase EC 1.6.4.3) (10 mg) is dissolved in 5 ml 50 mM Tris–HCl buffer, pH 8.0. Carrier protein may again be added to stabilize the enzyme which is then extensively dialysed against 20 mM sodium phosphate buffer, pH 7.2, at 4 °C.

After dialysis as described above, the two enzymes are mixed in the appropriate proportions and diluted tenfold in amplifier diluent. The optimium enzyme ratio, allowing for slight variations in individual enzyme activity, can be determined by preparing and assaying several different mixtures as

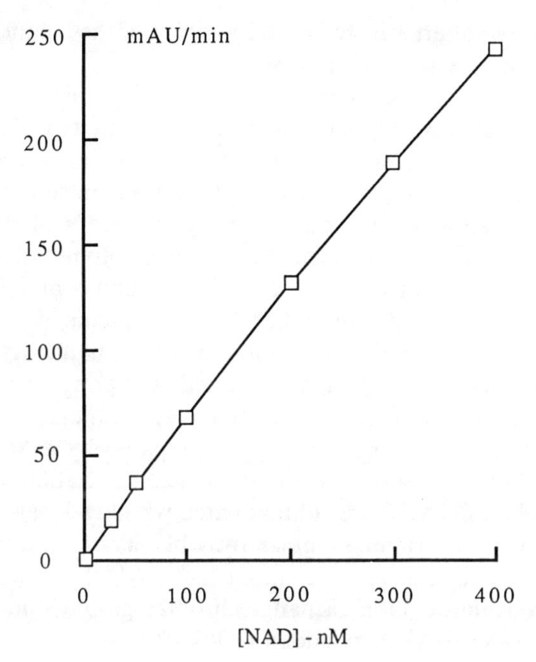

FIGURE 5 Cycling activity as a function of NAD concentration. The activity obtained with two-step amplifier is plotted against the final concentration of NAD in the assay

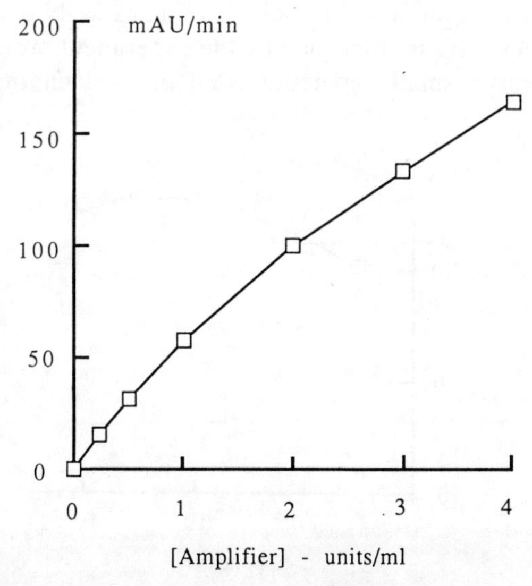

FIGURE 6 Amplifier activity as a function of amplifier enzyme concentration. The rate of increase in absorbance at 492 nm is plotted against the amplifier concentration as defined in the text

shown in Figure 4; alternatively, a 1 : 1 mixture of the solutions prepared as described should give satisfactory results.

For the purposes of determining and standardizing amplifier activity, the most convenient activity to monitor is the cycling activity with a fixed amount of NAD (100 nM). The amplifier activity is simply determined by mixing 2 volumes of amplifier and 1 volume of 300 nM NAD in substrate diluent and measuring the absorbance increase at 492 nm. Assays of this type can be performed in a conventional recording spectrophotometer or a microplate reader with kinetic reading facilities. An arbitrary unit of amplifier activity has been defined as the amount of amplifier that, in a volume of 1 ml, will produce an increase in absorbance of 67 mAU/min in a 1 cm pathlength at 25 °C. The absolute activity of the amplifier may be varied by altering the enzyme dilution, but an activity of 1 U/ml is suitable for most applications.

Working solutions of substrate may be prepared by dissolving the tetra-sodium salt of NADP in substrate diluent to a concentration of 0.1 mM. Even the purest grades of NADP are contaminated with low levels of NAD (< 0.1 per cent) which can contribute significantly to blank rates in amplified assays. If it proves desirable to remove trace amounts of NAD this is most conveniently achieved by conventional ion exchange chromatography on a suitable cation exchanger such as DEAE-Sephadex or Q-Sepharose.

Properties of amplifier

Two-step amplifier acts as a linear detection system for NAD so long as the concentration of the cycling intermediates remains well below the K_ms of the amplifier enzymes. This is confirmed by the experiment presented in Figure 5 which shows only a small departure from linearity in the rate of colour

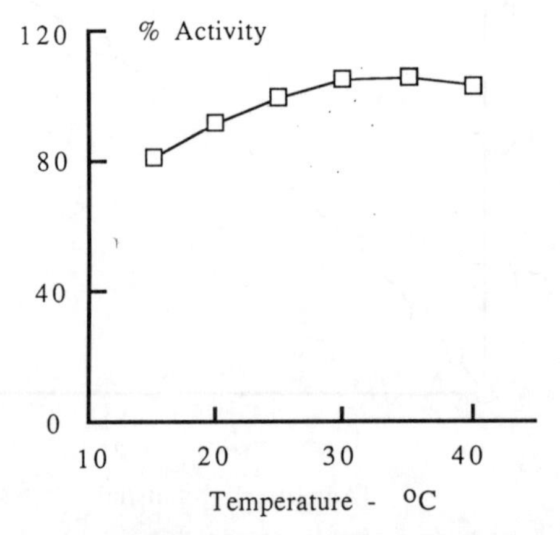

FIGURE 7 The effect of temperature. The amplifier activity with 100 nM NAD was determined over the temperature range indicated. The activity is expressed as a percentage of the activity at 25 °C

development as a function of NAD concentration up to 400 nM.

Over a limited range the amplifier activity is proportional to the concentration of the amplifier enzymes (Figure 6). As the enzyme concentration increases, however, the a·tivity saturates above about 4U/ml as predicted from the kinetic analysis. This behaviour is peculiar to an enzyme reaction where the enzyme concentration is much higher than the substrate (NAD and NADH) in contrast to the usual situation. Under these conditions, the maximum reaction velocity is limited by the substrate rather than the enzyme.

Another property attributable to the unusual enzyme/substrate ratio is the temperature response of the amplifier (Figure 7). The slope of the activity curve is abnormally shallow because the cycling activity is a function of both the V_{max} and K_m of the amplifier enzymes. These constants are temperature dependent but the ratio V_{max}/K_m may be almost invariant if both parameters change in a similar way.

AMPLIFIED ELISA

The simplest method for amplifying alkaline phosphatase activity uses the substrate and amplifier reagents described above in a two-step or sequential incubation with the colour development in the second step stopped by the addition of 0.5 M sulphuric acid. A simple protocol is outlined below. After the immunoincubation stage of the ELISA is complete and adequate washing has been done, substrate is pipetted into each well including a blank antibody-

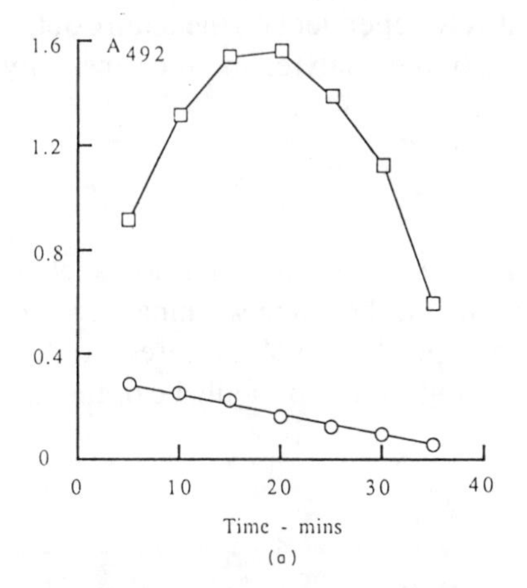

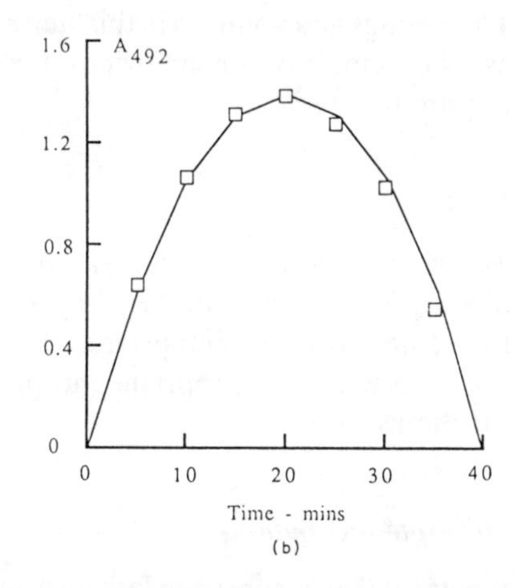

FIGURE 8 Two-step amplification. (a) In a two-step amplified alkaline phosphatase assay the duration of the substrate and amplifier incubations were varied keeping the total assay time constant at 40 min. The signals obtained with 3 pg alkaline phosphatase (□) and the reagent blank (○) are plotted against the time of the substrate incubation (first step) (b) When the reagent blank is subtracted, the same results (□) show a close fit to the theoretical curve obtained assuming each step to be a linear reaction. The maximum signal is obtained when the two incubations are of equal duration

coated well that has not received conjugate. During this first incubation, NADP is dephosphorylated to produce NAD which then catalyses the reduction of INT to the red INT-formazan during the amplifier incubation. Alkaline phosphatase is effectively turned off by the phosphate buffer (a powerful competitive inhibitor) in the amplifier, thus making the kinetics of this step linear. After the reaction has been stopped, the colour is stable for about 1 hr (kept away from bright light) and can be read in a variety of microtitre plate readers. The absorbance maximum for this dye is close to 492 nm.

The optimum timing of the individual incubation steps is mostly dependent upon the inherent background in the amplification reagents (the reagent blank as described above). Under the standard conditions described, blank rates due to contaminating NAD are in the range 5–10 mAU/min, but can be reduced to <1 mAU/min, by purification of the NADP. These rates are not influenced by the length of the substrate incubation and so lower blank rates are achieved if the amplifier incubation (colour development step) is kept relatively short. Although maximum signal is obtained when the two steps are equal (Figure 8), better signal to blank ratios are achieved by keeping the amplifier step to 10 min and increasing the substrate step to 20 or 30 min.

Two-step amplification protocol

Immunoincubation

The timings and volumes in this step are entirely dependent on the nature of the assay (competitive or sandwich), the nature of the analyte, and the sensitivity required.

Washing

To achieve the maximum assay performance it is essential that non-specific binding is minimized in the assay and careful and thorough washing steps are therefore critically important. This step typically involves three to five washings with an appropriate phosphate-free buffer which should be optimized for the assay.

Substrate incubation

Pipette 100 μl of substrate into each well, including blanks, in a timed sequence and, if possible, use a multichannel pipette. Incubate for 20 min at 25 °C. Keep away from temperature gradients and draughts.

Amplifier incubation

Pipette in the same time sequence 200 μl of amplifier into each well and incubate at 25 °C for 10 min or until the highest signal reaches 2 absorbance units. Stop the reaction at this point by adding 50 μl of 0.5 M sulphuric acid.

Data analysis

The absorbance in the wells should be determined in an appropriate reader within 1 hr of stopping the assay. In less sophisticated assays, results may be assessed visually and semi-quantitative results may be obtained provided the appropriate standards are run.

Variations

Reagent volumes

The reagent volumes described are suitable for immunoassays run in most commercially available microtitre plates and strip-plates which have well capacities of about 400 μl. Where appropriate, these volumes may be varied to accommodate smaller wells or to economize on reagents. Both substrate and amplifier volumes may be reduced in equal proportion or the amplifier may be reduced from 200 to 100 μl. The effect on the final signal depends on the resulting optical pathlength and the final amplifier concentration. Reducing the amplifier to 100 μl causes a loss of approximately 25 per cent in the signal.

Kinetic reading

Both the precision and dynamic range of an amplified ELISA can be improved by reading the colour development kinetically instead of in a stopped mode. A number of microplate readers are equipped to operate kinetically and the linear kinetics of two-step amplifiers (Figure 9) are readily analysed, either manually or by interfacing the spectrophotometer directly to a computer.

One-step amplification

One-step amplifiction, where both the alkaline phosphatase label and the amplifier enzymes are working simultaneously, can occur if a non-inhibitory buffer is used instead of phosphate. This is most simply achieved by replacing 20 mM sodium phosphate with 5 mM Tris–HCl, pH 7.2, throughout the methods described above. The following properties and modifications to the basic protocol apply only to a one-step amplifier.

(1) The addition of reagents may be simplified such that amplifier (200 μl) is added first to a well followed by substrate (100 μl) which initiates the colour development, or the appropriate volume may be pre-mixed and a single addition made (300 μl). However, once these reagents are mixed they are not stable.

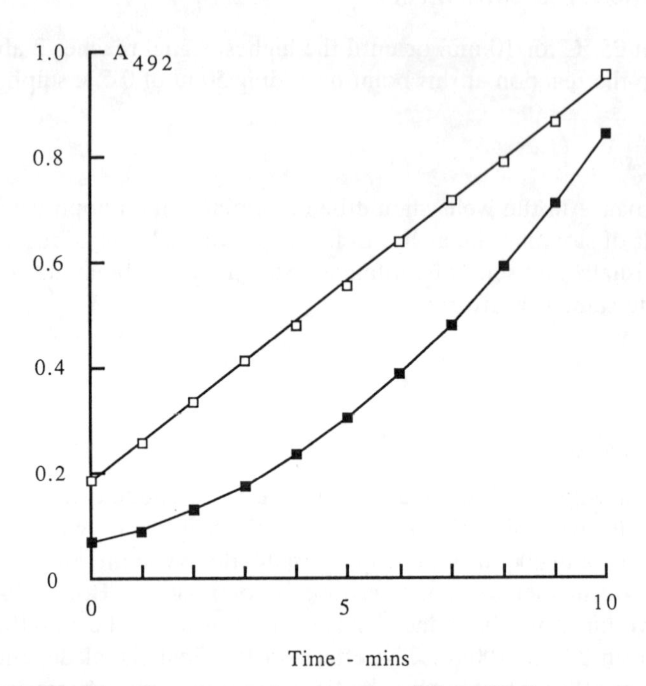

FIGURE 9 The kinetics of one-step and two-step amplifiers. The increase in absorbance at 492 nm is shown for a two-step amplifier with 100 nM NAD (□) and for a one-step amplifier with 30 pg alkaline phosphatase (■). A least-squares analysis gave a rate of 72.3 mAU/min for the two-step system and a quadratic rate of 5.88 mAU/min^2 for the one-step amplifier

(2) The kinetics of a one-step assay are quadratic as indicated above and are described by an equation of the type:

$$A = a + bt + ct^2$$

where A is absorbance, a, b, and c are the zero-, first-, and second-order coefficients respectively, and t is time. As with the two-step assay, the data may be analysed kinetically or as a stopped reaction (stopped in the same way). In the kinetic mode, the second-order coefficient is proportional to the concentration of the analyte, but in a one-step assay, where the label is immobilized on a solid phase, the kinetics are complicated by the local development of colour at the liquid/solid interface and some form of agitation is required to overcome this. With free enzyme, however, this complication does not arise and true quadratic kinetics are observed (Figure 9).

(3) Because NAD can be cycled as soon as it is produced, higher signals are obtained with a one-step amplifier than with a two-step amplifier. Ideally,

the signal would increase by a factor of two, but in practice the increase in blank rate also limits the signal to blank ratio.

(4) In a two-step amplifier the main contributor to the blank rate is NAD. In the one-step system, where alkaline phosphatase is not inhibited during the colour development, any contamination of the amplifier enzymes will add to the blank. Phosphatases are ubiquitous enzymes and it may be necessary to screen reagents, particularly enzymes, for this application.

ALKALINE PHOSPHATASE AND NAD-ACTIVATED ENZYME AMPLIFICATION

Alkaline phosphatase is one of the most suitable enzymes for use as a label in immunoassays, being readily available, quite stable, and having a relatively high specific activity. Its broad substrate specificity also enables it to be used with a number of artificial chromogenic substrates and, in the application decribed here, to dephosphorylate NADP.

The primary enzyme alkaline phosphatase and the enzymes of the amplifier cycle (see Figure 3) illustrate two contrasting modes in which an enzyme reaction can be exploited. Alkaline phosphatase is detected in minuscule quantities (attomoles), but works at high efficiency (close to its V_{max}), nevertheless producing only low concentrations of product (typically 10–500 nM NAD). The amplifier enzymes, however, although working far below their K_ms, are present at sufficiently high concentrations that high fluxes can be achieved and high product concentrations are reached (typically 3–150 μM). The two biochemical 'amplifiers' (dephosphorylation of NADP catalysed by alkaline phosphatase and reduction of INT catalysed by NAD) are thus analogous to an audioamplifier where the pre-amplifier normally has a high gain (ratio of NAD produced to alkaline phosphatase present) and low current (moles of NAD produced) whereas the power amplifier has a low gain (ratio of reduced INT to NAD present) but a high current (moles INT reduced).

Detection of alkaline phosphatase with a sensitivity of 3000 molecules

Alkaline phosphatase is commonly assayed using *p*-nitrophenyl phosphate (*p*-NPP) as a substrate and this is the natural comparison to make with the amplified assay (Figure 10). The specific activity of the enzyme used in these studies was found to be 1780 U/mg with *p*-NPP under standard conditions, whereas with NADP as substrate the specific activity was 547 U/mg, which translates to a turnover number of 76 600/min for a molecule of M_r 140 000. The lower activity with NADP is more than compensated for by the addition of the amplifier cycle which turns over at about 35 times per minute at standard activity. In a 30 min stopped assay (Figure 10) the slope of the dose–response curve for the amplified assay was 248 mAU/pg and for the p-NPP assay

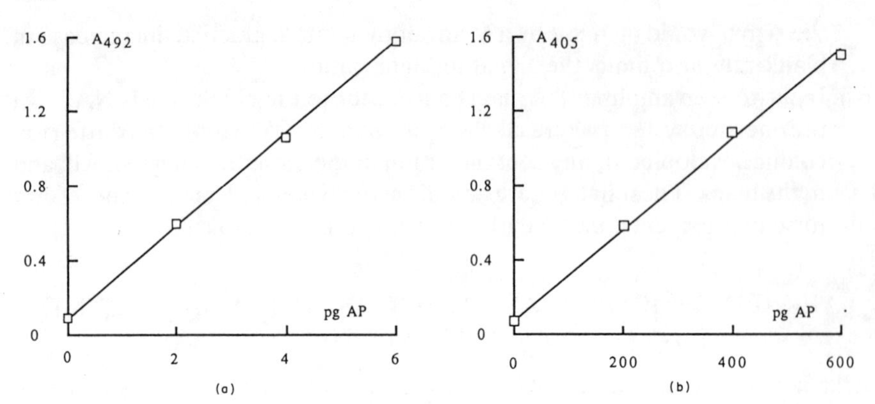

FIGURE 10 Comparison of enzyme-amplified and conventional detection of alkaline phosphatase. The amplified assay (a) followed the two-step protocol presented in the text with equal incubation times. The *p*-NPP assay (b) used 5 mM *p*-NPP in 1 M diethanolamine buffer, pH 9.8, containing 0.5 mM $MgCl_2$. The toal volume (150 μl) and total duration (30 min) were the same for each assay. The final absorbances, including reagent blanks, are plotted as a function of the amount of alkaline phosphatase per well

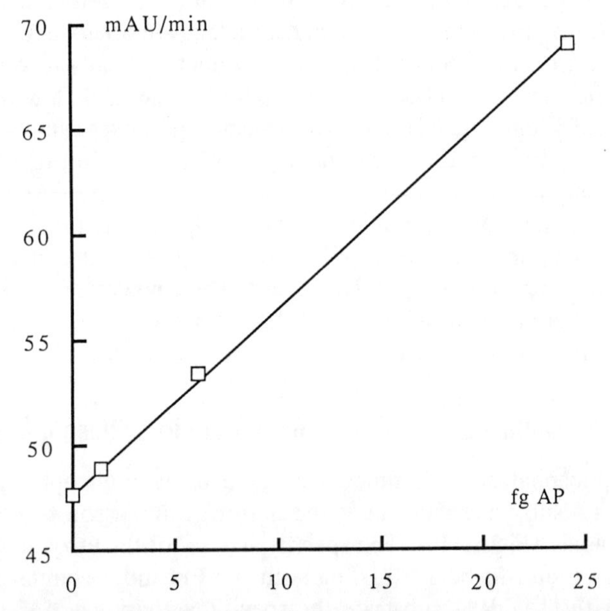

FIGURE 11 Enzyme-amplified detection of alkaline phosphatase. An alkaline phosphatase solution (567 752, BCL, Bell Lane, Lewes, East Sussex, BN7 1LG, UK) was diluted into substrate diluent containing 0.1 per cent bovine serum albumin (A 8022, Sigma, Fancy Road, Poole, Dorset BH17 7NH, UK) and dispensed in six replicates of 10 μl aliquots into the wells of a microplate and assayed as described in the text. The rate of increase in the absorbance at 492 nm is plotted against the amount of alkaline phosphatase per well

2.36 mAU/pg giving an amplification factor of 105-fold. By increasing the amplifier concentration to 6 U/ml and lengthening the substrate incubation step to 3 hr, the results shown in Figure 11 were obtained in a kinetic assay. The limit of sensitivity of the assay may be defined (Ekins, 1983) as the dose corresponding to a response one standard deviation from the mean of the zero. The coefficient of variation (CV) in six replicates of the zero was 1.26 per cent in this experiment, which gives a sensitivity of 0.69 fg. This is equivalent to 0.005 attomoles (10^{-18} moles) or approximately 3000 enzyme molecules.

Assay for TSH with a sensitivity of 0.0004 mIU/l

The assay was carried out in quintuplicate essentially as described by Stanley, Johannasson, and Self (1985), except the microtitre plate was incubated overnight at room temperature and an Fab′ conjugate (Ishikawa *et al.*, 1983) was used. The specific activity of the conjugate was 5.6×10^5 mAU/min per 50 μl of conjugate solution. A serum sample from a patient with a T3 suppressed TSH level was also assayed.

Figure 12 shows the response obtained in milliabsorbance units per minute as a function of the TSH concentration of the calibrators. In this experiment the CV of the response of singlicate determinations varied between 1.5 and 2.4 per cent except the zero calibrator which gave the higher CV of 5.7 per cent leading to a sensitivity (Ekins, 1983) of only (!) 0.0004 mIU/l. This is equivalent to less than 100 000 molecules or one-sixth of an attomole of TSH per well. The T3 suppressed serum sample measured 0.018 mIU/l. The non-specific binding of

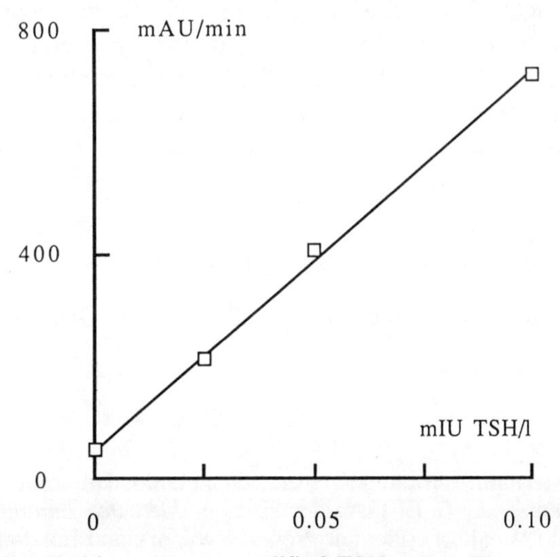

FIGURE 12 An ultrasensitive enzyme-amplified TSH assay. The rate of increase in absorbance at 492 nm is plotted as a function of the TSH concentration in the calibrators

TABLE 1 Applications of enzyme amplification in immunoassays

Assay type	Analyte	Reference
Capture	Placental alkaline phosphatase	Self (1985)
		Cooper *et al*. (1985)
	Alkaline phosphatase isoenzymes	Warren *et al*. (1985)
Competition	Progesterone	Stanley, Johannsson, and Self (1985)
		Stanley *et al*. (1986)
	CGRP	Self *et al*. (1985)
Sandwich	Prostatic acid phosphatase	Johannsson, Stanley, and Self (1985)
		Moss *et al*. (1985)
	TSH	Stanley, Johannsson, and Self (1985)
		Johannsson *et al*. (1986)
		Clark and Price (1986)
	Chlamydial antigen	Pugh *et al*. (1985)
		Caul and Paul (1985)
	Clostridium botulinum toxin type A	Shone *et al*. (1985)
	Hepatitis B surface Ag	Gatley (1986)

conjugate to the microtitre plate was not measurable. A non-specific binding equivalent to 3 mAU/min (the standard deviation of the response of the zero dose) would have corresponded to only 0.0005 per cent. It is important to restrict the non-specific binding to such low levels in order to take advantage of the high detectability of alkaline phosphatase.

These data together with the examples listed in Table 1 clearly show the wide applicability of enzyme amplification to immunoassays giving improved sensitivity and/or speed. The ease with which the endpoint colour can be determined using simple colorimetry or even visual inspection gives this technology an important advantage over many other emerging non-isotopic detection methods.

ACKNOWLEDGEMENT

The authors wish to thank their many colleagues at IQ(Bio) Limited for their contributions.

REFERENCES

Butler, J.E., Peterman, J.H., and Koertge, T.E. (1985). The amplified enzyme-linked immunosorbent assay (a-ELISA). In: *Enzyme-Mediated Immunoassay* (eds. T.T. Ngo, and H.M. Lenhoff). Plenum Press, New York and London, p. 241.

Caul, E.O., and Paul, I.D. (1985). Monoclonal antibody based ELISA for detecting *Chlamydia trachomatis*. *Lancet*, **i,** 277.

Clark, P.M.S., and Price, C.P. (1986). Enzyme-amplified immunoassays: A new ultransensitive assay of Thyrotropin evaluated. *Clin. Chem.*, **32,** 88.

Cooper, E.H., Pidcock, N.B., Jones, W.G., and Milford Ward, A. (1985). Evaluation of an amplified enzyme-linked immunoassay of placental alkaline phosphatase in testicular cancer. *Eur. J. Cancer Clin. Oncol.*, **21,** 525.

Ekins, R.P. (1983). The precision profile: its use in assay design, assessment and quality control. In: *Immunoassays for Clinical Chemistry* (eds. W.M. Hunter, and J.E.T. Corrie). Churchill Livingstone, London, pp. 76–105.

Gatley, S. (1986). A case for amplification. *Med. Lab. World*, Jan. 1986, 27.

Harper, J.R., and Orengo, A. (1981). The preparation of an immunoglobulin–amyloglucoside conjugate and its quantitation by an enzyme-cycling assay. *Anal. Biochem.*, **113,** 51.

Ishikawa, E., Imagawa, M., Hashida, S., Yoshitake, S., Hamaguchi, Y., and Ueno, T. (1983). Enzyme-labeling of antibodies and their fragments for enzyme immunoassay and immunohistochemical staining. *J. Immunoassay*, **4,** 209.

Jackson, T.M., and Ekins, R.P. (1986). Theoretical limitations on immunoassay sensitivity. Current practice and potential advantages of fluorescent Eu^{3+} chelates as non-radioisotopic tracers. *J. Imm. Methods*, **87,** 13.

Johannsson, A., Stanley, C.J., and Self, C.H. (1985). A fast highly sensitive colorimetric enzyme amplification system demonstrating benefits of enzyme amplification in clinical chemistry. *Clin.Chim.Acta*, **148,** 119.

Johannsson, A., Ellis, D.H., Bates, D.L., Plumb, A.M., and Stanley, C.J. (1986). Enzyme amplification for immunoassays. Detection limit of one-hundredth of an attomole. *J.Imm.Methods*, **87,** 7.

Moss, D.W., Self, C.H., Whitaker, K.B., Bailyes, E., Siddle, K., Johannsson, A., Stanley, C.J., and Cooper, E.H. (1985). An enzyme-amplified monoclonal immunoenzymometric assay for prostatic acid phosphatase. *Clin. Chim. Acta*, **152,** 85.

Pugh, S.F., Slack, R.C.B., Caul, E.O., Paul, I.D., Appleton, P.N., and Gatley, S. (1985). Enzyme amplified immunoassay: A novel technique applied to direct detection of Chlamydial specimens. *J.Clin.Path.*, **38,** 1139.

Rotman, B. (1961). Measurement of activity of single molecules of β-D-galactosidase. *Proc.Natl.Acad.Sci.USA*, **47,** 1981.

Self, C.H. (1982). European patent application no. 82301170.5 (EP60123), Abstracted in *Chemical Abstracts*, **97,** 212066D.

Self, C.H. (1985). Enzyme amplification – a general method applied to provide an immunoassisted assay for placental alkaline phosphatase. *J. Imm. Methods*, **76,** 389.

Self, C.H., Wimalawansa, A., Johannsson, A., Bates, D.L., Girgis, S.I., and MacIntyre, I. (1985). A new sensitive and fast peptide immunoassay based on enzyme amplification used in the determination of CGRP and the demonstration of its presence in the thyroid. *Peptides*, **6,** 627.

Shone, C., Wilton-Smith, P., Appleton, P.N., Hambleton, N., Modi, N., Gatley, S., and Melling, J. (1985). Monoclonal antibody-based immunoassay for type A Clostridium botulinum toxin is comparable to the mouse bioassay. *Appl. Env. Microbiol.*, **50,** 63.

Stanley, C.J., Johannsson, A., and Self, C.H. (1985). Enzyme amplification can enhance both the speed and the sensitivity of immunoassays. *J. Imm. Methods,* **83,** 89.

Stanley, C.J., Paris, F., Webb, A.E., Heap, R.B., Ellis, S.T., Hamon, M., Worsfold,

A., and Booth, J.M. (1986). Use of a new and rapid milk progesterone assay to monitor reproductive activity in the cow. *Vet. Record*, **118,** 664.
Warren, R.C., McKenzie, C.F., Rodeck, C.H., Moscoso, G., Brock, D.J.H., and Barron, L. (1985). First trimester diagnosis of hypophosphatasia with a monoclonal antibody to the liver/bone/kidney isoenzyme of alkaline phosphatase. *Lancet*, **ii,** 856.

ELISA and Other Solid Phase Immunoassays
Edited by D.M. Kemeny and S.J. Challacombe

CHAPTER 5

The Amplified ELISA (a-ELISA): Immunochemistry and Applications

J.E. Butler
University of Iowa, Iowa, USA

CONTENTS

BASIC PRINCIPLES

Basic principles of antigen-specific ELISAs

The term antigen-specific ELISA, as used here, denotes an immunochemical configuration in which the goal is to measure antibody (Ab) activity in serum, secretions, or culture fluids to a particular antigen (Ag). Typically, such systems are constructed in the form of so-called heterogeneous immunoassays in which the Ag of interest is attached to the solid phase. Solid phases can be agarose, dextran, acrylamide, or latex particles or they may be the walls of the reaction vessel; initially plastic tubes were used but currently microtitre wells have come into popular use as solid phases. Antigens may be absorbed on to these solid phases (Catt and Treager, 1967; Brash and Lyman, 1969; Engvall and Perlmann, 1972), covalently attached (Maiolini and Masseyeff, 1975), covalently bonded to particles which are in turn adsorbed to hydrophobic membranes (Lambden and Watt, 1978), or attached non-convalently to an adsorbed 'carrier' (Suter and Butler, 1986). Simple adsorption of the Ag or its carrier is most common when microtitre plates are employed. Although convenient, this method can have undesirable effects (see below).

The Ag-coated solid phase is incubated with the potential Ab-containing specimen. During this step, efforts must be made to prevent Ab and particularly other immunoglobulins (Igs), from adsorbing directly on the solid phase while permitting specific Abs to bind the Ag. This is typically accomplished by adding non-ionic detergents (Triton X-100; Tween 20, Tween 80 or NP-40) or abundant proteins like albumin or gelatine to the reaction buffer. Because the detergent Tween 20 in phosphate-buffered saline is most commonly used, the reaction buffer[a] is abbreviated PBS–T. The primary antigen–antibody reaction is typically allowed to proceed for 2–3 hr at RT or 37 °C, or overnight at 4 °C. In most microtitre systems, >75 per cent of maximal binding will have occurred after a 3 hr, 37 °C incubation period[b].

After the Ab to be measured has been allowed to bind the immobilized Ag, the unbound protein is removed and a detection system is added. The detection of the primary antibody through use of a variety of enzyme-labelled secondary Abs, is the principle which distinguishes ELISA from other solid phase Ag-specific assays which utilize fluorescent- or radioisotope-labelled Abs. The amplified ELISA (a-ELISA; Butler, McGivern, and Swanson, 1978a) described in this chapter is an 'indirect' system which employs an enzyme immune complex (EIC) as the final, macromolecular reaction step (Figure 1 and below). The term 'indirect' is defined as any detection system in which the enzyme, fluor- or radio-isotope is not part of the secondary Ab. The experience gained in more than a decade with this 'workhorse assay', has permitted a great deal to be learned about the immunochemistry of Ag-specific ELISAs in general.

Basic principles of the a-ELISA

The a-ELISA was developed for measuring antibodies to proteins, haptens, viruses, and even bacteria, which could be stably adsorbed on polystyrene. Polystyrene microtitre plates from Dynatech or NUNC are currently used in our laboratory. Typical of most Ag-specific ELISAs, the adsorbed Ag is then incubated with the primary Ab-containing specimen in the presence of PBS–T.

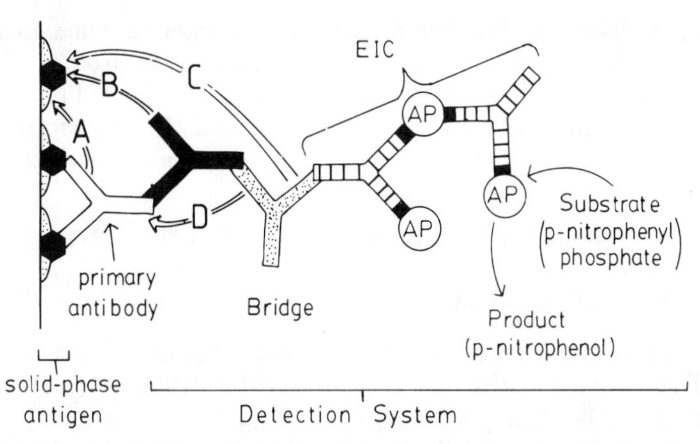

FIGURE 1 The reaction sequence of the a-ELISA. The various reactants are labelled on the figure with exception of the isotype-specific antiglobulin (black antibody between primary and bridge). The various arrows, A–D, illustrate the potential short-circuts which can occur in this system. The stoichiometry of the EIC is described elsewhere (Koertge, Butler and Dierks, 1985). The circular arrows at the right illustrate the substrate conversion which provides the detection signal in the a-ELISA

Following this incubation, the plates are washed and an isotype or sub-isotype-specific antiserum is added. Next a so-called bridging reagent (goat anti-rabbit IgG; Figure 1) is added and eventually, after incubation and washing, the EIC, composed of alkaline phosphatase (AP) and anti-alkaline phosphatase, is added (Figure 1). The reaction is finally terminated by washing away excess EIC and by adding an excess of the substrate *p*-nitrophenyl phosphate. The amount of antibody bound is indirectly determined by measuring the extent of enzymic reaction by the absorbancy at 405 nm of the yellow product, *p*-nitrophenol.

The stoichiometry (Figure 1) of the a-ELISA theoretically permits three moles of AP to bind per mole of the antibody being measured. This 'amplified' indirect detection system further benefits from the fact that the enzymic activity of the AP is not altered by the non-covalent complex which it forms with anti-AP (Koertge, Butler, and Dierks, 1985). This Ag-specific ELISA is capable of detecting as little as 30 pg of specific Ab (Dierks, Butler and Richerson, 1986). Like other indirect detection systems, the a-ELISA is highly adaptable for measuring Abs in any species and for any isotype or sub-isotype. The isotype-specific secondary Ab (or antiglobulin, Figure 1) can be readily exchanged for another in the reaction sequence of the a-ELISA without changing any other components. This permits Abs of various isotypes to be more comparatively measured than when separate antibody–enzyme conjugates are used for each isotype. The a-ELISA configuration illustrated in

Figure 1 has been employed to measure Abs in humans, mice, rats, swine, cattle, and marsupials to a variety of Ags. A separate version of the a-ELISA is used to measure rabbit Abs that utilizes guinea pig antiglobulins to rabbit Ig isotypes and guinea pig anti-AP to form the EIC (Butler, McGivern, and Swanson, 1978a; Butler *et al.*, 1982; Dierks, Butler and Richerson, 1986).

SOURCES AND SPECIFICITY OF REAGENTS

Nature of the solid phase antigen

The a-ELISA is currently performed in our laboratory by adsorbing Ag, diluted in coupling buffer[c], to the surface of polystyrene microtitre wells (Immulon 2, Dynatech, Alexandria, VA). The Ags commonly used are ovalbumin, serum albumin, *Brucella*-soluble antigen (BASA), staphylococcal α-toxins, viruses, oral bacteria, various milk proteins, various proteineous extracts of allergens, and various hapten–protein conjugates. Those protein–hapten conjugates most commonly used are phosphorylcholine–gelatine, dinitrophenyl (DNP)–albumin, and fluorescein (FLU)–gelatine. The effect of pH on adsorption, although not rigorously studied in this laboratory, does not appear to be critical. Neither ovalbumin nor FLU–gelatin differs significantly in adsorption as a function of pH from 5 to 10 (Dierks, 1985; Peterman, 1986). We have successfully adsorbed corona, parvo, and herpes viruses to plastic in the same manner as for proteins but have experienced unstable adsorption when the same method was applied to bacteria (Sloan, 1975; Sloan and Butler, 1978). The use of methyl glyoxal as described by Czerkinsky *et al.* (1983) has permitted the successful adsorption of sonicates of *Streptococcus mutans, Actinomyces viscosus*, and *Escherichia coli* to Immulon 2 (see Chapter 15). The use of glutaraldehyde pre-treatment (Suter, 1982) has in our experience not significantly improved the preparation of solid phase Ags on microtitre plates (Koertge, 1984).

Initial studies in our laboratory (Cantarero, Butler, and Osbourne, 1980) indicated that a linear relationship existed between the amount of protein added and the amount bound. This 'region of independent binding' extended to an upper limit of about 1000 ng of added Ag/6.5 cm^2 of conventional polystyrene tubes (Falcon, Oxnard, CA; Sarstedt, Princeton, NJ). Studies using Immulon 2, a specifically treated plastic, indicate a similar phenomenon and an upper limit for the region of independence of *c*. 150 ng/1.41 cm^2[c]. The affinity of Immulon 1 plates for proteins appears to be less than for Immulon 2 and NUNC certified.

Within the region of independent binding, the percentage of protein which binds is a characteristic of the protein (Cantarero, Butler and Osbourne, 1980). Above the upper limits of the region of independence, a greater amount, but a progressively smaller percentage, of added Ag becomes adsorbed. For FLU–gelatine adsorbed on Immulon 2 microtitre wells, the percentage which binds decreases with the amount added, e.g. 50 per cent of 100 ng; 25 per cent of 1000 ng, and 16 per cent of 2000 ng (Peterman, 1986). It has been proposed that the Ag which adsorbs at high concentration of added Ag, i.e. above the upper limits of the region of independent binding, may bind to sites on plastic of lower affinity or may in fact represent protein–protein adsorption (Cantarero, Butler and Obsbourne, 1980).

A significant factor in the behaviour of adsorbed Ags is their stability. Desorption has been reported (Lehtonen and Viljanen, 1980) and recent studies show that repetitive washing up to ten times results in a small but continual and constant release of Ags of *c.* 2.5 per cent (Peterman, 1986). The desorption of adsorbed Ag can be a significant factor in the performance of ELISA because desorbed Ag can serve as a competitive inhibitor for free Ab in the system, e.g. when Ag desorption is high and the amount of added Ab is low, desorption can significantly influence the ELISA titration plot and the estimation of Ab activity (Peterman, 1986). A partial but practical means of reducing the effect of Ag desorption is to incubate washed, Ag-coated plates for 3 hr at 37 °C in PBS–T or wash solution[d] (Peterman, 1986). For proteins, optimal conditions are usually obtained using Ag solutions at 5 μg/ml to coat Immulon 2 plastic.

The data on the adsorptive and desorptive behaviour of Immulon 2, cited above, is based on studies with proteins and hapten–protein conjugates. No comparative data are available on the behaviour of viruses and bacteria.

Studies using scanning electron microscopy at the 40 A° level indicate that the surface of Immulon 2 is relatively smooth. Hence, when 150 ng of IgG/ 1.41 cm^2 is stably bound to plastic, 80 per cent of the plastic surface contains no adsorbed protein (Peterman, 1986). This means that a relatively large area is potentially available for the adsorption of proteins in the subsequent reaction steps. Reducing the concentration of adsorbed Ag increases non-specific adsorption of such protein (Butler, Peterman, and Koertge, 1985; Peterman 1986). Hence, maximizing the amount of stably adsorbed protein plus the use of Tween-20, albumin or gelatine in the reaction buffer, are important for good performance in conventional microtitre ELISAs.

The adsorption of proteins on solid surfaces is known to cause conformational changes in the adsorbed protein (Bull, 1956); Burghardt and Axelrod, 1983). Such changes could mean that epitopes are buried or lost. Studies by Kennel (1982) and Dierks, Butler and Richerson (1986) have shown that such surface adsorption can significantly alter antigenicity. In the latter study, only a very small percentage of the rabbit anti-OA Abs capable of binding OA in solution are able to stably bind OA adsorbed on plastic. Aggregation of OA using glutaraldehyde significantly improves its antigenic behaviour when later adsorbed on plastic (see Figure 7 this chapter and Dierks, Butler, and Richerson, 1986). Hence, either the polymerization or the non-denaturing covalent attachment of Ags to plastic, as suggested by Nimmo *et al.* (1984), seems advisable.

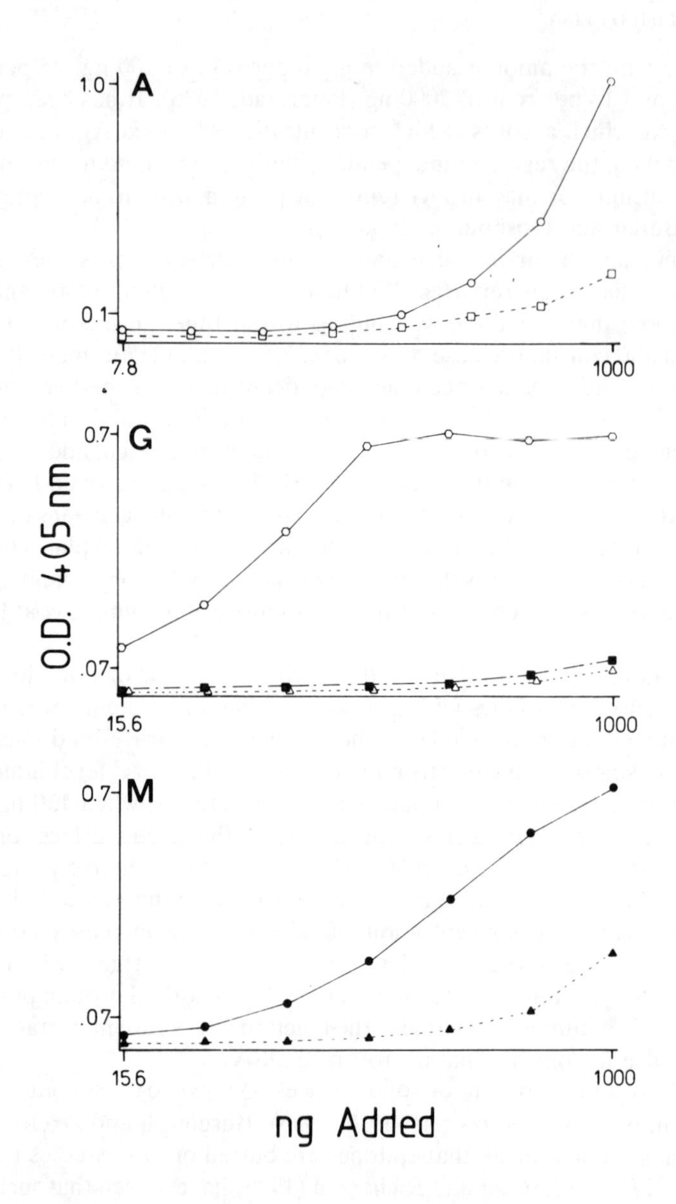

FIGURE 2 The specificity of commercial anti-human Ig tested in the Adsorbed Antigen Activity Assay (see (Figure 3). A, G, and M denote the isotype of the purified Ig adsorbed on the plastic. IgA (top) is tested against anti-α (○–○) and anti-μ (□–□). IgG (middle), is tested against anti-γ (◇–◇), anti-α (■–■), and anti-ε (△–△). IgM (bottom) is tested against anti-μ (●–●) and anti-α (▲–▲). Most commercial anti-γ reagents do not cross-react with IgM and IgA and hence no data are shown

Source, specificity, and titration of anti-Ig sera

The specificity of commercially available polyclonal anti-Ig sera for use in ELISA cannot be assumed. Proof of specificity currently rests with the consumer, not the supplier, such that each one must be evaluated prior to its use in the a-ELISA. Anti-μ α-, and ϵ- reagents from, e.g. DAKO (Copenhagen, Demark) require additional adsorption on human IgA, IgM, and IgG affinity columns respectively, before they are specific for use in the a-ELISA. Similar results have been obtained with the reagents from other suppliers, especially those directed against the Igs of animals of veterinary importance (Miles Laboratory, Naperville, IL; Cappel Labs, Cooper Biomedical Westchester, PA). Most common is the reciprocal cross-reactivity between anti-μ and anti-α reagents (Figure 2). Until commercial reagents are available which do not require absorption prior to use (Butler, Peterman, and Koertge, 1985), the preparation of one's own reagents can sometimes be more practical if a routine long-range testing programme is planned. Currently, with the exception of anti-IgE- and IgG-subclass-specific monoclonals, antisera against human Igs used in our laboratory are prepared in our own laboratory.

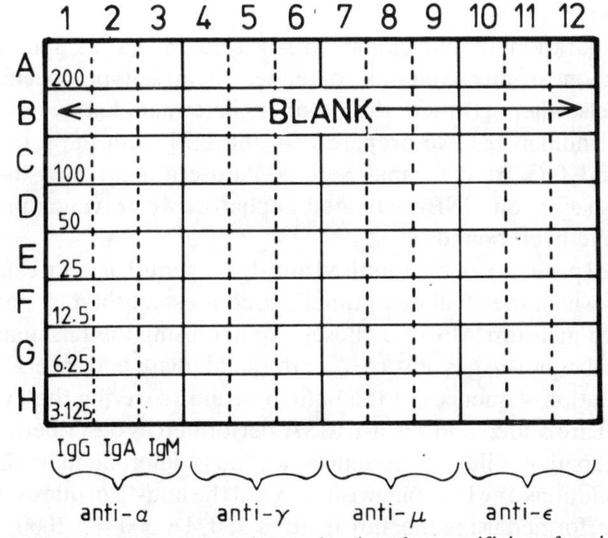

FIGURE 3 The microtitre test format for evaluating the specificity of anti-Ig reagents using the Adsorbed Antigen Activity Assay (A_4). The format shown is for testing the specificity of rabbit anti-α (columns 1–3), anti-γ (columns 4–6), anti-μ (columns 7–9), and anti-ϵ (columns 10–12) against added concentrations of purified IgG (columns 1, 4, 7, 10). IgA (columns 2, 5, 8, 11) and IgM (columns 3, 6, 9, 12) ranging from 3.125–200 ng/well

The specificity of anti-Ig reagents used in our work is tested in two ways. The first method utilizes the principle of the Absorbed Antigen Activity Assay (formerly referred to as EADA, Butler, 1981[e]), which was originally performed in individual polystyrene tubes (Butler *et al.*, 1980a). In the case of purified human anti-Igs, IgA, IgM, IgG[f] are adsorbed on microtitre wells at added protein concentrations ranging from 3.12–200 ng/well (Figure 3). The specificity, i.e. cross-reactivity, of each anti-Ig is tested using the a-ELISA against each different adsorbed Ig. Initial use of the reagent at a low dilution (1: 250) is a highly stringent assay for cross-reactivity (Figure 3). The optimal dilution at which the antiglobulin will be used in routine assays is determined only after its specificity has been ascertained at low dilution. Provided each purified Ig is known by appropriate criteria to be pure, any significant cross-reactivity would indicate the need to absorb further the reagent (see below). Should further adsorption not alter the results of such specificity tests, it is most likely that the purity of the adsorbed Ig and not the specificity of the anti-Ig, is in question. Due to the potential loss of antigenicity by adsorbed Igs (Dierks, Butler, and Richerson, 1986) an antiglobulin should be used in the a-ELISA at one two fold dilution greater than the dilution by which it was judged specific in the Adsorbed Antigen Activity Assay. The specificity of four 'ELISA quality' commercial anti-Igs is illustrated in Figure 2 before additional absorption.

The preparation of polyclonal anti-Ig sera in varous species and their absorption on affinity columns to render them isotype specific, has been described elsewhere (Butler, 1983; Butler, Peterman, and Koertge, 1985). In the case of human Igs, we prepare IgA and IgM from the sera of myeloma patients and IgG from normal Serum. Antisera are raised in rabbits and rendered specific on CNBr-activated Sepharose 4B to which purified Igs or F(ab)'2 have been bound.

Once the specificity of the required anti-Ig reagents has been established, the dilution at which they will be optimally used is established by performing the a-ELISA to measure Abs to a chosen Ag and using the terminal reagents in excess (see below). Ag is adsorbed to the solid phase and several, seven-well, twofold dilution sequences of the patient serum (covering the expected titration range) are added and the a-ELISA performed as described. The number of identical patient dilution sequences needed is determined by the number of different dilutions at which one wishes to test the anti-Ig in question. Typically, tests are performed using the anti-Ig at 1 : 250, 1 : 500, 1 : 1000, and 1 : 2000 dilutions. The optimal dilution is selected on the basis of: (a) the linearity and range of the titration plot obtained; (b) the maximum OD obtained; (c) the time required for the enzymic reaction to proceed such that a complete titration is obtained; and (d) reagent economy. Most polyclonal anti-Igs will produce suitable titrations in 1 hr at dilutions of 1 : 1000 or greater, while very 'weak' anti-Igs will require 2 hr when used even at a 1 : 250 dilution.

Terminal detection reagents of the a-ELISA

The terminal detection reagents of this assay include the 'bridging' antiserum and the EIC (Figure 1). The preparation of these reagents has been described elsewhere (Butler, 1981; Butler, McGivern, and Swanson, 1978a; Butler, Peterman, and Koertge, 1985; Butler *et al.*, 1980a) and is briefly reviewed below.

The bridging antiserum used in the a-ELISA configuration illustrated in Figure 1, is prepared by repeatedly immunizing goats with 5 mg of rabbit IgG; the initial immunization is given intramuscularly with Freund's complete adjuvant while subsequent immunizations are given with incomplete adjuvant. For use in the measurement of swine and bovine Abs, such bridging antisera require no absorption. However, in human and rats, such reagents must be absorbed either on a human or rat IgG affinity column to prevent 'short-circuiting' (Figure 1) when used in the a-ELISA. Bridging antisera for use in the a-ELISA for measuring rabbit Abs, are prepared in guinea pigs (Butler, McGivern and Swanson, 1978a).

The EIC of the a-ELISA is prepared with rabbit (or guinea pig) anti-AP raised by toe-pad immunization of rabbits or guinea pigs with 1–2 mg of AP (calf intestinal mucosa, type VII) in Freund's complete adjuvant. Animals are boostered i.v. with a similar amount of AP 30 days later and exsanguinated. The equivalence precipitin titre of such anti-AP reagents is determined first by immunodiffusion (Butler, 1980) and then using nephelometry (Leone, 1968). Typically 1 ml of rabbit anti-AP is incubated at 37 °C for 1 hr and then overnight with a serologically equivalent amount of AP (typically 0.5 mg) in PBS. The immune precipitate is washed twice and then resolubilized with a ninefold excess of AP at 4 °C with overnight stirring. The resultant soluble EIC is then stored at −20 °C in 50 per cent glycerol in aliquots of a size necessary to perform ELISAs in one or two plates. Each batch of EIC is titred as described previously (Butler, Peterman, and Koertge, 1985) and is typically used at a final dilution of 1 : 2000. In 50 per cent glycerol at −20 °C, the EIC is stable for at least 5 years (Koertge, Butler and Dierks, 1985).

The substrate used for AP is *p*-nitrophenyl phosphate and is purchased from Sigma as No. 104. It is stored at −20 °C until used. Substrate solution is freshly prepared in 50 mM carbonate substrate buffer[g]. Activities have been reported to increase if a diethanolamine buffer is used. We have not found this necessary because the sensitivity of the a-ELISA is not limited by the enzyme signal, but by non-specific and other immunochemical phenomena (see later).

Titration of reagents, short-circuitry, and storage

The titration of the anti-Ig reagents was described when their specificity was tested (see above). In principle, all reagents, except the primary Ab to be measured, must be used in excess so that the limiting step in the assay is the concentration of primary Abs, not the detection system. This should be empirically demonstrated by the investigator. Typically anti-Ig reagents should be used at dilutions of 1 : 500 – 1 : 2000, bridging antisera at dilution of 1 : 5000 – 1 : 4000, and the EIC at a 1 : 2000 dilution. Increasing the concentration of the antiglobulin or any of the terminal reagents may increase the rate of reaction but should not alter the characteristics of the final titration plot. Should this be the case, use of the higher dilution is recommended to conserve reagent even though the substrate reaction time will be increased.

A potential problem in any multi-Ab assay is short-circuitry, e.g. the secondary antiglobulin recognizes the Ag plus the primary Ab or the bridge recognizes the primary Ab and/or Ag as well as the secondary antibody. These various possibilities are illustrated in Figure 1. Hence, the specificity of each reagent needs to be evaluated and this can be done using the Adsorbed Antigen Activity Assay described above. When short-circuitry is a problem, appropriate absorption of the troublesome reagent, or a search for an alternative reagent with a lower degree of cross-reactivity, is necessary.

We have routinely stored all secondary and bridging antisera at −20 °C in aliquots of useful size containing 0.02 per cent azide, 1 mM EDTA and 25 mM ϵ-amino caproic acid. Most recently, 10 mM benzamide–HCl has been used in place of ϵ-amino caproic acid. We find the use of 1.0 ml plastic tubes from Sarstedt, which have rubber-seal caps, to be ideal for the storage of such aliquots[h]. Self-defrosting freezers should be avoided to reduce ice crystal formation.

MEASUREMENT OF ANTIGEN-SPECIFIC ANTIBODIES USING THE a-ELISA

Establishment of the optimal assay procedure

The routine use of the a-ELISA should be such that sensitivity, convenience, and economy of reagents are optimized. The multi-step nature of the a-ELISA means it is best performed over a two-day period which is adjusted for convenience of the technician and the maximum sensitivity of the assay. Three sample protocols are presented in Table 1 and the general procedure is described below.

Optimal conditions for preparation of the solid phase Ag are obtained by allowing proteins to adsorb overnight at RT at a concentration of 5 μg/ml. After the plates have been emptied and washed[i], they are incubated for 2 hr at 37 °C in either wash solution or PBS–T, i.e. 'postwash' treatment, to encourage the desorption of weakly bound protein (Peterman, 1986). Adsorption of viruses is performed in the same manner while bacteria are adsorbed in the presence of methyl glyoxyl (see above). A postwash treatment is also useful when bacteria are used.

TABLE 1 Conditions employed when measuring three different antibodies* using the a-ELISA

Reaction step of a-ELISA	Incubation time			Sample or reagent dilution		
	Human IgA anti *S.mutans*	Human IgE anti-ragweed	Bovine IgG_1 anti-FLU	Human IgA anti *S.mutans*	Human IgE anti-ragweed	Bovine IgG_1 anti-FLU†
Antigen adsorption	overnight RT	overnight 4 °C	overnight 4 °C	1×10^9‡ methyl glyoxal	5 μg/ml whole ragweed extract	5 μg/ml FLU-gelatine
Postwash treatment		2 hr 37 °C	2 hr 37 °C		2 hr 37 °C	2 hr 37 °C
Primary antibody	overnight 4 °C	overnight 4 °C	overnight 4 °C	1:25 initial dilution§ of saliva	1:4 initial dilution§ of *S.aureus*-treated serum«	1:10 000 dilution§ of bovine serum
Secondary antibody (antiglobulin)	2 hr RT on shaker	2 hr RT on shaker	2 hr RT on shaker	1:1000 rabbit anti-α	1:250 rabbit anti-ε	1:500 rabbit anti-γ_1
Goat anti-rabbit IgG 'bridge'	2 hr RT on shaker	2 hr RT on shaker	2 hr RT on shaker	1:4000 #129»	1:4000 #129»	1:500 #107»
EIC	overnight	overnight	overnight	1:2000	1:2000	1:2000
Substrate	1–2 hr RT	2–3 hr RT	0.5–1 hr RT	1 mg/ml	1 mg/ml	1 mg/ml

* Examples chosen are those for which the optimal condition and nature of the samples tested are quite different.
† FLU = fluorescein hapten.
‡ Performed with modification of the method of Czerkinsky *et al.* (1983).
§ All samples are tested over a range of dilutions, the value given is the initial dilution and hightest concentration.
« Human serum must first be partially depleted of IgG anti-ragweed using *S.aureus* (Cowan I) to allow the IgE, which binds protein-A poorly under the conditions used, to be detected without competition from IgG.
» Goat #129 has been absorbed with human IgG to prevent a short-circuit in the a-ELISA. Goat 107 requires no adsorption when used with bovine antibodies.

The Ag absorption procedure is typically timed so that the microtitre plates are ready for the addition of the test sample and the reference standard at *c.* 4:00 p.m. This allows the technician adequate time in the early afternoon to prepare all dilutions of the test sample in PBS–T; these dilutions are then applied at 4:00 p.m. In our laboratory we use either Pipetmans (Rainin Manufacturing Company, Emeryville, CA) or Titerteks (FLOW Laboratories, McLean, VA) to add samples. The plates are covered with Parafilm (American Can Co., Greenwich, CT) and incubated overnight at 4 °C. We have found by experience that the primary Ag–Ab interaction is slow to reach equilibrium and only approaches it after 16–24 hr (Koertge, 1984; Peterman, 1986[b]). Hence, we permit this step of the assay to be incubated overnight for scientific as well as logistical reasons.

On the following day, the plates are brought to RT, emptied and washed and a dilution of the antiglobulin in PBS–T of the appropriate specificity, is added. The plates are again covered and placed on a Minishaker (Dynatech Labs, Alexandria, VA) for 2 hr. The plates are again emptied and washed and the bridging antiserum added at appropriate dilution in PBS–T. Plates are incubated as for the antiglobulin.

The plates are again emptied, washed, and the EIC, diluted appropriately in PBS–T, is added. For convenience, we typically incubate the plates overnight at 4 °C. If there is reason for the experiment to be completed exactly at 5:00 p.m., incubation can be done for 3 hr at RT on a Minishaker.

On the following day, the substrate solution is prepared and allowed to come to room temperature for 1 hr prior to its use. The substrate buffer is typically stored in the cold to prevent microbial growth. *p*-Nitro-phenyl phosphate is stored at −20 °C, and when weighing out the solid substrate, users should allow the frozen bottle to warm somewhat before opening to prevent rapid condensation of moisture. We prepare the substrate solution immediately prior to use to avoid the photolysis which can occur with *p*-nitrophenyl phosphate.

The emptied and washed plates are then treated with the substrate solution and allowed to develop at RT. For convenience, we use a BioTek EL310 plate reader which can be programmed to monitor the reaction of a specific well, typically the well containing the highest concentration of the standard. When the OD_{405} of this well reaches 1.2, the plate is automatically read and recorded (see below).

The apparent discrepancy among incubation times for the various antibody–antibody reaction steps of the a-ELISA are probably related to several factors. First, the primary Ab combines with adsorbed Ag in close proximity to the plastic solid phase while subsequent reactions can occur at a greater distance from the support (Figure 1). This is proposed because it does not seem possible to explain the affinity-dependence of the microtitre ELISA (Butler *et al.*, 1978; Lew, 1984; Lehtonen and Eerola, 1982; Peterman, 1986) merely on the ratio of Ab to solid phase Ag[j]. Hence solid phase antigen–antibody interaction must be different from that in free solution. The greater enzymic signal generated by the EIC when incubated for longer periods is probably a diffusion effect resulting from its much larger size (Figure 1).

In establishing the optimal conditions for using the a-ELISA in a new system, variables most likely to require adjustment are: (a) the dilution of the anti-Ig if the animal species and/or Ab isotype is different from that of previous tests; (b) the titration range of the standard or representative unknown specimens, and (c) a test for potential short-circuits which can occur between anti-Ig or any terminal reagent and the new Ag being used. Ag absorption, dilution of bridge and the EIC remain constant. The tests for new antiglobulin reagents were described above. The optimal titration ranges of the standard or test specimens are those which produce a log–log linear titration of sufficient length in a reasonable time period, i.e. 1–3 hr (Figure 5).

Plate format and standards for routine testing

The routine use of the a-ELISA involves the testing of dilutions of unknown samples and the reference standard on the same plate. We have found the use of four twofold or threefold dilutions of each unknown sample tested on a plate together with a six-dilution sequence of the reference standard, to be an optimal configuration for such routine testing. This format is illustrated in Figure 4. The format allows 20 unknown samples to be tested on a single plate and for the background correction to be made using the mean OD_{405} of four control wells. These control wells are filled with PBS–T instead of the primary Ab, but receive all other reagents. Alternatively, if little is known about the titre of antibodies in the test samples, a plate format testing seven dilutions of ten different unknowns can also be employed (Peterman and Butler, 1987).

The reference standard method for routine measurement of Ab titres is widely used, and while it is not quantitative for absolute gravimetric amounts of Abs (see below and Figure 6), it is conveniently reproducible. Preferably a serum specimen is chosen which is: (a) available in a large quantity (10–50 ml depending on the scope of the study) and (b) has a relatively high titre against the antigen in question. Serum makes a practical reference standard because, stored at −20 °C, it retains its Ab activity probably by virtue of the stable natural environment of proteins in serum and the minimal proteolytic activity for Abs in serum. Serum is clear and can be readily sterile-filtered into preservable aliquots. In the situation in which the titre of the isotype of specific Ab selected is too low, e.g. some IgA Abs, secretions rather than serum, can be used as a reference standard. The secretion should be stored in 50 per cent glycerol at either −20 °C or −70 °C. We typically store reference standards in 200 μl/ml aliquots in plastic vials with rubber seal caps to prevent dehydration.

A six-dilution sequence (in independent duplicate) is prepared from the reference standard each day the assay is performed. It is highly recommended

DILUTION SEQUENCES

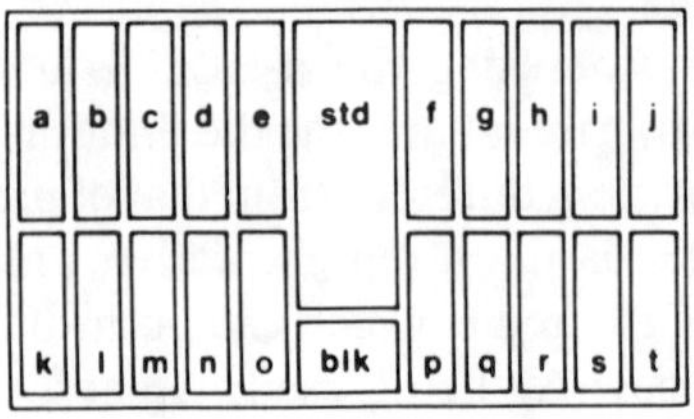

a b c d e std f g h i j
k l m n o blk p q r s t

If all the dilution values are correct, press ENTER, otherwise, enter a letter for the sample to be changed. Use Z for the standard.

Std.	Starting Dilution	Sequence Factor
Std.	20000	2
a	1000	3
b	300	3
c	300	3
d	50	3
e	300	3
f	300	3
g	300	3
h	300	3
i	300	3
j	300	3
k	300	3
l	300	3
m	1000	3
n	300	3
o	300	3
p	300	3
q	300	3
r	300	3
s	10000	3
t	300	3

FIGURE 4 Microtitre test format for ELISANALYSIS I as displayed to the operator on the computer monitor. The letters (a–t) represent 20 different patient sera that are tested at four different dilutions. In the example, the initial dilution (highest conc.) tested is indicated on the right. The number to the right of each initial dilution indicates the dilution factor. The example shows that a 1/2 dilution sequence was used for the standard while 1/3 dilution sequences were used for each test sample. Std = six-well dilution sequence in duplicate, of the reference standard; blk = assay blanks, PBS–T instead of patient sera is added

that the same pipetting device be used for the preparation of the dilution of the reference standard and the test sequences[k]. We prepared all dilutions in 10 × 75 mm disposable glass test tubes. Such tubes are actually re-usable depending on the cleanliness of the test-tube washing procedure. The dilutions are then transferred to the wells of the antigen-coated microtitre plates.

Dilutions of the test samples are prepared in the same manner as for the standard in glass tubes and later transferred to the appropriate wells of microtitre plates. Provided the highest dilution is at first transferred, the same pipette tip may be used to transfer sequentially the lower dilutions of the test samples (and reference standard). Automatic diluting machines, such as the Cetus Propipet or Dynatech Autodilutor III, were found to be impractical because the greatest amount of time needed for the preparation of dilutions was consumed in the preparation of the initial dilution and this must be done manually even if an autodiluter is available! When done carefully, dilutions can be prepared directly in microtitre wells using a multipipetter.

Certain sera or secretions cannot be tested without pre-treatment. For example, the measurement of IgE Abs in serum in the presence of a normally much higher concentration of IgG Abs of the same specificity, requires that the IgG Abs be depleted to prevent them from inhibiting binding by blocking all antigenic sites. We have calculated that if, in a normal serum, 10 per cent of the IgG and 25 per cent of the IgE were anti-allergen, the upper limits of accurate detection of the IgE would be 1.25 ng/ml (of course most IgG Ab is in the order of 1 per cent of total IgG). At lower serum dilutions, the higher IgE concentration would be inhibited by IgG because of the Ag-limiting nature of microtitre Ag-specific ELISAs (see Figure 6B). Hence, we have used Cowan I strain *Staphylococcus aureus* to deplete about 50 per cent of the IgG from patient sera prior to assay for IgE antigen-specific Abs.

Competition among Ab isotypes is also possible in other situations. In cattle we have observed that IgM Abs inhibit IgA Abs in serum (Butler *et al.*, 1980b) and that IgG Abs inhibit IgA Abs in rabbit serum (Butler *et al.*, 1980a). The latter presumably occurs by the same mechanism as inhibition of IgE while inhibition of IgA by IgM probably occurs due to the slow off-rate of the polyvalent IgM. This type of inhibition is usually apparent from the characteristics of the ELISA titration plots (Butler *et al.*, 1980b).

Data acquisition and analysis

The fact that 96 data points are rapidly generated from each microtitre plate dictates that it is desirable, if not necessary, that a computer-based data acquisition and analysis system be employed. Three manufacturers currently market automated microtitre plate readers (Dynatech MR series, FLOW Multiscan, and Biotek EL310). Each is capable of reading and recording the optical densities of all 96 wells of a standard microtitre plate in about 60 sec. Each of these is readily interfaced with a personal computer. In our laboratory, an EL310 is interfaced with an IBM-PCjr. The latter is inexpensive and its use for such data acquisition does not interrupt the operation of more sophisticated laboratory computers. It is also adequate for data analysis although the latter can be done at a subsequent time and the data stored on diskettes by the interfaced IBM-PCjr.

Various hardware manufacturers offer simplified programs for analysis of ELISA data. In almost every case, these fail to consider ELISA immunochemistry, contain no in-built means for comparing the slope of the standard and specimen titration plots, and are too often designed only to treat 'endpoint' determinations. ELISANALYSIS (Peterman and Butler, 1987) is designed to overcome these disadvantages and to provide user-friendly programs for correct analysis of ELISA data. ELISANALYSIS I is a simplified program designed to collect and analyse data obtained when the ELISA test format of Figure 4 is used. ELISANALYSIS II is a more flexible program for experimental studies, and will not be described here.

```
-+-+-+-+-+-+-+-+-+-+-+-+-+-+-+-+-+-+-+-+-+-+-+-+-+-+-+-+-+-+-+-+-+-+-+-+-+-+-+-+-+-+-+-+-

060186C IgA ABS TO S. MUTANS(c)

LN/LN plot
Equation of the line:     Y = -1.035756   * X +  5.456356
Correlation Coef.= .9969003 , for N= 11

Background set at  .17025

For Dilution Range  200  to  6400
```

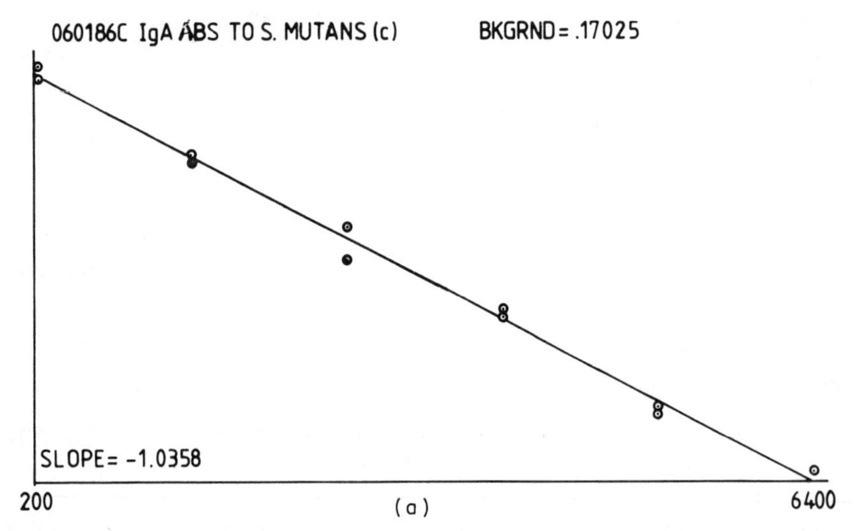

FIGURE 5 Computer printout of the measurement of IgA antibodies to *S. mutans* using the test format described in Figure 4 and the graphic analyses presented in Figure 6D. (a) The log–log regression analysis, with its equation (p.122). (b) The data obtained with ELISANALYSIS for four patients (p.123). The values obtained with the reference standard (tested in columns 6 and 7; see Figure 4) are tabled. At extreme left is the actual OD_{405}, the next column indicates the dilution of the reference standard tested and even further to the right, the computer-determined actual dilutions (equiv. dilution) are presented. At the extreme right of this table, the calaculated EU/ml at each dilution is given. ELISANALYSIS corrects for dilution so that the mean of these values should be 100 EU/ml of the reference standard. The same format is used as for the reference standard. Values which fall above or below the reference standard plot are appropriately 'flagged'. When the values in EU/ml are independent of sample dilution, the investigator can be sure that the test sample is titrating with the same slope as the reference standard (see Figure 6D). When three or more values fall within the reference standard curve and give values indicating that they titrate with the same slope, ELISANALYSIS computes the mean and appropriate statistics for such samples (exemplified in a box)

ELISANALYSIS I receives input OD_{405} values for the six duplicate dilutions of the reference standard and uses pre-programmed information on the initial

-+-

060186C IgA ABS TO S. MUTANS(c)

#	O.D.	D.F.	Equiv. D.F.	E.U./ml	On Curve?
1	1.132	200	201.472	99.26959	
2	1.228	200	183.789	108.8205	
3	.651	400	393.514	101.6481	
4	.617	400	422.392	94.69876	
5	.421	800	737.701	108.445	
6	.361	800	960.631	83.27863	
7	.285	1600	1569.094	101.9697	
8	.294	1600	1458.775	109.6811	
9	.219	3200	3585.855	89.23952	
10	.223	3200	3322.974	96.29928	
11	.2	6400	5776.65	110.7909	
12	.178	6400	21168.715	30.2333	Below

-+-

060186C IgA ABS TO S. MUTANS(c)

#	O.D.	D.F.	Equiv. D.F.	E.U./ml	On Curve?
25	1.187	100	190.939	52.37266	
26	.727	200	341.523	58.5612	
27	.564	400	477.162	83.82891	
28	.436	800	697.46	114.702	

-+-

060186C IgA ABS TO S. MUTANS(c)

#	O.D.	D.F.	Equiv. D.F.	E.U./ml	On Curve?
45	.582	1000	457.007	218.8148	
46	.427	2000	721.05	277.3733	
47	.292	4000	1481.905	269.9228	
48	.189	8000	9020.707	88.68484	Below

218.8148
277.3733
269.9228
Mean= 2.55E+02 for 3 points
Variance(N-1)= 1.02E+03
Std Error= 1.84E+01
Coeficient of Variation= 10.2%

-+-

060186C IgA ABS TO S. MUTANS(c)

#	O.D.	D.F.	Equiv. D.F.	E.U./ml	On Curve?
65	1.368	400	163.005	245.392	Above
66	.743	800	332.308	240.7409	
67	.44	1600	687.472	232.7368	
68	.332	3200	1126.43	284.0834	

240.7409
232.7368
284.0834
Mean= 2.53E+02 for 3 points
Variance(N-1)= 7.63E+02
Std Error= 1.59E+01
Coeficient of Variation= 8.9%

-+-

060186C IgA ABS TO S. MUTANS(c)

#	O.D.	D.F.	Equiv. D.F.	E.U./ml	On Curve?
77	.599	20	439.501	4.550617	
78	.434	40	702.565	5.693421	
79	.293	80	1470.247	5.441262	
80	.229	160	2994.724	5.34273	

4.550617
5.693421
5.441262
5.34273
Mean= 5.26E+00 for 4 points
Variance(N-1)= 2.44E-01
Std Error= 2.47E-01
Coeficient of Variation= 8.1%

(b)

dilution and the nature of the dilution sequence, i.e. twofold, threefold, etc. (Figure 4). The system is programmed to calculate a mean from the four control wells (wells G 5, 6 and H 5, 6) and to perform a linear regression analysis using data from the reference standard. The reference standard is assigned an activity of 100 ELISA Units (EU) per undiluted ml. Hence for each dilution, ELISANALYSIS corrects for dilution and background and prints out a value for EU/ml (Figure 5b). Theoretically this should be 100 EU/ml ± experimental error at each dilution at which the reference standard was tested. ELISANALYSIS then uses pre-programmed information on the dilution of the control samples together with OD_{405} data for each unknown dilution and those of the control wells, to calculate a value in EU/ml for each dilution of each unknown. If the EUs/ml at three of the four dilutions of the same unknown are the same (we currently use a thumb role of accepting means with CV $<$ 20 per cent) and show no dilution-related trend, the slope of the linear regression plot of the unknown is considered to be the same as that of the standard. If not, the test sample is judged as titrating with a different slope than the standard either for reasons of dilution error or immunochemistry (see below). Also, ELISANALYSIS 'flags' any values of the unknowns which are above or below that of the standard, and labels them as such on the printout (Figure 5b). When more than two of the four values of the test sample are above or below the range of the reference standard, the sample is re-assayed using a corrected initial dilution. The ELISANALYSIS format shown in Figure 4 accommodates the testing of 20 unknowns, but another version of ELISANALYSIS I permit eight different dilutions of ten different samples to be tested. This version can be useful when very little is known about the expected titre of an unknown sample.

The immunochemical basis as to why certain test samples titrate differently from the standard (e.g. Figure 6) can be: (a) competition of isotypes for the same antigen; (b) excessive non-specific binding of the specimen which raises its true background value significantly above the four-well background mean; or (c) a combination of isotype competition and epitope heterogeneity when multivalent antigens are used. It is noteworthy that affinity does not influence the slope of such log–log plots, only the Y-intercept (Peterman, 1986; Peterman and Butler, 1988; Figures 6B and 6C).

ELISANALYSIS also can be asked to compute a mean EU/ml from all dilutions of each unknown which titrate in the manner of the standard. These are printed out together with their corresponding statistics (Figure 5b).

IMMUNOCHEMISTRY OF ANTIGEN-SPECIFIC ELISAs

Mass Law considerations in the binding of antibodies to solid phase antigens

The issues concerning the *in vivo* significance of Ab measured using microtitre ELISAs warrant discussion. Two of these have directly to do with Mass Law considerations and are described here. A third is related to solid phase antigen

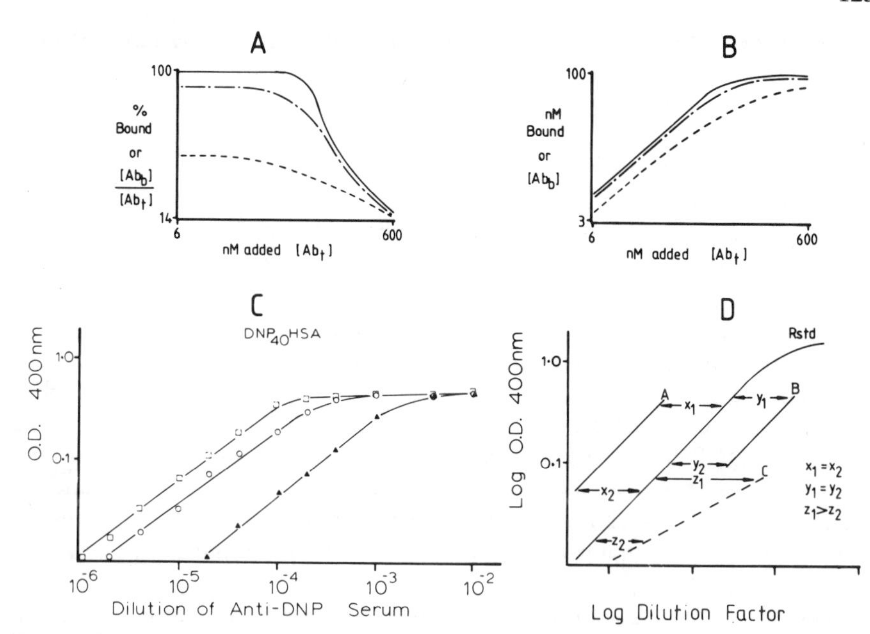

FIGURE 6 Titration plots for antigen-specific ELISAs and their interpretations. (A) The theoretical log–log binding plots for antibodies of three different affinities when tested against antigen with a fixed epitope density; data expressed as per cent of added Ab which can bind as a function of affinity. (B) Same as A but data expressed as the amount of antibody bound. (Data from Peterman, 1986). (C) The log–log ELISA titration plots for rat sera containing different amounts of anti-DNP antibodies and possessing different affinities. □ = 3.25 mg/ml; $K^a = 1.1 \times 10^8$ *l*/m. ● = 0.075 mg/ml; $K^a = 7.8 \times 10^7$ *l*/m. ▲ = 0.27 mg/ml; $K^a = 2.6 \times 10^6$ *l*/m. (Data from Butler *et al.*, 1978b and Butler, 1981). (D) Graphic illustration of the titration behaviour of three test sera in an antigen-specific ELISA compared to that obtained with the reference standard (Rstd). Note that only in those situations in which test samples titrate with the same slope as the Rstd, can valid estimations of antibody activity be made

and is discussed below. First, what percentage of the Abs in a dilution becomes bound to the solid phase Ag and is subsequently detected? This depends on the K^a of the Ab and a theoretical model explaining this relationship is shown in Figure 6A. As can be seen, for Abs with K^a of $>10^8$, >90 per cent of the added Abs will bind in the region of excess Ag (plateau region of Figure 6A, linear region of plot 6B). Hence affinity influences the detectability of Abs using the ELISA. This agrees with observations we made when ELISA was employed to measure anti-DNP in rats (Butler *et al.*, 1978b; Figure 6C). These observations have been corroborated by the work of others (Lehtonen and Eerola, 1982; Lew, 1984; Nimmo *et al.*, 1984). Furthermore, affinity influences the intercept of the log–log titration plots but not their slope (Figure 6B); this principle has also been empirically demonstrated (Butler, 1981; Peterman, 1986; Figure 6C). Further evidence for the affinity of ELISA is supported by data on the influence of epitope density in hapten–carrier systems (Butler, 1981; Lew, 1984; Nimmo *et al.*, 1984; Peterman, 1986). Increasing epitope density preferentially improves the binding of lower affinity antibodies (Peterman, 1986).

Secondly, the theoretical predictions of Figures 6A and B combined with their empirical demonstration (Peterman, 1986; Figure 6C) mean that: (a) the slopes of the log–log plots for all ELISAs should be constant, ideally 1.0; and (b) because the intercepts of such plots are affinity dependent, the ELISA cannot be used as an assay to quantitate antibodies in absolute units unless the affinity of each test sample is the same as the affinity of the reference standard. The latter is biologically unlikely, so that ELISAs, including the a-ELISA, in which the reference standard method is used, cannot measure Ab in absolute units. Figure 6C illustrates empirically how the Y-intercepts obtained with three anti-DNP antibodies correlate better with their affinity than their antibody concentrations. Finally, there is no guarantee that the *in situ* bound Ab measured by ELISA represents the total amount of Ab in the sample tested, only the proportion that can bind under the conditions of the assay (see above and Figure 6A).

Therefore, the calculation of data in EU/ml as the read-out of ELISA-NALYSIS, is a practical and utilitarian system which rather than providing data in absolute units, provides data as units of 'activity', i.e. EU/ml. This activity reflects both theoretical and empirical arguments that it is simplistically the product of Ab concentration and its K^a.

Characteristics of solid-phase antigens

The phenomenon of protein adsorption on plastic has been previously discussed. The early studies of this phenomenon were done with little or no control for the antigenic properties of the adsorbed antigens. It was later observed that adsorption of ovalbumin results in a significant loss of antigenic activity (Kennel, 1982; Dierks, Butler, and Richerson, 1986); this finding is consistent with earlier studies that indicated protein denaturation occurred during adsorption (Burghardt and Axelrod, 1983). In our laboratory, adsorption of FLU–gelatine at 37 °C for 3 hr results in a greater amount of more stably bound Ag than when adsorbed overnight at 4 °C. However, Ag bound at 4 °C is more antigenically active.

Previously published data on the nature of the protein–plastic bond (Brash and Lyman, 1969) are totally consistent with the loss of antigenicity observed as a result of surface adsorption. It would appear that surface adsorption results in a progressive unfolding of protein to expose more hydrophobic residues to the plastic. This results in more stable binding (as we observed at 37 °C) but simultaneously results in alteration of Ag epitopes. Hence, the exact 'functional' nature of surface-adsorbed proteins is, in practice, unknown and its concentration is difficult to assess because the true reaction volume in solid phase immunoassays is difficult to measure[j]. Therefore, the behaviour of an Ab for the solid phase antigen used to measure it *in vitro*, and its behaviour for antigen *in vivo*, could be quite different.

The immobilization of Ag on plastic surface also sterically hinders subsequent antigen–antibody interactions. Empirical data show that steric hindrance prevents the detection of some of the bound Abs by the antibody–enzyme conjugates and complexes used in ELISAs (Koertge and Butler, 1985). Data from EM scanning studies reveal that Immulon 2 surfaces are regular, and that adsorbed antigen occupies <25 per cent of the surface (Peterman, 1986). Hence, the detection system, but not the primary Ab, would most likely be sterically hindered; this was shown to be the case using conjugates of different sizes (Koertge and Butler, 1985).

Stoichiometry of antigen-specific ELISAs: factors affecting the titration plot

The molecular stoichiometry of antigen-specific ELISA, including the a-ELISA, is variable depending on what point in the titration plot is analysed. The failure to maintain a constant stoichiometry throughout the linear titration range (Figure 7A) means that the titration plot prescribed by the enzymic reaction is not a direct quantitative measure of the amount of primary antibody bound. This phenomenon is caused by two factors acting alone or together. First, the detection system is progressively hindered from binding the primary Ab with constant stoichiometry as the concentration of the latter increases. The degree and range over which this hindrance occurs is directly related to the size of the detection system (Koertge and Butler, 1985). Second, deviations between the binding of the primary Ab and the extent of the enzyme reaction at high dilution of Ab, can result from the underestimation of the background value, i.e. an underestimation of the degree of non-specific binding by the detection system in the absence of the primary antiserum. There is no practical, empirical means of obtaining a 'correct' background value because it would require selectively and totally removing all the specific Abs from the same sample and then re-testing it comparatively with the unaltered sample. Not only is such a method impractical, it would be open to the criticism that the non-specific binding factor had also been removed during depletion of the primary Ab! In addition, the other proteins present in the primary specimen but absent when PBS–T is used in the control wells, could lower the true non-specific background by blocking sites on the Ag-coated well which could otherwise non-specifically bind components of the detection system. It is not uncommon, especially when testing samples with high protein content but low antibody content, for the 'background values' of control wells treated with PBS–T to have absorbancies higher than the highest dilution of the test sample. The devil's circle created by this dilemma makes empirical determination of

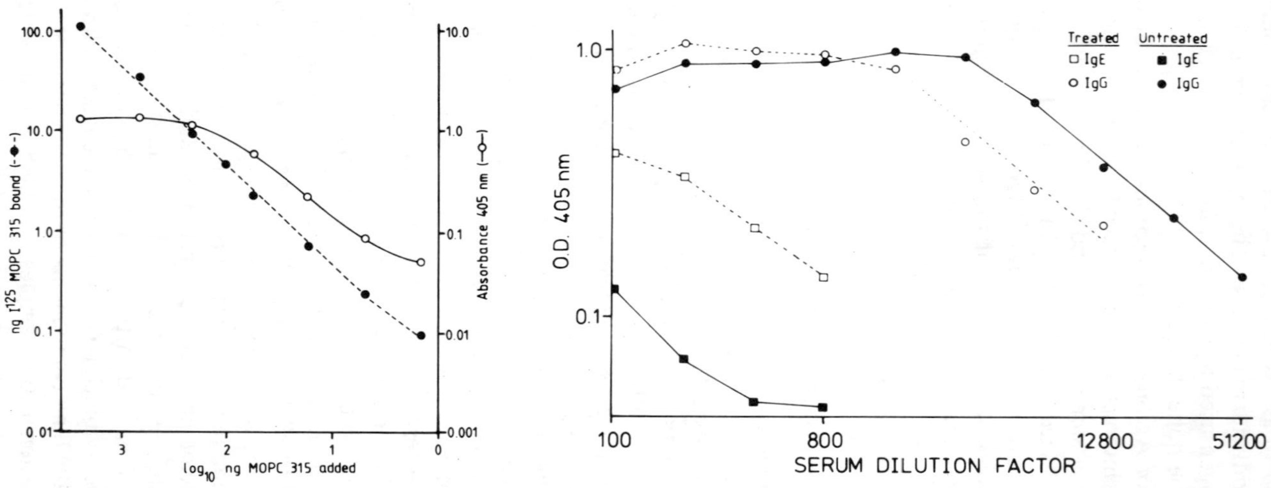

FIGURE 7 Steric hindrance and antibody competition in the performance and interpretation of antigen-specific microtitre ELISAs. *Left*: The relationship between the actual binding of M315 (left-hand axis) as determined by direct measurement of ^{125}I M315 bound versus the indirect measure of the amount bound on the same wells using the detection system of the a-ELISA (right-hand axis expressed as OD_{405}). (From Koertge and Butler, 1985). *Right*: The competitive inhibition of IgE antibodies to ragweed by IgG antibodies directed to epitopes on the same allergen. Titration plots generated using the a-ELISA for IgE and IgG before and after treatment of the serum with 1×10^{11} Cowan I strain *S. aureus*

background values all but impossible. An alternative but time-consuming method for establishing the correct background is to titrate the test samples to 'infinite' dilution rather than using just four dilutions as we described here for routine assays (Figure 4). By this technique, the background is selected as the highest dilution of the sample which shows no further significant decrease in absorbancy (Figure 7A). Finally, experienced investigators may estimate their background values for systems with which they are familiar. While such a procedure may seem 'unscientific', it may in fact be more correct than any empirical method suitable for use in routine assay!

The inability to obtain a constant stoichiometry in ELISA, most noticeably that resulting from steric hindrance of the detection system (Koertge and Butler, 1985) means that absolute quantitation of even *in situ* bound antibodies by stoichiometric calculations is not possible (Koertge, Butler, and Dierks, 1985).

Another factor which can influence the nature of Ag-specific ELISA titration plots is competition of two or more Abs of different isotypes, for the same antigenic determinant or adjacent determinants, e.g. IgE and IgG (Figure 7B). As stated earlier this can produce a titration plot, no portion of which is linear or, if linear, then only over a range too short to be usable or with a different slope than the reference standard (Figure 6D). After removal of some of the competing Abs, a more usable titration which also yields higher estimates of activity for the isotype in question, e.g. IgE, is obtained (Figure 7B).

SUMMARY

The discussions presented in this section provide the basis for the establishment of the routine test format described above and ELISANALYSIS I. The rationale for the read-out in Ab activity, i.e. EU/ml, should now be apparent. The importance, but difficulty in establishing a 'correct' background value, is the basis for allotting four wells to serve as controls; they receive no primary antibody but all other reagents and a background value is calculated from the mean of their ODs. The questionable validity of this means of establishing background values is the weakest part of the routine a-ELISA design and analysis. Such a determination is important as it may very well influence the comparative slope differences between test samples and the reference standard. When efficiency and precision have to be considered, the procedure and format we described in this chapter for the a-ELISA is an optimal system which sacrifices little in the way of scientific accuracy while permitting 20 test samples to be analysed per microtitre plate.

A major weakness in all solid phase ELISAs in which Ags are adsorbed to plastic is the effect of the consequent antigen alteration. In the case of hapten–protein conjugates and whole bacteria this effect may be minimal, but such denaturation could significantly alter the measurement of antibodies to

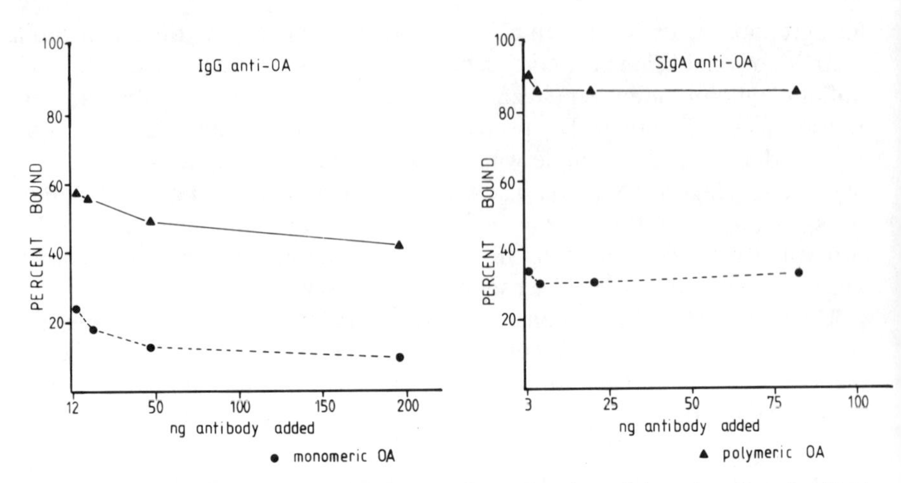

FIGURE 8 The influence of aggregation on the antigenic activity of ovalbumin (OA) absorbed on Immulon 2 microtitre wells. *Left*: Relationship between nanograms of IgG anti-OA added and the percentage bound to either adsorbed monomeric OA (●) or glutaraldehyde-polymerized OA (▲). *Right*: Same as on left but system tests the ability of monomeric or polymerized OA to bind SIgA anti-OA

protein antigens (Dierks, Butler, and Richerson, 1986). The difference observed between the binding of Ab to Ag in free solution versus Ag adsorbed on plastic suggests either: (a) antigenic alteration or (b) steric hindrance. One technique which appears to improve the antigenicity of adsorbed antigen is to use polymerized antigen (Figure 8). Therefore, aggregation or other means of preventing adsorption-induced denaturation of antigenic determinants will have to be developed to allow solid phase Ag-specific ELISAs to reach their full potential in immunology and medical science.

ACKNOWLEDGEMENTS

The author wishes to acknowledge Ms Jill Spradling, Dr J.H. Peterman, and especially Ms Patricia McGivern for their contributions to the data presented. This review is dedicated to the memory of Patricia McGivern who recently succumbed to malignant adenocarcinoma. Patricia McGivern was an inspiration to all laboratory workers for her diligence, concern for precision and detail, and overall awareness of the implications of her work to medical science. Her tireless contribution to ELISA biotechnology, even in the wake of terminal cancer, will not be forgotten.

NOTES

(a) Reaction buffer is 0.01 M sodium phosphate, pH 7.1, containing 0.15 M NaCl and 0.05 per cent Tween 20. Addition of 0.02 per cent sodium azide is optional and this

depends on how rapidly the reaction buffer is used and replaced. Reaction buffer in routine use is stored at RT.
(b) The kinetics of Ab–Ag interactions, when Ag is immobilized on plastic, are slower than that observed in free solution (Koertge, 1984). These altered kinetics are due in part to the rate of diffusion (Peterman, 1986) but appear also to reflect true differences in the nature of solid phase Ag–Ab compared to those in free solution.
(c) Coupling buffer is 0.1 M sodium carbonate, pH 9.6. This is stored at 4 °C and freshly prepared every 2–3 weeks. As cited above, adsorption can be done using a variety of detergent-free buffers. The value of 150ng/1·41 cm^2 applies to most proteins but may be higher or lower for proteins of higher or lower affinity for plastic, respectively.
(d) Wash solution is 0.15 M NaCl containing 0.05 per cent Tween 20. The solution is prepared approximately every 2 days in 16 l volumes. Wash solution is stored in chromic-acid-clean glass carboys.
(e) Adsorbed Antigen Activity Assay: this designation reflects our recognition that Ags can undergo denaturation during adsorption so that the activity measured is that of the adsorbed Ag, not the native Ag in solution.
(f) In practice, purified IgA and IgM are myeloma proteins while IgG is purified from pooled human serum.
(g) Substrate buffer is 50 mM carbonate, pH 9.8, containing 1 mM $MgCl_2$. It is stored at 4 °C and prepared fresh at least monthly.
(h) The adsorption of protein on plastic should not be ignored in laboratory procedures. Storage of samples at high protein concentration, i.e. >5 mg/ml, presents no problem although storage of dilute solutions should be avoided.
(i) Microtitre plates are emptied by inverting them over a sink, typically combined with a sharp snapping action of the wrist. They are washed with wash solution (note d) using a gravity feed device developed in our laboratory in collaboration with K. Breier, Institute of Animal Husbandary, FRG (Butler, Peterman, and Koertge, 1985). The device washes 16 wells simultaneously for approximately 10 seconds. Wash solution passes through a 8 μm Millipore in-line filter before entering the washer.
(j) It is currently difficult to calculate precisely the concentration of the solid phase Ag because Ag is not uniformly distributed within the 200 μl volume in which the Ag–Ab occurs. Rather, Ag is located on the surface and its concentration at the point of interaction with Ag may be much higher. As the Mass Law depends on Ag concentration, uncertainly exists as to the appropriate means for its calculation. If Ag concentration is based on a 200 A° region from the plastic surface, most solid phase immunoassay should not be affinity dependent.
(k) Air displacement 'autopipets' are widely used in clinical and experimental laboratories of immunology. The accuracy of such devices is influenced by sample viscosity and by their condition. Pipetmans from Rainin require periodic calibration which should be done only after replacement of 'o' rings and other expendible parts.

REFERENCES

Brash, J.L., and Lyman, D.J. (1969). Adsorption of plasma protein in solution to uncharged hydrophobic polymer surfaces. *J. Biomed. Mat. Res.*, **3,** 175–89.
Bull, H.B. (1956). Adsorption of bovine serum albumin on glass. *Biochim. Biophys. Acta*, **19,** 464–71.
Burghardt, T.P., and Axelrod, D. (1983). Total internal reflection fluorescence study of energy transfer in surface-adsorbed and dissolved bovine serum albumin. *Biochemistry*, **22,** 979–85.
Butler, J.E. (1980). The nature of the antibody-antigen and antibody-hapten

interaction. In: *Enzyme Immunoassay; Principles and Practice* (ed. E. Maggio). CRC Press. Chapter 2. pp. 5–52.

Butler, J.E. (1981). The amplified ELISA: Principles of and application for the comparative quantitation of class and subclass antibodies and the distribution of antibodies and antigens in biochemical separates. In: *Methods in Enzymology* (eds. H.J. Vunakis and J.J. Lagone). Vol. 73, pp. 482–523.

Butler, J.E. (1983). Bovine immunoglobulins: An augmented review. *Vet. Immunol. Immunopath.*, **4,** 43–152.

Butler, J.E., McGivern, P.L., and Swanson, P. (1978a). Amplification of the enzyme-linked immunosorbent assay (ELISA) for the detection of antibodies. *J. Immunol. Meth.*, **20,** 365–83.

Butler, J.E., Peterman, J.H., and Koertge, T.E. (1985). The amplified enzyme-linked immunosorbent assay (a-ELISA). In: *Enzyme Mediated Immunoassay* (eds. T.T. Ngo and H.M. Lenhoff). Plenum Press, NY, pp. 241–76.

Butler, J.E., Feldbush, T.L., McGivern, P.L., and Stewart, N. (1978b). The enzyme-linked immunosorbent assay (ELISA): A measure of antibody concentration or affinity? *Immunochemistry*, **15,** 131–6.

Butler, J.E., Cantarero, L., Swanson, P., and McGivern, P.L. (1980a). The amplified ELISA: Principles, modifications and applications. In: *Enzyme Immunoassay* (ed. E. Maggio). C.R.C. Press, pp. 197–212.

Butler, J.E., McGivern, P.L., Cantarero, L.A., and Peterson, L. (1980b). Application of the amplified enzyme-linked immunosorbent assay: Comparative quantitative of bovine serum IgG1, IgG2, IgA and IgM antibodies. *Am. J. Vet, Res.*, **41,** 1479–91.

Butler, J.E., Swanson, P.A., Richerson, H.B., Ratajczak, H.V., Richards, D.W., and Suelzer, M.T. (1982). The local and systemic IgA and IgG antibody responses of rabbits to a soluble inhaled antigen. Measurement of responses in a model of acute hypersensitivity pneumonitis. *Am. Rev. Resp. Dis.*, **126,** 80–5.

Butler, J.E., Peterman, J.H., Suter, M., and Dierks, S.E. (1987). The immunochemistry of solid-phrase sandwich enzyme-linked immunosorbent assays. *Fed. Proc.* **46**, 2548–56.

Cantarero, L.A., Butler, J.E., and Osbourne, J.W. (1980). The binding characteristic of six proteins to polystyrene and its implications to solid-phase immunoassays using polystyrene tubes. *Anal. Biochem.*, **105,** 375–82.

Catt, K., and Tregear, G.W. (1967). Solid-phase radioimmunoassay in antibody-coated tubes. *Science*, **158,** 1570–2.

Czerkinsky, C., Rees, A.S., Bergmeir, L.A., and Challacombe, S.J. (1983). The detection and specificity of class specific antibodies to whole bacterial cells using a solid phase radioimmunoassay. *Clin. Exp. Immunol.*, **53,** 192–200.

Dierks, S.E. (1985). Differential recognition of solid phase immunoglobulin and antibody bound to solid phase antigen: A method of antibody quantitation. M.S. Thesis, University of Iowa.

Dierks, S.E., Butler, J.E., and Richerson, H.B. (1986). Altered recognition of surface-adsorbed compared to antigen-bound antibodies in the ELISA. *Molec. Immunol.*, **23,** 403–11.

Engvall, E., and Perlmann, P. (1972). Enzymes-linked immunosorbent assay, ELISA. III. Quantitation of specific antibodies by enzyme-labeled anti-immunoglobulin in antigen-coated tubes. *J. Immunol.*, **109,** 129–35.

Kennel, S. (1982). Binding of monoclonal antibody to protein antigen in fluid phase and bound to solid supports. *J. Imm. Methods.*, **55,** 1–12.

Koertge, T.E. 1984. A study of the quantitative capability of the enzyme-linked immunosorbent assay (ELISA) to study the transport of antibodies into secretions. Ph.D. Thesis, University of Iowa.

Koertge, T.E., and Butler, J.E. (1985). The relationship between the binding of primary antibody to solid-phase antigen in microtitre plates and its detection by the ELISA. *J. Imm. Methods*, **83,** 283–99.

Koertge, T.E., Butler, J.E., and Dierks, S.E. (1985). Characterization of the soluble immune comples (EIC) of the amplified enzyme-linked immunosorbent assay (a-ELISA) and an evaluation of this assay for quantitation by reaction stoichiometry. *J. Immunoassay*, **6,** 371–90.

Lambden, P.R., and Watt, P.J. (1978). A solid phase radioimmunoassay on hydrophobic membrane filters: detection of antibodies to gonococcal surface antigens. *J. Imm. Methods*, **20,** 277–86.

Lehtonen, O.P., and Eerola, E. (1982). The effect of different antibody affinities on ELISA absorbance and titer. *J. Imm. Methods*, **54,** 233–40.

Lehtonen, O.P., and Viljanen, M.K. (1980). Antigen attachment in ELISA. *J. Imm. Methods*, **341,** 61–70.

Leone, C.A. (1968). Turbidimetric assay methods: Application to antigen-antibody reactions. In: *Methods in Immunology and Immunochemistry* (eds. C.A. Williams and M.W. Chase). Academic Press, NY, pp. 174–81.

Lew, A.M. (1984). The effect of epitope density and antibody affinity on the ELISA as analysed by monoclonal antibodies. *J. Imm. Methods*, **72,** 171–6.

Maiolini, R., and Masseyeff, R. (1975). A sandwich method of enzymoimmunoassay. I. Application to rat and human alpha-fetoprotein. *J. Imm. Methods*, **8,** 223–4.

Nimmo, G.R., Lew, A.M., Stanley, C.M., and Steward, M.W. (1984). Influence of antibody affinity on the performance of different antibody assays. *J. Imm. Methods*, **72,** 177–87.

Peterman, J.H. (1986). Factors which influence the binding of antibody to solid phase antigens: Theoretical and experimental investigations. Ph.D. Thesis, University of Iowa.

Peterman, J.H., and Butler, J.E. (1988). ELISANALYSIS: A data acquisition and analysis system for antigen-specific and sandwich ELISAs which considers their immunochemistry. *J. Imm. Methods* (in press).

Peterman, J.H., Voss, Jr., E.W., and Butler, J.E. (1985). Antibody affinity and antibody–antigen interactions: Theoretical considerations and applications to solid-phase binding assays. *Fed. Proc.*, **44,** 1870.

Sloan, G.J. (1975). Enzyme-linked immunosorbent assay (ELISA): Application to the quantitation by subclass of bovine antibodies against *Staphylococcus auerus*. M.S. Thesis, University of Iowa.

Sloan, G.J., and Butler, J.E. (1978). Evaluation of enzyme-linked immunosorbent assay for quantitation by subclass for bovine antibodies. *Am. J. Vet. Res.*, **39,** 935–41.

Suter, M. (1982). A modified ELISA technique for anti-hapten antibodies. *J. Imm. Methods*, **53,** 103–8.

Suter, M., and Butler, J.E. (1986). The immunochemistry of sandwich ELISAs. II. A novel capture system which prevents the adsorption-induced denaturation of capture antibodies. *Immunol. Lett.*, **13,** 313–16.

ELISA and Other Solid Phase Immunoassays
Edited by D.M. Kemeny and S.J. Challacombe

CHAPTER 6

The Role of Antibody Affinity in the Performance of Solid Phase Assays

M.E. Devey and **M.W. Steward**

Immunology Unit, London School of Hygiene and Tropical Medicine, London

CONTENTS

INTRODUCTION

The requirement for rapid, specific, and sensitive means to detect antibodies, antigens, and other substances of biological importance has led to the development of a wide range of different immunoassays. In particular, the solid phase assays, especially those utilizing enzyme-linked conjugates, have become widely used in laboratories throughout the world. The ability of such assays to measure absolute antibody concentration has been questioned (Ahlstedt, Holmgren and Hanson, 1974; Butler *et al.*, 1978) and it has become apparent that antibody affinity may significantly influence the results of solid phase and other assays. In addition, although antibody responses are often considered solely in terms of concentration, it is important to realize that antibody affinity may play a critical role in determining biological activities of antibodies (Steward and Steensgaard, 1983). Therefore, it is necessary not only to be able to characterize an antibody response in terms of affinity as well as magnitude but also to understand how antibody affinity may affect a particular assay. In this chapter the measurement of antibody affinity will be discussed and the role of affinity in the performance of various antibody assays assessed, with particular emphasis on enzyme-liked assays (ELISA).

ANTIBODY AFFINITY

Antibody affinity describes the strength of the primary interaction between a single antibody combining site and its homologous antigenic determinant or epitope, which has its most precise application in monovalent hapten–anti-hapten systems.

The term avidity has not been defined precisely in thermodynamic terms and has acquired a number of different meanings. It is sometimes used synonymously with affinity but most often it is used to describe the ability of antibodies to form stable complexes with antigen. Avidity is largely determined by the affinity but also depends upon the valence of both the antibody and the antigen as well as on other properties, such as antibody heterogeneity, net charge and hinge region flexibility, which are independent of the primary binding reaction. The term avidity is therefore best avoided and, as suggested by Karush (1978), it is preferable to use the term *intrinsic affinity* to describe the primary binding energy of a single antibody combining site with a single antigenic determinant, while the term *functional affinity* should be used to describe reactions between multivalent antibodies and complex antigens (Figure 1). Multivalent binding thus leads to a considerable increase in the equilibrum constant compared to monovalent binding. For example, IgM antibodies may have a high functional affinity, due to their ability to interact multivalently with antigen, despite having a low intrinsic affinity.

The interaction between antibody and antigen is reversible:

$$(\mathrm{Ab}) + (\mathrm{Ag}) \underset{k_\mathrm{d}}{\overset{k_\mathrm{a}}{\rightleftharpoons}} (\mathrm{AbAg}) \qquad [1]$$

where (Ab) represents free antibody, (Ag) free antigen, (AbAg) the antigen–antibody complex and k_a and k_d the association and dissociation constants respectively. At equilibrium, the rate of formation of the complex will equal its rate of dissociation and the equilibrium constant (K) may be calculated from the ratio of the complexed to free reactants. Applying the Law of Mass Action:

$$K = \frac{(\mathrm{AbAg})}{(\mathrm{Ab})\,(\mathrm{Ag})} = \frac{k_\mathrm{a}}{k_\mathrm{d}} \qquad [2]$$

K, therefore, is a measure of affinity and its value will increase with increasing complex formation, a situation that will occur with high affinity antibodies.

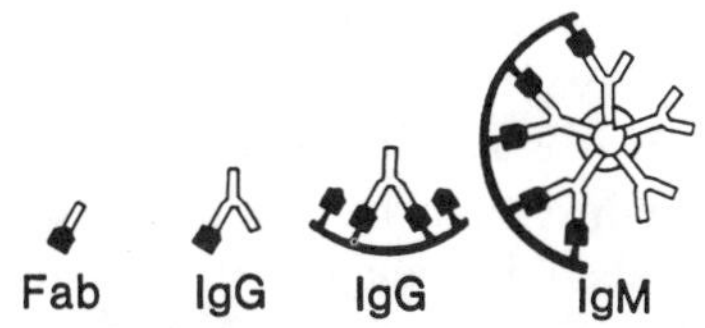

ANTIBODY	Fab	IgG	IgG	IgM
effective antibody valence	1	1	2	up to 10
antigen valence	1	1	n	n
equilibrium constant	10^4	10^4	10^7	10^{11}
advantage of multivalence	-	-	10^3-fold	10^7-fold
definition of binding	affinity	affinity	avidity	avidity
	intrinsic affinity		functional affinity	

FIGURE 1 Antibody affinity and avidity

Low affinity antibodies, on the other hand, have a greater tendency for dissociation and the value of K, accordingly, will be lower.

BIOLOGICAL SIGNIFICANCE OF ANTIBODY AFFINITY

It has been known for many years that the amount of antibody increases with time after immunization, and in addition that there is a progressive change in the quality of the antibody. Thus, the ability of antibodies produced late in the immune response to react more rapidly with antigen and form less dissociable complexes can be explained, at least in part, by the progressive increase in antibody affinity (affinity maturation) that occurs with time.

There is experimental evidence from a number of studies showing that high affinity antibodies mediate a variety of different biological effector functions more effectively than low affinity antibodies (reviewed by Steward and Steensgaard, 1983; Devey, 1986). In many cases only small differences in affinity were able to produce significant changes in biological reactivity (Fauci, Frank, and Johnson, 1970). In addition, the difference in biological reactivity between high and low affinity antibodies became less marked for certain reactions (complement fixation and passive cutaneous anaphlaxis) when high antigen concentrations were present. Thus, it seems possible that the disadvantage of low affinity responses may be overcome if large amounts of antibody can be produced or, in some cases, with high concentrations of antigen. Nevertheless, it is likely that the production of high affinity antibody responses is generally advantageous to the host and that a continued low affinity response may have immunopathological consequences (reviewed by Steward and Devey, 1981). Therefore, in assessing antibody responses, particularly after vaccination (Brown *et al.*, 1984), the importance of antibody affinity should be recognized. Furthermore, it is also necessary to recognize situations where low affinity antibody responses may be produced, such as early in the immune response, in the malnourished, in the aged, in the immunosuppressed, and in patients with multiple infections, so that appropriate tests capable of detecting low affinity antibodies can be used.

MEASUREMENT OF ANTIBODY AFFINITY

Equilibrium dialysis

The basis of many methods for determining antibody affinity is the measurement of the relative amounts of bound and free antigen at equilibrium over a range of antigen concentrations. A number of techniques have been used to separate bound antigen from free antigen. The classical method is equilibrium dialysis in which a dialysis membrane, permeable to the antigen (usually a radiolabelled hapten) but not to the antibody, is used to separate antigen from antibody. At equilibrium, the amount of *free* antigen will be the same in both compartments but any additional antigen in the antibody compartment will represent *bound* antigen and can be measured. Several mathematical approaches can be used to analyse data generated from such experiments and one of three forms of equation is generally used:

1. *The Scatchard equation*:

$$\frac{r}{(\mathrm{Ag})} = nK - rK \qquad [3]$$

were r = moles of antigen bound per moles of antibody present and n = antibody valence.

A plot of $r/(\mathrm{Ag})$ versus r over a range of antigen concentrations allows calculation of n (from the intercept on the abscissa) and K (from the slope, which is equal to $-K$).

2. *The Langmuir equation*:

$$\frac{1}{r} = \frac{1}{n} \times \frac{1}{(\mathrm{Ag})} \times \frac{1}{K} + \frac{1}{n} \qquad [4]$$

A plot of $1/r$ versus $1/(\mathrm{Ag})$ again allows both n and K to be determined. The intercept on the ordinate = $1/n$ and the slope, $1/nK$.

3. *The Sips equation:*

$$\log \frac{r}{n-r} = a \log K + a \log (\mathrm{Ag}) \qquad [5]$$

where a = heterogeneity index

A plot of log $r/n-r$ versus log (Ag) can be used to calculate K. Thus, where log $r/n-r = 0$ then $K = 1/(\mathrm{Ag})$. Furthermore, the heterogeneity index, a, is given by the slope of the lines.

Limitations of equilibrium dialysis

The mathematics of affinity calculations and the relative merits of the different approaches have been discussed in detail by Steward and Steensgaard (1983) and Steward (1986). Ideally, measurement of antibody affinity should be performed using homogeneous reactants in terms of both antigen and antibody, for example a monoclonal antibody and a monovalent hapten. Measurement of the affinity of antibodies in whole serum is complicated by the heterogeneity of the normal antibody response in terms of amount, isotype, and affinity. Furthermore, the fact that most antigens are large, complex, and multivalent poses further problems for the assessment of affinity. Under such circumstances, and particularly if large numbers of serum samples need to be evaluated, equilibrium dialysis would certainly not be an appropriate method.

Radioimmunoprecipitation

A relatively simple technique for measuring antibody affinity in whole serum is a modification of the Farr assay in which ammonium sulphate precipitation is used to separate bound from free antigen (Steward and Petty, 1972; Gaze, West, and Steward, 1973). This utilizes the fact that immunoglobulins are insoluble in 50 per cent saturated ammonium sulphate (SAS) and depends upon the requirement that the equilibrium is not disturbed by the precipitation step (Farr, 1958). Antibody is mixed with radiolabelled antigen over a range of antigen concentrations. The reaction is carried out in antigen excess in order to achieve binding site saturation of both the high and low affinity subpopulations.

The amounts of bound and free antigen are determined after SAS precipitation by counting the amount of radioactivity in the precipitate and supernatant and, using the Steward–Petty modification of the Langmuir equation (equation 4), a binding curve is constructed by plotting the reciprocal of bound antigen (1/b) against the reciprocal of free antigen (1/c) at each antigen concentration (Figure 2). The slope of the line is dependent on antibody affinity and by calculating the reciprocal of the free antigen concentration when

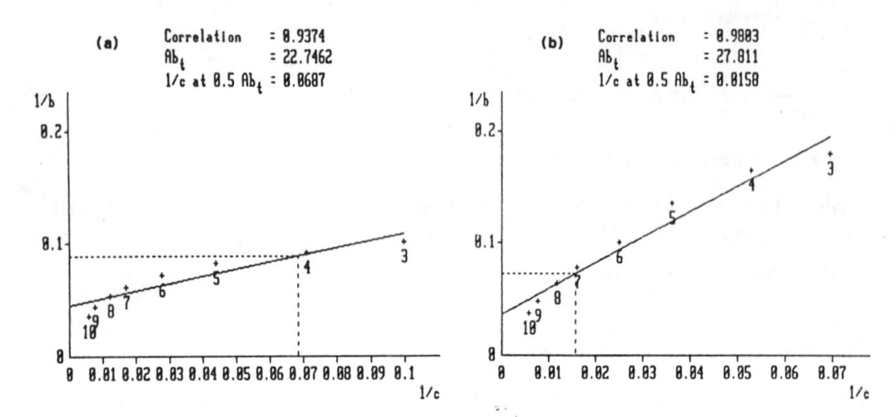

FIGURE 2 Steward–Petty plot of sera containing similar levels of high affinity antibodies (a) and low affinity antibodies (b) to HSA

half the antibody binding sites are occupied by antigen, a value for *K* is obtained. The total antibody binding sites (i.e. antibody concentration, Ab_t) can be found by extrapolating the straight line to the intercept (where the reciprocal of bound antigen = 0).

A limitation of the method is that the antigen must be soluble in 50 per cent SAS, although with antigens that are not, the problem may be overcome by using other selective precipitating agents such as polyethylene glycol or a second antibody. This method is ideal for the rapid estimation of the overall affinity of an antibody response in whole serum and has been shown to rank antibodies, in affinity terms, in the same order as equilibrium dialysis (Stanley, Lew, and Steward, 1983), although it does not give any information about the distribution of different affinity populations, i.e. affinity heterogeneity. Such information can only be gained by analysis of data by the Sips equation (equation 5).

Enzyme-linked immunosorbent assays

Analysis of dose–response curves

The finding by some authors that ELISA may be dependent on antibody affinity has led to suggestions that it may be used to measure affinity (Lehtonen and Eerola, 1982; Gripenberg and Gripenberg, 1983). These methods are based on the analysis of the shape of the dose–response curve when optical density is plotted against dilution of antibody. At low dilutions of sera (antibody excess) it would be expected, from the Law of Mass Action, that high affinity antibody would bind preferentially to antigen on the solid phase,

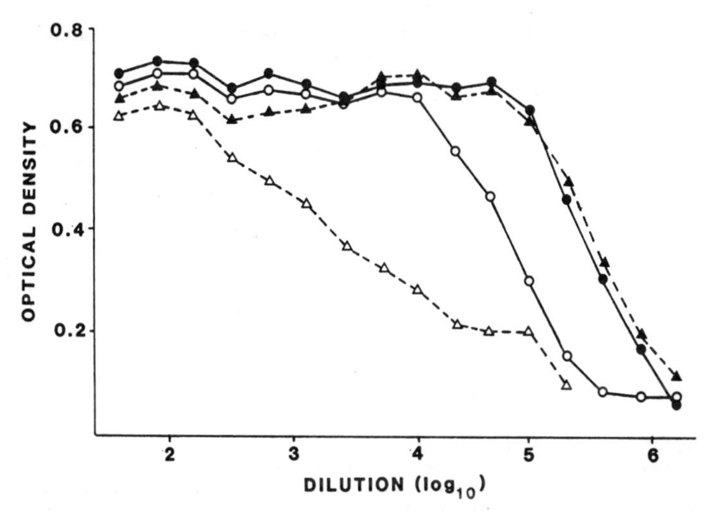

FIGURE 3 ELISA binding curve of a high affinity monoclonal antibody to DNP with $DNP_{40}BSA$ (●–●) and $DNP_{11}BSA$ (○–○) and a low affinity monoclonal antibody with $DNP_{40}BSA$ (▲--▲) and DNP_{11} BSA (△--△). (Redrawn from Nimmo *et al.*, 1984 BSA)

whereas at high dilutions of sera (antigen excess) low affinity antibody should also bind. Therefore, affinity may influence the slope of the linear part of the curve, which should be steep with high affinity antibodies and gradual with low affinity antibodies. Experimentally, this has been shown to be the case using monoclonal antibodies of defined affinity to haptens (Figure 3), but marked differences between high and low affinity antibodies were only observed when the antigen on the solid phases was at a low epitope density (Nimmo *et al.*, 1984; Lew, 1984). However, with sera containing polyclonal antibodies, the relative amounts of the different affinity populations will also influence the shape of the curve. In addition, there are theoretical and experimental grounds to suggest that antibody binding to solid-phase-immobilized antigen is not governed by a dynamic equilibrium (Nygren, Czerkinsky, and Stenburg, 1985) and it is likely that conformational changes of immobilized antigen will also affect antibody binding. Therefore, whether analysis of ELISA dose–responses curves can be utilized as a reproducible, reliable, and sensitive method for measuring antibody affinity in whole serum has not yet been established and available evidence suggests that it may not be useful.

Dissociation assays

An alternative approach, which may have greater practical application, has been applied to antibodies to viral antigens (Inouye *et al.*, 1984). This is based

TABLE 1 Relative affinity of polyclonal anti-HSA antibodies measured by SAS precipitation and DEA–ELISA (data from A. Falconar and P.R. Young, unpublished)

Serum	SAS precipitation* $K \times 10^6 M^{-1}$	DEA–ELISA† Fold reduction‡ (mean ± SD)	$\frac{1}{\log_{10} \text{reduction}}$
33	2.15	14.7 ± 3.4	0.86
11	1.50	19.0 ± 2.7	0.78
44	1.49	19.6 ± 2.5	0.77
83	0.83	62.8 ± 15.0	0.56
96	0.50	90.4 ± 12.3	0.51
95	0.45	77.7 ± 10.3	0.53

* Measured by the method of Gaze, West, and Steward (1973) using ^{125}I–HSA
† Using HSA as the substrate
‡ Fold reduction in ELISA at 50 per cent of the maximum optical density (OD)
Kendall's rank correlation = 0.87 between affinities measured by the two methods

on ELISA performed in the presence and absence of a mild protein denaturant (0.5 M guanidine hydrochloride) in the serum diluent. The presence of this denaturing agent was shown to have a differential effect on the dose–reponse curve, by interfering with the antibody–antigen interaction and producing a marked shift with antibodies produced early in the immune response (generally accepted as being of low affinity) but having less effect with sera taken later (containing higher affinity antibodies). The extent of the shift, which was a measure of the susceptibility of the antibodies to dissociation, was taken to be an indication of the relative affinity.

A similar assay, using chaotropic thiocyanate ions, has been used to demonstrate affinity maturation after immunization against rubella (Pullen, Fitzgerald and Hosking, 1986). A. Falconar and P.R. Young (personal communication), using 0.05 M diethylamine (DEA) to inhibit antibody binding to the solid phase antigen, have demonstrated a significant correlation between the affinities of polyclonal anti–HSA antibodies measured by DEA-ELISA and by SAS precipitation (Table 1) suggesting that this assay may be a simple and convenient way of assessing functional affinity (avidity) of antibodies to complex antigens which cannot be measured by conventional affinity assays.

Isotype-specific inhibition assay

This assay has been used to analyse the relationship between the isotype and the affinity of specific antibodies produced after immunization in man. Low affinity antibody responses to tetanus toxoid have been shown to correlate with high levels of IgG_4 antibodies whereas high affinity responses correlated with a predominantly IgG_1 response (Devey *et al.*, 1985). As most normal individuals produce antibodies of both isotypes, the question arises as to whether IgG_4

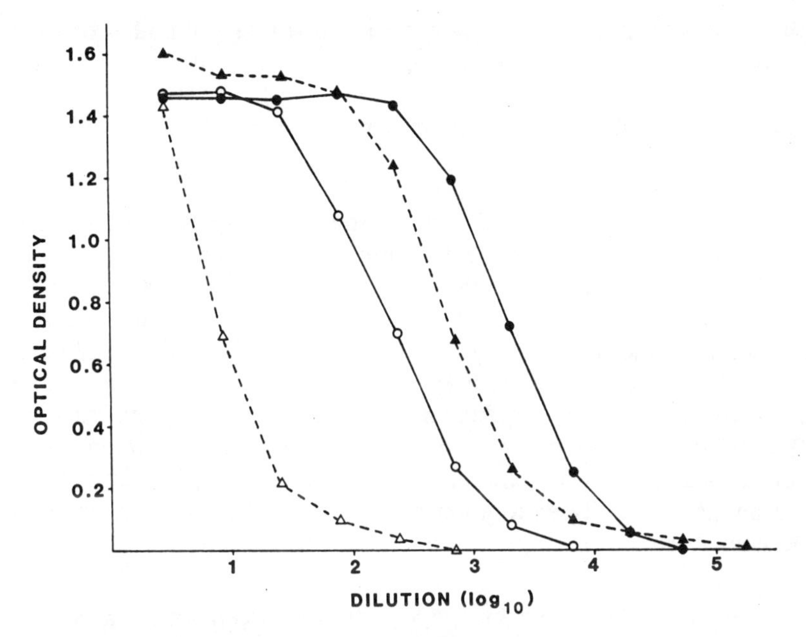

FIGURE 4 ELISA dissociation assay of antibodies to tetanus toxoid in a single serum sample using 20 mM DEA and monoclonal antibodies to IgG_1 and IgG_4. IgG_1 binding curve with (○–○) and without (●–●) DEA, IgG_4 binding curve with (△--△) and without (▲--▲) DEA. (Data from Devey *et al.*, 1988)

TABLE 2 Functional affinities of IgG_1 and IgG_4 antibodies to tetanus toxoid in normal individuals measured by a subclass-specific DEA–ELISA (data from M.E. Devey, K. Bleasdale, and S. Rath, unpublished)

Serum	IgG_1	IgG_4
	1/Shift log_{10}*	1/Shift log_{10}*
1	1.05	0.53
2	0.95	0.33
3	0.91	0.69
4	0.83	0.52
5	0.61	0.45
6	0.61	0.45
7	0.57	0.38
8	0.55	0.49
MEAN + SD	0.78 + 0.19	0.51 + 0.13†

* Reciprocal Shift with 20 mM DEA at 50% maximum OD
† $t = 3.27$, $p = .006$

antibodies are intrinsically of lower affinity than IgG_1 antibodies or whether individuals with high IgG_4 responses produce low affinity antibodies of both isotypes. Since the isolation of specific IgG_1 and IgG_4 antibody populations may present a number of technical problems, a subclass-specific DEA-ELISA has been used to measure the functional affinities of IgG_1 and IgG_4 antibodies without the need for purification (Figure 4).

Preliminary results (Devey, Bleasdale, and Rath, unpublished) have shown that, in 8/8 normal individuals, IgG_4 antibodies to tetanus toxoid were more easily inhibited from binding to the solid phase antigen than IgG_1 antibodies in the same serum (Table 2), suggesting that IgG_4 had a lower functional affinity than IgG_1. Furthermore, analysis of the affinity of purified antibodies to hepatitis B surface antigen (HBsAg) of different IgG subclass in hepatitis B vaccine recipients, using a radioimmunoprecipitation assay, has shown that IgG_4 anti-HBsAg antibodies were of lower affinity than were antibodies of the other subclasses (Brown *et al.*, unpublished observations) suggesting that, for some antigens, affinity maturation proceeds at different rates for different subclasses.

THE EFFECT OF AFFINITY ON ANTIBODY ASSAYS

Enzyme-linked immunosorbent assays

ELISA is a discontinous solid phase assay usually consisting of two cycles of three washes and a second antibody step. From first principles it might be expected that low affinity antibodies may not be detected due to dissociation from the solid phase antigen during the washing steps. Indeed, there has been considerable controversy as to whether ELISA is primarily a measure of antibody concentration (Engvall and Perlmann, 1972; Svenson and Larsen, 1977) or affinity (Ahlstedt, Holmgren, and Hanson, 1974; Butler *et al.*, 1978). This discrepancy can be explained by the different ways in which ELISA results can be expressed: either as an endpoint titre (the maximum serum dilution giving a standard optical density (OD) reading above a control value) or as a direct absorbance reading at a single serum dilution which can be related to a standard serum. The latter approach assumes that the dose–response curve of the test serum is identical to that of the standard, which may not be the case (see below). Those reports which indicated that ELISA measured antibody concentration used endpoint titration whereas those claiming that it measured affinity used the direct absorbance approach.

Thus, ELISA appears to correlate well with high affinity antibodies when the results are expressed as a direct OD reading at a low serum dilution rather than when an endpoint titre at a low absorbance is used (Lethonen and Eerola, 1982). This was confirmed by Peterfy, Kuusela, and Makela (1983) using a panel of monoclonal antibodies to show that the maximum OD value (i.e. the binding curve plateau) correlated with affinity while the endpoint titre at a low

OD correlated with antibody concentration. Both studies showed that ELISA was poor at detecting low affinity antibodies, with suggested threshold levels of 6.3×10^4 mol^{-1} (Lethonen and Eerola, 1982) and 3×10^5 mol^{-1} for IgG and 2×10^5 mol^{-1} for IgM (Peterfy, Kuusela, and Makela, 1983).

ELISA, affinity, and epitope density

The study of the effect of affinity on *in vitro* antibody assays has, until recently, depended on the use of immune sera raised *in vivo*. These sera contain a heterogeneous population of antibodies of various isotypes and affinities and therefore interpretation of the results has depended on an estimate of the 'average' affinity. However, as there are both experimental and theoretical grounds for believing that affinity distribution is skewed or bimodal and does not necessarily follow a 'normal' distribution (Werblin and Siskind, 1972), the use of 'average' affinity values may be misleading. Monoclonal antibodies have provided a good opportunity to study homogeneous antibody populations of known isotype, affinity, and concentration in order to determine the effect of these parameters on *in vitro* antibody assays.

The performance in ELISA of a panel of monoclonal antibodies of different isotypes and affinities for the dinitrophenyl (DNP) hapten (Stanley, Lew, and Steward 1983) has been investigated (Nimmo *et al.*, 1984; Lew, 1984). The results of these studies illustrate the difficulties currently encountered in the interpretation of such assays. Nimmo *et al*, (1984) used six monoclonal antibodies of the IgG_1, IgG_{2a} and IgG_{2b} isotypes, with affinities in the range $0.056–0.713 \times 10^6$ M^{-1} (measured by a polyethylene glycol precipitation assay), in ELISA. Plates were coated with $DNP_{40}BSA$ or $DNP_{11}BSA$ in order to highlight any differences between divalent interactions at high epitope density (i.e. functional affinity) and monovalent interactions at low epitope density (which should reflect intrinsic affinity).

Table 3 shows a comparison of the results at the two epitope densities taking the endpoint titre at OD_{492} as either 0.5 or 0.2. The differences in titre obtained using the two different epitope densities were expressed as the ratio of endpoint titre $DNP_{40}BSA/DNP_{11}BSA$ (Table 4). The greatest difference was seen with low affinity antibodies particularly at an endpoint OD of 0.5, suggesting that both epitope density and choice of endpoint OD may influence ELISA titre and that this effect may be most marked with low affinity antibodies.

The relative antibody concentrations of the various monoclonal antibodies were obtained by dividing the antibody concentration (mg/ml) by the corresponding titre obtained by ELISA at OD 0.5 and OD 0.2 (Table 5). The minimum antibody concentration detected by ELISA was lower with $DNP_{40}BSA$ than with $DNP_{11}BSA$ at both endpoints, particularly for low affinity antibodies. For example, the minimum antibody concentration detected in ELISA at an endpoint of 0.2 using $DNP_{11}BSA$ was 2.7×10^{-5} mg/ml

TABLE 3 Comparison of ELISA endpoint titres (EPT) of monoclonal anti-DNP antibodies of different isotypes and affinities (data from Nimmo *et al.*, 1984)

MAB	Isotype	Affinity ($\times 10^6 M^{-1}$)	EPT OD_{492} 0.5		EPT OD_{492} 0.2	
			$DNP_{11}BSA$	$DNP_{40}BSA$	$DNP_{11}BSA$	$DNP_{40}BSA$
A	IgG_1	0.056	320	81920	20480	327680
B	IgG_1	0.216	20	—	160	40
C	IgG_1	0.347	160	5120	5120	20480
D	IgG_{2a}	0.366	1280	10240	10240	40960
E	IgG_{2b}	0.459	20480	81920	81920	327680
F	IgG_1	0.713	2560	40960	20480	163840

TABLE 4 Ratio of ELISA endpoint titre (EPT) using $DNP_{40}BSA$ to EPT using $DNP_{11}BSA$ with monoclonal antibodies of different affinities (data from Nimmo *et al.*, 1984)

MAB	Ratio EPT OD_{492} DNP_{40}/DNP_{11}	
	OD_{492} 0.5	OD_{492} 0.2
A	256	16
B	—	0.25
C	32	4
D	8	4
E	4	4
F	16	8

for a high affinity antibody (E) compared to 5.1×10^{-3} mg/ml for a low affinity antibody (B), a sensitivity difference of 190-fold. The poor performance of monoclonal antibody B in ELISA may have been due to dissociation during the washing procedure. Surprisingly, antibody A, which was of even lower affinity although of the same isotype, behaved as well as the higher affinity antibodies.

The binding curve in ELISA at a low epitope density ($DNP_{11}BSA$) was significantly different for the high and low affinity monoclonal antibodies (see Figure 3). The high affinity curve had a long plateau phase with a steep fall starting at a high dilution whereas the low affinity curve had a gradual fall starting at a much lower dilution. With a high epitope density ($DNP_{40}BSA$) the binding curves for both these antibodies were remarkably similar, although the low affinity antibody started to fall at a lower dilution. This indicates that taking the OD reading of a test serum at a single dilution to compare with a standard serum may be highly misleading, especially if a high epitope density has not been achieved, and highlights the need for careful analysis of dose–response curves in ELISA (see Chapter 3).

TABLE 5 Relative antibody concentrations measured by ELISA at endpoint OD 0.5 and 0.2 with $DNP_{40}BSA$ and $DNP_{11}BSA$ with monoclonal antibodies of different affinities (data from Nimmo *et al.*, 1984)

MAB	EPT OD_{492} 0.5		EPT OD_{492} 0.2	
	$DNP_{11}BSA$	$DNP_{40}BSA$	$DNP_{11}BSA$	$DNP_{40}BSA$
A	1.9×10^{-2}	7.5×10^{-5}	3.0×10^{-4}	1.8×10^{-5}
B	4.1×10^{-2}	—	5.1×10^{-3}	2.0×10^{-2}
C	2.8×10^{-3}	8.8×10^{-5}	8.8×10^{-5}	4.9×10^{-5}
D	5.2×10^{-4}	6.5×10^{-5}	6.5×10^{-5}	1.6×10^{-5}
E	1.1×10^{-4}	2.7×10^{-5}	2.7×10^{-5}	6.9×10^{-6}
F	7.0×10^{-4}	4.4×10^{-5}	8.8×10^{-5}	1.1×10^{-5}

— = no data available

In additional studies using monoclonal antibodies, Lew (1984) showed that the maxiumum OD obtained in ELISA (i.e. at the plateau phase) increased as higher epitope densities were used (Figure 5), confirming the findings of Nimmo *et al.* (1984) of the affinity dependence of low affinity antibodies, particularly at low epitope density. At high epitope density, affinity dependence was not so evident, presumably because antibodies were able to bind

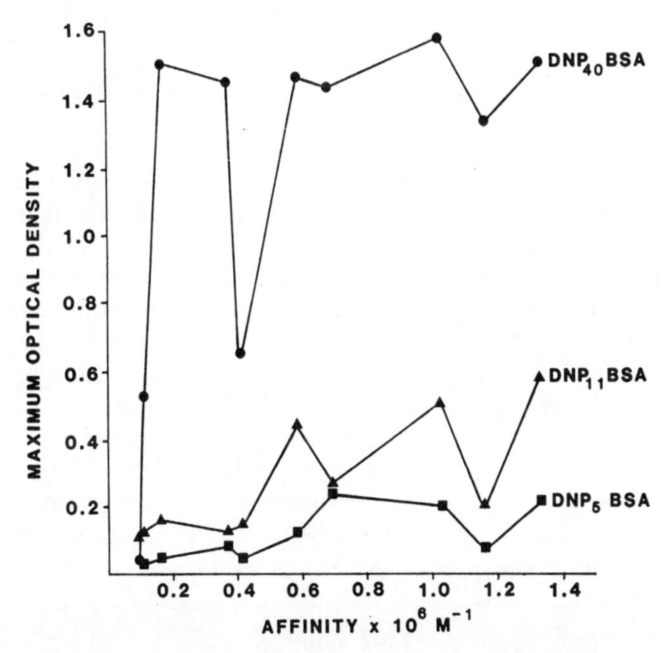

FIGURE 5 Effect on affinity and epitope density on the maximum optical density obtained in ELISA with IgG_1 monoclonal antibodies to DNP. (Data from Lew, 1984)

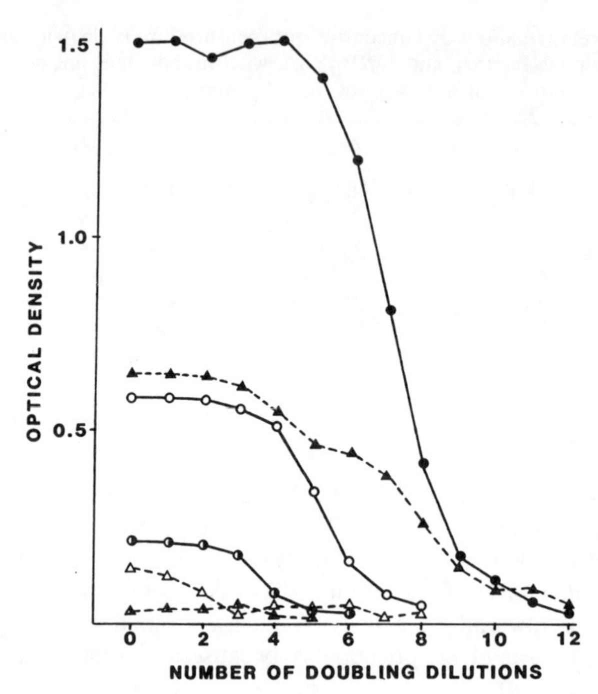

FIGURE 6 ELISA binding curves of a high affinity monoclonal antibody to DNP (starting concentration 1 µg/ml) with $DNP_{40}BSA$ (●–●), $DNP_{11}BSA$ (○–○), and DNP_5BSA (◑–◑) and a low affinity monoclonal antibody (starting concentration 2 mg/ml) with $DNP_{40}BSA$ (▲--▲), $DNP_{11}BSA$ (△--△), and DNP_5BSA (▲--▲) (Data from Lew, 1984)

divalently. Comparison of the dose–response curves of a high affinity and a low affinity antibody showed a steeper dose–response curve for the high affinity antibody compared with the low affinity antibody at all three (DNP_5, DNP_{11}, and DNP_{40}) epitope densities (Figure 6). Taking the endpoint OD as 0.1, the difference in ELISA sensitivity between the high and low affinity antibodies

TABLE 6 Sensitivity of ELISA for high and low affinity monoclonal antibodies at three different epitope densities (data from Lew, 1984)

Antigen	Sensitivity* (ng/ml)	
	Low affinity ($0.41 \times 10^6 M^{-1}$)	High affinity ($1.33 \times 10^6 M^{-1}$)
DNP_5BSA	not detectable	63 –125
$DNP_{11}BSA$	$0.5–1 \times 10^6$	8 –16
$DNP_{40}BSA$	$2.0–4 \times 10^3$	0.5–1

* Minimum antibody concentration required for detection at endpoint titre at OD = 0.1

TABLE 7 Relative antibody concentrations measured by SP–RIA with $DNP_{40}BSA$ and $DNP_{11}BSA$ with monoclonal antibodies of different affinities (data from Nimmo *et al.*, 1984)

MAB	Isotype	Affinity ($\times 10^6 M^{-1}$)	$DNP_{11}BSA$	$DNP_{40}BSA$
A	IgG_1	0.056	6.2×10^{-2}	3.0×10^{-5}
B	IgG_1	0.216	1.0×10^{-1}	5.2×10^{-2}
C	IgG_1	0.347	9.3×10^{-3}	7.2×10^{-5}
D	IgG_{2a}	0.366	8.8×10^{-4}	2.2×10^{-4}
E	IgG_{2b}	0.459	1.8×10^{-4}	5.5×10^{-6}
F	IgG_1	0.713	1.4×10^{-4}	1.8×10^{-5}

was 6×10^4-fold at a low epitope density and 4×10^3-fold at a high epitope density (Table 6).

Solid phase radioimmunoassay

As with ELISA, solid phase radioimmunoassay (SP-RIA) involves a number of washing steps which will clearly affect the measurement of low affinity antibodies. Peterfy, Kuusela and Makela (1983) found that the performance of high and low affinity antibodies in SP-RIA was similar to that in ELISA, with threshold affinity of $3 \times 10^5 M^{-1}$ for IgG. Nimmo *et al.* (1984) showed that the sensitivity of SP-RIA decreased with descreasing antibody affinity with an antigen of low epitope density but was not so dependent on affinity when an antigen of high epitope density was used (Table 7).

Other antibody assays

Peterfy, Kuusala, and Makela (1983) found that four other commonly used

TABLE 8 Threshold affinity and sensitivity of six commonly used antibody assays using monoclonal antibodies (data from Peterfy, Kuusela, and Makela, 1983)

Assay	IgG		IgM	
	Threshold affinity ($10^6 M^{-1}$)	Sensitivity (ng/ml)	Threshold affinity ($10^6 M^{-1}$)	Sensitivity (ng/ml)
SP-RIA	0.3	7	0.06	6
ELISA	0.3	14	0.2	18
Haemagglutination	0.5	400	0.09	70
Haemolysis	2.0	700	0.2	230
Farr assay	1.0	1 100	0.2	1 100
Precipitation	2.0	60 000	0.2	60 000

antibody assays (haemagglutination, complement haemolysis, precipitation, and the Farr assay) were even more dependent on antibody affinity than ELISA and SP-RIA and defined a 'threshold affinity', above which affinity did not affect the sensitivity of the assay (Table 8). The results of Nimmo *et al.* (1984) were broadly in agreement with this, except that they did not find any correlation between affinity and the ability of monoclonal antibodies to precipitate antigen, which is not inconsistent with the nature of the precipitation reaction which is known to depend on other factors in addition to the primary binding reaction. In addition, an unpublished observation from this laboratory, that the Farr assay was more sensitive in the detection of low affinity antibodies than ELISA using DNP_4BSA as the antigen, would argue that the Farr assay is less dependent on affinity then ELISA. A comparison of the performance of various affinity assays (Stanley, Lew, and Steward, 1983) showed that, although both equilibrium dialysis and the Farr assay were able to rank antibodies in the same order of affinity, equilibrium dialysis was more efficient in detecting both the highest and lowest affinity antibodies. The Farr assay tended to minimize the differences between the highest and lowest affinity antibodies.

CONCLUSIONS

Early work by Minden, Ried, and Farr (1966) showed that several commonly used tests for antibody produced very different results when applied to a panel of human sera. All the sera had detectable antibody when primary binding tests, such as ammonium sulphate precipitation and radioimmunoelectrophoresis, were used. However, secondary tests (precipitation and haemagglutination) and tertiary tests (passive cutaneous anaphylaxis and Prausnitz–Kustner) did not consistently give positive results nor did they correlate with the amount of antibody present. This clearly illustrated the importance of the quality, as well as the quantity, of antibody in performance of both the secondary and tertiary tests. Recently, Peterfy, Kussela, and Makela (1983) and Nimmo *et al.* (1984) have shown that all antibody tests are affinity dependent and that this may lead to problems in assessing antibody concentration. Although some solid phase assays, such as ELISA, are less dependent on antibody affinity than others, they are still poor at detecting low affinity antibodies. In addition, the effect of affinity on dose–response curves may lead to erroneous comparisons with standards if sera are compared at a single dilution at a low antibody concentration. Thus, the most reliable way to estimate antibody content by ELISA may be by determination of endpoint titre at a low optical density although this part of the dilution curve may be subject to the greatest variation (see Chapter 3). It should be remembered, also, that affinity differences between monoclonal or polyclonal isotype-specific antibodies may lead to incorrect assumptions as to the relative concentrations of different isotypes in double antibody assays (Seppala *et al.*, 1984).

Nimmo *et al.* (1984) and Lew (1984) have shown that a high epitope density minimizes the influence of affinity in ELISA and SP-RIA. This is presumably because, at high epitope density, antibodies are able to interact bivalently with the ligand. Nygren, Czerkinsky, and Stenberg (1985), using $TNP_{25}OA$ as the ligand, showed that dissociation of whole antibodies was very slow compared to that of monovalent Fab fragments, suggesting that bivalence was an important factor in maintaining stability. The effect of monovalent versus bivalent interaction on the equilibrium in solid phase systems has also been examined by Horejsi and Matousek (1985) who suggest that, even at high concentrations of ligand, bivalent interactions do not occur frequently due to unsuitable orientation of epitopes on the solid phase. Obviously this would not be a problem with antigens which have repeating epitopes, such as DNA or highly haptenated proteins. However, for many antigens epitope density cannot be easily increased, although one possible way may be to couple the antigen covalently to the solid phase rather than by simple absorption. Antigen adsorption and desorption from the solid phase has been studied by Nieto *et al.* (1986) and antigen desorption was found not to be affected by either the concentration or affinity of the test serum.

The increased awareness of the importance of antibody affinity, not only in biological and functional terms but also in its effect on the performance of immunoassays, has highlighted the need for careful appraisal of this parameter in the choice of antibody assay. This is of particular importance in situations, discussed earlier, where low affinity responses are likely to occur, so that the presence of such antibodies does not go undetected.

REFERENCES

Ahlstedt S., Holmgren J., and Hanson L.A. (1974). Protective capacity of antibodies against *E. Coli* O antigen with special reference to the avidity. *Int. Arch. Allergy*, **46**, 470.

Brown S.E., Howard C.R., Zuckerman A.J., and Steward M.W. (1984). Affinity of antibody responses in man to hepatitis B vaccine determined with sythetic peptides. *Lancet*, **ii**, 184.

Butler J.E., Feldbush T.L., McGivern P.L., and Stewart N. (1978). The enzyme-linked immunosorbent assay (ELISA): a measure of antibody concentration or affinity? *Immunochemistry*, **15**, 131.

Devey M.E. (1986). The biological and pathological significance of antibody affinity. In: *Immunoglobulins in Health and Disease* (ed. M.A.H. French). MTP Press, Lancaster, pp. 55–73.

Devey M.E., Bleasdale K.M., French M.A.H., and Harrison G. (1985). The IgG4 subclass is associated with a low affinity antibody response to tetanus toxoid in man. *Immunology*, **55**, 565.

Devey M.E., Bleasdale K., Lee S., and Rath S. (1988). Determination of the functional affinity of IgG1 and IgG4 antibodies to tetanus toxoid by isotype-specific solid-phase assays. *J. Imm. Methods*, **106**, 119.

Engvall E., and Perlmann P. (1972). Enzyme-linked immunosorbent assay, ELISA. III. Quantitation of specific antibodies by enzyme-labelled anti-immunoglobulin in antigen-coated tubes. *J. Immunol.*, **109**, 129.
Farr R.S., (1958). A quantitative immunochemical measure of the primary reaction between I* BSA and antibody. *J. Infect. Dis.*, **103**, 239.
Fauci A.S., Frank M.M., and Johnson J.S. (1970). The relationship between antibody affinity and the efficiency of complement fixation. *J. Immunol.*, **105**, 215.
Gaze S., West N.J., and Steward M.W. (1973). The use of a double isotope method in the determination of antibody affinity. *J. Imm. Methods*, **3**, 357.
Gripenberg M., and Gripenberg G. (1983). Expression of antibody activity measured by ELISA. Anti-ssDNA antibody activity characterized by the dose response curve. *J. Imm. Methods*, **62**, 315.
Horejsi, V., and Matousek V. (1985). Equilibrium in the protein-immobilized-ligand–soluble-ligand system: estimation of dissociation constants of protein-soluble-ligand complexes from binding inhibition data. *Molec. Immunol.*, **22**, 125.
Inouye S., Hasgawa A., Matsuno S., and Katow S. (1984). Changes in antibody avidity after virus infections: denaturation by an immunosorbent assay in which a mild protein denaturing agent is employed. *J. Clin. Microbiol.*, **20**, 525.
Karush F. (1978). The affinity of antibody: range, variability, and the role of multivalence. In: *Comprehensive Immunology*, Vol. 5 (eds. G.W. Litman and R.A. Good). Plenum Press, New York, p. 85.
Lehtonen O-P., and Eerola E. (1982). The effect of different antibody affinities on ELISA absorbance and titre. *J. Imm. Methods*, **54**, 233.
Lew A.M. (1984). The effect of epitope density and antibody affinity on ELISA as analysed by monoclonal antibodies. *J. Imm. Methods*, **72**, 171.
Minden P., Ried R.T., and Farr R.S. (1966). A comparison of some commonly used methods for detecting antibodies to bovine albumin in human serum. *J. Immunol.*, **96**, 180.
Nieto A., Gaya A., Moreno C., Jansa M., and Vives J. (1986). Adsorption–desorption of antigen to polystyrene plates used in ELISA. *Ann. Inst. Pasteur*, **137 C**, 161.
Nimmo G.R., Lew A.M., Stanley C.M., and Steward M.W. (1984). Influence of antibody affinity on the performance of different antibody assays. *J. Imm. Methods.*, **72**, 177.
Nygren H., Czerkinsky C., and Stenberg M. (1985). Dissociation of antibodies bound to surface-immobilized antigen. *J. Imm. Methods*, **85**, 87.
Peterfy F., Kuusela P., and Makela O. (1983). Affinity requirements for antibody assays mapped for monoclonal antibodies. *J. Immunol.*, **130**, 1809.
Pullen G.R., Fitzgerald M.G., and Hosking C.S. (1986). Antibody avidity determination by ELISA using thiocyanate elution. *J. Imm. Methods*, **86**, 83.
Seppala I.J.T., Routonen N., Sarnesto A., Mattila P.A., and Makela O. (1984). The percentages of six immunoglobulin isotypes in human antibodies to tetanus toxoid: standardization of isotype-specific second antibodies in solid-phase assay. *Eur. J. Immunol.*, **14**, 868.
Stanley C., Lew A.M., and Steward M.W. (1983). The measurement of antibody affinity: a comparison of five techniques utilizing a panel of monoclonal anti-DNP antibodies and the effect of high affinity antibody on the measurement of low affinity antibody. *J. Imm. Methods*, **64**, 119.
Steward M.W. (1986). Overview: Introduction to methods used to study the affinity and kinetics of antibody–antigen reactions. In: *Handbook of Experimental Immunology*, Vol. 1 (ed. Weir D.M.). Blackwell Scientific Publications, Oxford, Chap. 25, p. 1.
Steward M.W., and Devey M.E. (1981). Antigen–antibody complexes: their nature and role in animal models of antigen–antibody complex disease. In: *Immunological*

Aspects of Rheumatology (ed. W. Carson Dick). MTP, Lancaster, pp. 63–91.
Steward M.W., and Petty R.E. (1972). The use of ammonium sulphate globulin precipitation for determination of antibody affinity of anti-protein antibodies in mouse serum. *Immunology*, **22**, 747.
Steward M.W., and Steensgaard J. (1983). *Antibody Affinity: thermodynamic aspects and biological significance.* CRC Press, Boca Raton, Florida.
Svenson S.B., and Larsen K. (1977). An enzyme-linked immunosorbent assay (ELISA) for the determination of diphtheria toxin antibody. *J. Imm. Methods*, **17**, 249.
Werblin T.P., and Siskind G.W. (1972). Distribution of antibody affinities: technique of measurement *Immunochemistry*, **9**, 987.

ELISA and Other Solid Phase Immunoassays
Edited by D.M. Kemeny and S.J. Challacombe

CHAPTER 7

The Immunochemistry of Sandwich ELISAs: Principles and Applications for the Quantitative Determination of Immunoglobulins

J. E. Butler
University of Iowa, Iowa, USA

CONTENTS

BASIC PRINCIPLES

Sandwich immunoassay

The term sandwich immunoassay refers to a technique whereby the entity to be measured is 'sandwiched' between two molecules or layers of molecules, which recognize either the same epitope (symmetrical sandwich) or different epitopes

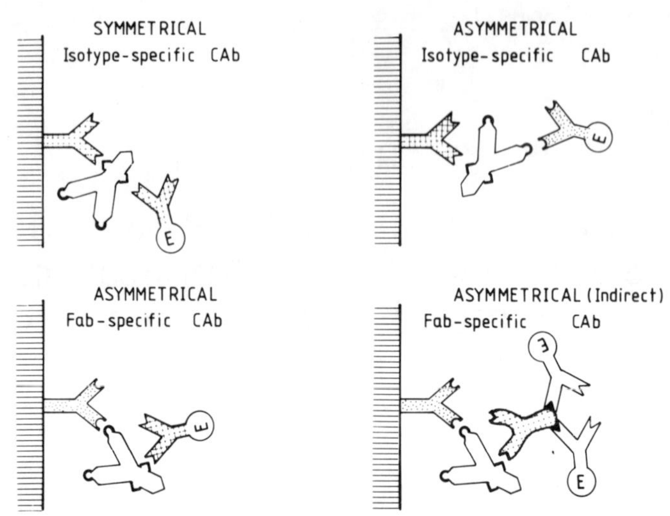

FIGURE 1 Types of configurations used in sandwich ELISAs for quantitation of immunoglobulins. Lined, vertical bar depicts plastic surface. Isotype-specific capture antibodies (CAb) or detection antibodies are chequered. Anti-Fab CAb or detection antibodies are stippled. The unshaded immunoglobulin is the one being measured. An encircled 'E' represents an antibody–enzyme conjugate. The unshaded antibodies attached to 'E' on the lower right are specific for the isotype-specific antibodies in an indirect or 'amplified' sandwich system. Enlargments on the Fc-region of the Ig being measured represent an isotype-specific determinant

(asymmetrical sandwich; Figure 1) on the molecule (normally the antigen) being measured. This configuration dictates that the molecule to be measured must be *multivalent*. Hence sandwich assays are restricted to the quantitation of multivalent antigens (Ags) such as proteins or polysaccharides. Monovalent Ags are typically of low molecular weight, e.g. certain hormones, drugs and experimental haptens, and must be measured by classical competitive immunoassays (Berson and Yalow, 1959).

The form of sandwich assay in widest use is one in which a so-called 'capture antibody' (CAb) is immobilized on a solid phase. While many solid phases are available, the use of microtitre plates and simple surface adsorption (Catt, Niall, and Tregear, 1966) has become increasingly popular (e.g. see this volume Chapter 5). The solid phase CAb is then incubated with a test sample or reference standard (Rstd) containing the Ag to be measured. As in other solid phase immunoassays, the test sample is diluted in a buffer such as PSB–T[a] which contains a detergent or a high concentration of inert protein, e.g. gelatine, to prevent the test Ag from adsorbing to the solid phase while simultaneously permitting the Ag to complex specifically with the CAb. Thereafter, the solid phase is washed and a detection reagent is added which recognizes exposed antigenic determinants on the captured Ag. The detection reagent is an Ab conjugated to a radioisotope, fluorescent compound or, in sandwich ELISAs, an enzyme. After incubation with the detection system, the solid phase is washed and the extent to which the detection system became bound is measured by radioactive counting, fluorescence intensity or extent of substrate conversion.

We have used the sandwich immunoassay to measure haemoglobin in infant stools, albumin and lactoferrin in dilute biological fluids, and experimentally, to study the immunoassay itself by using enzymes and hapten–protein conjugates as Ags. In this chapter, the application of the sandwich immunoassay to measure immunoglobulins (Igs) using ELISA detection systems, will be discussed.

Sandwich ELISA for measuring immunoglobulins

Sandwich ELISAs for quantitation of Igs are especially valuable when the concentration of Igs is low such as in exocrine body fluids, foetal sera, and tissue culture supernatants. Sandwich ELISAs, while unnecessary for quantitation of major serum Igs, i.e. IgG, IgA, and IgM, are valuable for the measurement of serum IgE. In addition, sandwich ELISAs are valuable when quantitation of subpopulations expressing subclass, allotypic or even idiotypic determinants, is desired. Igs are macromolecular Ags with multiple repeating epitopes and are therefore well suited for measurement by sandwich immunoassay. In addition, Igs of different isotypes: (a) share common epitopes, e.g. light chains, and (b) have their common epitopes, e.g. Fab determinants and idiotypic determinants[b], in a region of their structure which is physically separated from their isotype- and most allotype-specific epitopes (Figure 1). Although the large size of Igs makes it possible to use a symmetrical configuration to measure them (Figure 1; upper left), asymmetrical configurations (Figure 1; all other examples) permit a single Ab, e.g. anti-L-chain or anti-Fab, to be used either as a universal CAb (reverse asymmetrical configuration) or in a universal Ab-enzyme detection system to detect all classes of Igs. The performance characteristics of the different configurations shown in Figure 1 for sandwich immunoassays will be described in this chapter.

Ab specific for the various epitopes of Igs may be either polyclonal (PoAb) or monoclonal (MoAb). The unique specificity of the latter theoretically allows extremely small isotypic, allotypic, or idiotypic differences to be recognized and the Ig possessing them to be quantitatively measured. PoAbs are available from many commercial suppliers and an expanding library of MoAbs is available privately and commercially.

Important factors in the favourable performance of sandwich ELISAs are: (a) the Ag capture capacity (AgCC)[c] of the CAb; (b) the characteristics of the detection system; and (c) the equipment available for doing the assays. The latter requirement appears to be largely satisfied by microtitre ELISA systems which are simple to use and readily automated. Hence, we choose to base our sandwich ELISAs for Igs on the microtitre system. Further improvements in solid phase chemistry will no doubt improve the performance of sandwich ELISAs in microtitre systems (see later).

The AgCC of the CAbs and the ability to stably immobilize them at sufficient concentration and in a native configuration, appears essential. PoAbs are often used in the form of the globulin fraction of the antiserum. The inability to immobilize proteins at high concentration using current microtitre technology[d], means that the 'Ab abundance'[e] of the globulin fractions of such PoAbs as well as their AgCC, is critical. Consequently, PoAbs of low antibody content may require affinity purification to increase their Ab abundance. This topic is discussed below. MoAbs raised either *in vitro* or *in vivo* can be obtained in high Ab abundance without affinity purification. However, MoAbs characteristically have lower affinity than PoAbs, such that their AgCC may be inadequate; this is further complicated by their propensity towards denaturation upon adsorption (Suter and Butler, 1986; Suter, Butler, and Peterman, 1988; see below). It should be emphasized that it is the performance of the CAb which determines the suitability of using a sandwich ELISA for multivalent antigen; in cases where the CAb 'fails the requirements', a competitive immunoassay must be used.

The characteristics of detection systems which influence their performance in sandwich ELISAs include their: (a) complexity, (b) antibody abundance, and (c) specificity relative to that of the CAb. In antigen-specific ELISAs, increasing complexity is related to increasing sensitivity, i.e. 'amplified', indirect detection systems are more sensitive than direct detection systems (Koertge and Butler, 1985; see also this volume, Chapter 5). Such amplified systems involve multiple Abs and their enzymatic signal becomes therefore 'indirect'. Figure 1 illustrates an indirect, amplified configuration in a sandwich ELISA; this configuration increases the chance of steric hindrance and short-circuiting (see Figure 6). Simple conjugates are therefore desirable in sandwich ELISA and their performance will be determined by their Ab abundance, affinity, and the configuration in which they are used, i.e. asymmetrical versus symmetrical. These considerations will be discussed below.

IMMUNOCHEMISTRY OF SANDWICH ELISAs

Empirical studies on sandwich ELISA immunochemistry

The use of iodinated Igs (Figure 2, top) permits the immunochemistry of sandwich ELISAs to be studied in terms of: (a) the relationship between Ag (Ig) binding and its detection based on substrate conversion; (b) the amount and proportion of Ig which binds as a function of the Ag : CAb ratio; (c) the influence of the concentration of solid phase CAb; (d) the differential activity

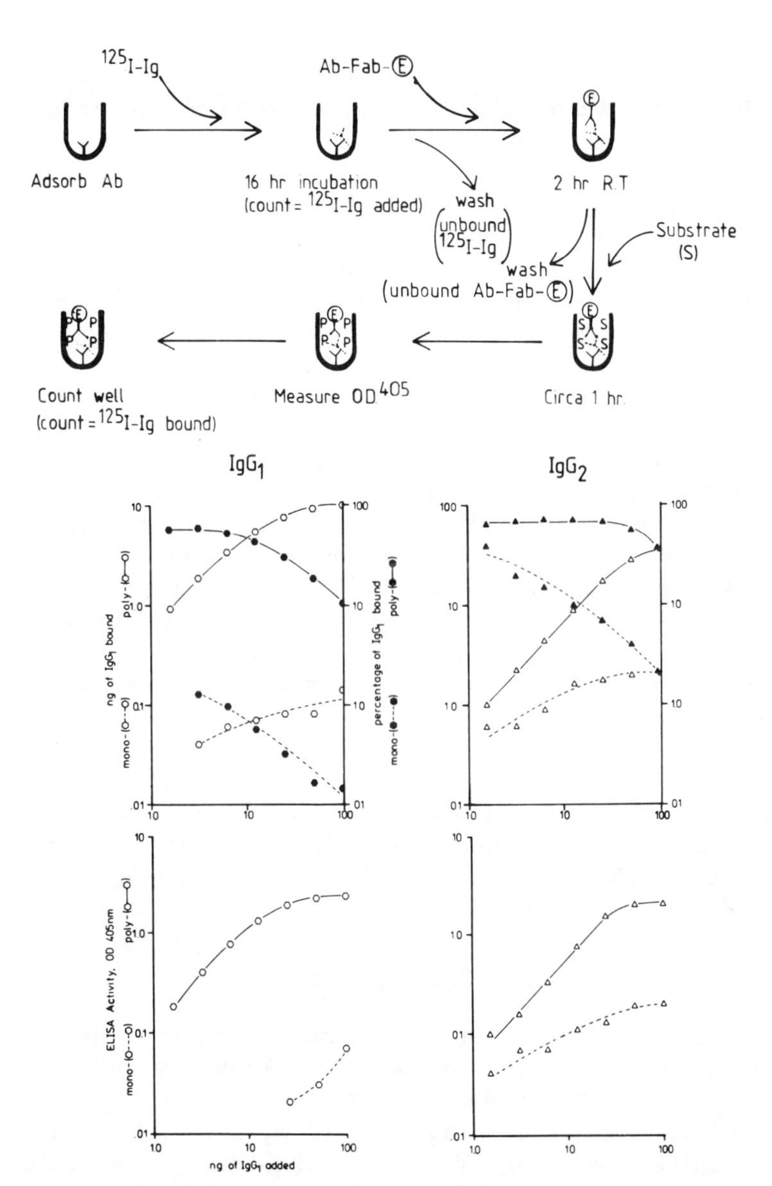

FIGURE 2 *Top*: experimental design. Method used to study the immunochemistry of sandwich ELISAs. The diagram depicts the immobilization of the CAb in a microtitre well (upper left), the capture of iodinated Ig (interrupted Y) and its detection by an asymmetrical anti-Fab enzyme conjugate. S = substrate, P = coloured product. The actual amount of captured Ig is eventually determined by counting the microtitre well. *Bottom*: 'titration plots'. The relationship between the amount of IgG_1 and IgG_2 bound to their respective CAbs and the enzymatic signal produced as a function of the amount of each added. Data were collected as illustrated above. Data are presented for affinity-purified polyclonal (solid lines) and monoclonal (dotted lines) CAbs adsorbed on Immunlon 2® at 5 μg/ml. The upper plot illustrates the actual amount and percentage of added IgG_1 or IgG_2 bound, as a function of the amount of IgG_2 or IgG_1 added. The left and right axes are as indicated for IgG_1. The lower plots are the enzymatic signals generated when an asymmetrical (anti-Fab detection system, see Figure 1) system is used to measure the amount of ^{125}I-IgG_1 or $-IgG_2$ bound to their corresponding CAbs

of different CAbs, especially PoAbs and MoAbs specific for the same Ig.

The experimental design illustrated in Figure 2 was used to study the immunochemistry of sandwich ELISAs using affinity-purified MoAbs and PoAbs specific for bovine IgG_1 and IgG_2 as CAbs in an asymmetrical configuration. These CAbs were adsorbed on plastic at the same concentration and ^{125}I-IgG_1 or -IgG_2 was added over a range of concentrations to determine the AgCC of the CAbs (see Figure 2 — 'titration plots'). The upper graph for each subclass determination shows that the PoAb has a greater AgCC than its corresponding MoAb; this is apparent whether AgCC is measured as the percentage or the actual amount of ^{125}I-IgG_1 or -IgG_2 which becomes bound. These diagrams also show that over the range in which the log–log titration plot (amount bound) is linear, the proportion of Ig which is 'captured' remains constant; the lack of a region in which a constant proportion of the added IgG_2 becomes bound to the MoAbs is paralleled by the lack of a linear log–log titration plot for these CAbs. The titrations in Figure 2 show that the enzymic signal generated using an asymmetrical configuration (anti-Fab conjugate) parallels the actual amount of ^{125}I-IgG_1 or -IgG_2 bound. The latter result sharply contrasts with that seen in antigen-specific ELISA in which the actual binding of the entity being measured (Ab in the case of Ag-specific ELISA) does not parallel the enzymatic signal generated due to steric hindrance of the detection system (Koertge and Butler, 1985; see also this volume Chapter 5). Figure 3 illustrates the influence of CAb concentration in a sandwich ELISA system. Reducing the CAb concentration results in a reduction in the region over which a constant proportion of the added Ig binds (above) which is in turn paralleled by a decrease in the linear region of the log–log titration plot (below).

Mass Law considerations for sandwich ELISAs

The observation presented in Figure 3 can be compared with theoretical predictions made for the binding of Ag to CAb in situations in which the affinity and concentration of the CAb are known and can be mathematically manipulated. A quadratic equation derived from the simple Mass Law by J.H. Peterman (Peterman, 1986), was used to develop computer-generated titration plots illustrating the effect of these variables in the performance of sandwich ELISAs (Figure 4). These data show that a reduction in affinity, i.e. K_a, results in: (a) a lowered proportion of Ig binding, i.e. a lowered AgCC (Figure 4; left), (b) a decrease in the log–log linear region of the titration plot (Figure 4; right), and (c) a decrease in the Y-intercept for log–log titration plots (Figure 4; right).

The theoretical data presented in Figure 4, and the empirical data of Figures 2 and 3, are consistent in showing that: (a) the AgCC of the CAb is paramont to the favourable performance of sandwich ELISAs, and (b) the concentration of CAb primarily determines the range of linearity of the titration plot but at very

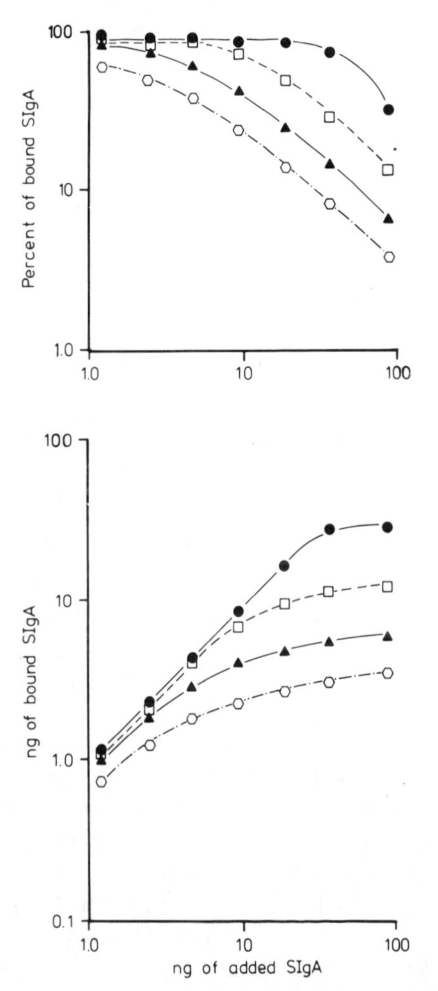

FIGURE 3 The influence of the concentration of solid phase capture antibody on the binding behaviour of iodinated SIgA. Each symbol shows data obtained with a different amount of added, solid phase anti-SIgA. ● = 5 μg/ml, □ = 2.5 μg/ml, ▲ = 1.25 μg/ml, and ○ = 0.6255 μg/ml. Upper curves show the percentage of SIgA bound while the lower plots show the ng of SIgA bound at each concentration of added SIgA. Experiments were conducted as illustrated at the top of Figure 2 but without use of the enzyme-labelled detection system. (From Butler *et al.*, 1986)

low concentration also lowers AgCC. When the results presented in Figure 2 are interpreted in terms of the theoretical data of Figure 4, the MoAbs to IgG_1 and IgG_2 behave with lower AgCC, i.e., lower affinity and lower functional concentration[c], than their PoAb equivalent. The same result has been observed with every PoAb–MoAb pair so far tested (Butler *et al.*, 1986; Suter, Butler, and Peterman, 1988).

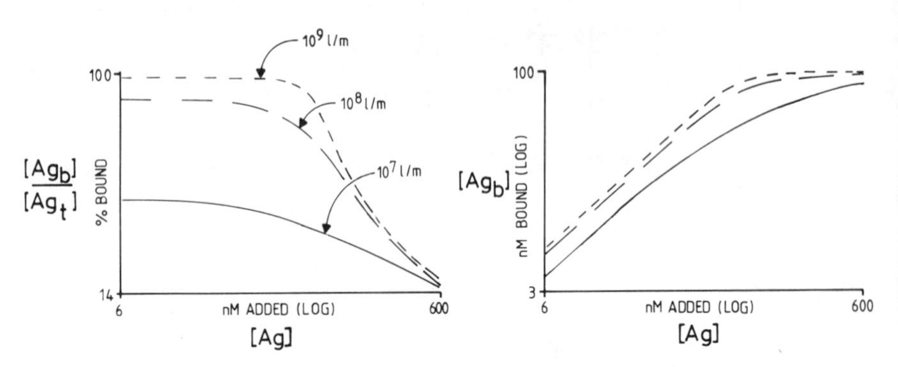

FIGURE 4 The theoretical influence of the affinity of the capture antibody (CAb) on the nature of the sandwich ELiSA titration plot. The equilibrium constants of the different CAbs are indicated. On the left, data are expressed in terms of the proportion of added Ig ($[Ag_b]/[Ag_t]$) which binds while, on the right, data are expressed in ng bound ($[Ag_b]$). (Modified from Butler *et al.*, 1986)

Effect of surface adsorption on the antigen capture capacity of CAb

The possibility that the lower AgCC of MoAbs could result from adsorption-induced denaturation was investigated using a model system in which MoAbs to fluorescein (FLU) were either immobilized by simple adsorption or immobilized using the protein–avidin–biotin capture (PABC) system (Suter and Butler, 1986). This configuration involves the immobilization of MoAbs or PoAbs biotinylated with the *N*-hydroxysuccinimide ester of biotin, to a biotinylated carrier protein adsorbed on the plastic. The intermediate linkage is formed by streptavidin (Butler *et al.*, 1987; Suter, Butler, and Peterman, 1988). The results of studies with this system revealed significant differences in the fluorescein binding capacity of MoAbs immobilized by PABC versus those directly adsorbed on plastic. When equal amounts of CAbs were comparatively tested (determined using ^{125}I-CAb), anti-FLU MoAbs immobilized by the PABC system showed a 5–400-fold greater AgCC than when adsorbed directly on plastic (Suter and Butler, 1986).

When the PABC system was employed to study anti-Ig MoAbs, the data shown in Figure 5a were obtained. The much lower per cent of swine IgG bound by the anti-swine IgG MoAbs when adsorbed on plastic versus when immobilized using the PABC system, indicates a substantial loss in AgCC due to direct adsorption. The lower Y-intercept is consistent with a lower affinity for the surface-adsorbed MoAb (Figure 5a; bottom). However, when a PoAb to swine Ig was studied in the two systems, little loss of AgCC was observed due to direct adsorption and no difference in affinity was observed (Figure 5b). Furthermore, the performance of this PoAb in the PABC after affinity purification, compared to its performance either when affinity-purified and directly adsorbed or adsorbed merely as the globulin fraction of PoAb serum, was not substantially different (Figure 5c).

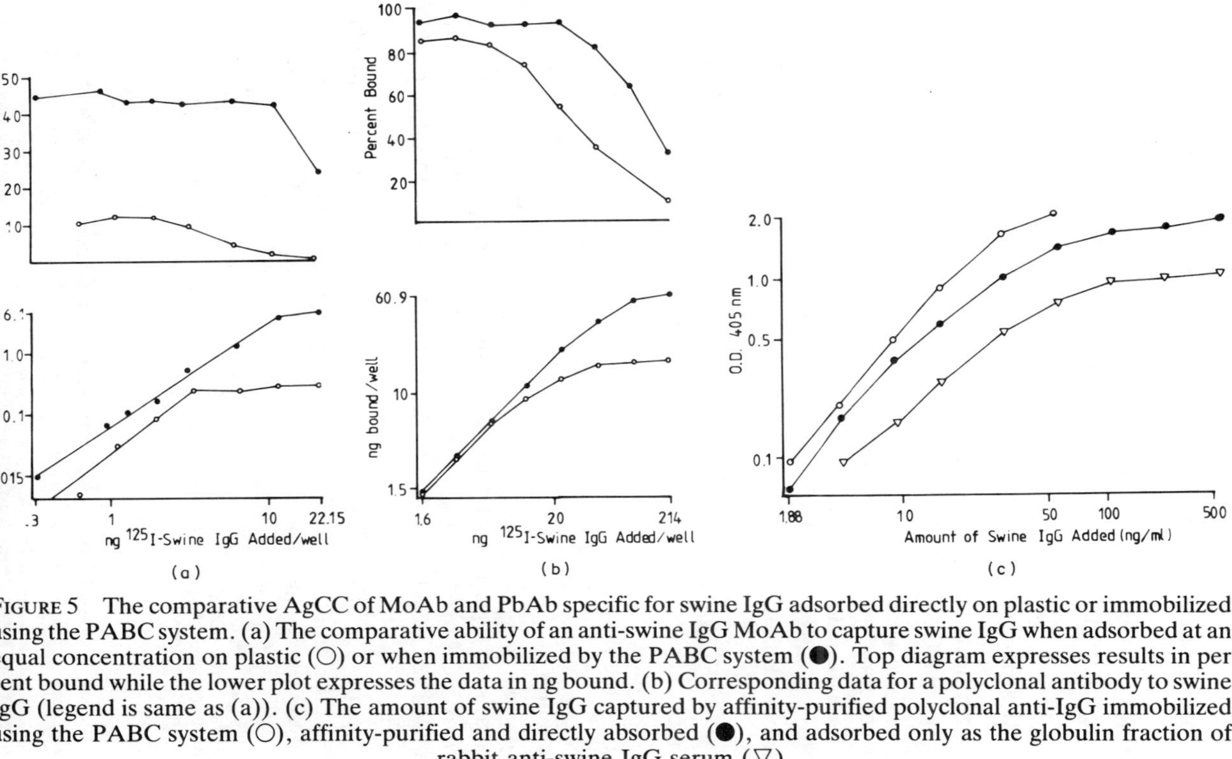

FIGURE 5 The comparative AgCC of MoAb and PbAb specific for swine IgG adsorbed directly on plastic or immobilized using the PABC system. (a) The comparative ability of an anti-swine IgG MoAb to capture swine IgG when adsorbed at an equal concentration on plastic (○) or when immobilized by the PABC system (●). Top diagram expresses results in per cent bound while the lower plot expresses the data in ng bound. (b) Corresponding data for a polyclonal antibody to swine IgG (legend is same as (a)). (c) The amount of swine IgG captured by affinity-purified polyclonal anti-IgG immobilized using the PABC system (○), affinity-purified and directly absorbed (●), and adsorbed only as the globulin fraction of rabbit anti-swine IgG serum (▽)

The results of Suter and Butler (1986) and of those presented in Figures 5a and 5b suggest that the differences seen between the anti-IgG_2 MoAb and PoAb (Figure 2) could reflect adsorption-induced denaturation and that MoAbs are especially affected. The basis for this difference is not easily explained but the practical implications are obvious. Hence, the unique specificity of MoAbs appears to be offset not only by their lowered K_a, but also by their propensity for denaturation when adsorbed on plastic.

Sandwich ELISA immunochemistry: a summary

The empirical and theoretical data presented above support the introductory remarks made concerning the importance of the AgCC[c] of CAb for favourable performance in sandwich ELISAs when Igs are being quantitated. Investigators developing such assays need to consider the efficacy of PoAbs versus MoAbs in the design of their assays and then take the necessary measures to ensure that the most useful assay is developed. The PABC system, although improving the AgCC of MoAb, suffers from the fact that only small amounts of CAb can be immobilized by this method. Hence, future improvements in sandwich ELISAs for Igs will in part depend on technical advances which permit larger amounts of CAb to be immobilized by non-denaturing methods.

The data in Figure 3 illustrate the value of having a high concentration of CAb on the solid phase while those in Figure 4 demonstrate the value of using CAbs of high 'function' affinity[f]. Hence, methods which can increase Ab abundance of moderate to high affinity Ab will also increase the linear log–log titration range of the assay. Furthermore, the data of Figures 2–4 illustrate that the nature of the sandwich ELISA titration plot reflects the AgCC of the CAb in an Ab (CAb)-limiting system. The parallelism between the amount of Ag captured and its detection (Figure 2) means that the proportional binding of Ag (Ig in this chapter), as determined by enzyme activity, is a direct result of the binding between the CAb and the Ig and is not influenced by steric hindrance. Hence, the influences of AgCC and concentration are predictable and reproducible and, as will be discussed later, suitable for transformation to a logit–log plot which effectively increases the linear region of the log–log titration plot (Figure 10). The immunochemical aspects demonstrated and discussed above form the basis for the procedural and technical aspects of sandwich ELISA discussed below.

SOURCES AND SPECIFICITY OF REAGENTS

Types of anti-immunoglobulin reagents

PoAbs and MoAbs can be used in sandwich ELISAs; regardless of their origin and nature there is no substitute for CAbs with high AgCC. It was shown earlier that AgCC can be determined from values for the percentage of Ig bound in the region of the sandwich ELISA titration plots where antibody is in excess. As reviewed above, the AgCC of PoAbs appears to be higher than that for MoAbs (Butler *et al.*, 1986). Furthermore, the concentration of the solid phase CAb is also an important factor in determining the useful titration range of sandwich assays (Figure 3). This is a practical concern when PoAbs are to be employed (see below).

When the globulin fractions of PoAbs have low Ab abundance, affinity purification is desirable if not absolutely required. Most likely candidates are PoAbs specific for sub-isotypes and allotypes, i.e. 'weak' antisera. Affinity purification is not a 'cure-all'; little improvement will be seen when Abs with low AbCC are affinity purified despite increasing Ab abundance (see Figure 4). Affinity purification of CAb for sandwich ELISA can have several disadvantages. Although discussed in greater detail below, two are worthy of repetition. First, affinity purification is time consuming and the decision to do it must be weighed against preparing or selecting another reagent with greater Ab abundance. Second, affinity-purified Ab can be contaminated with ligand

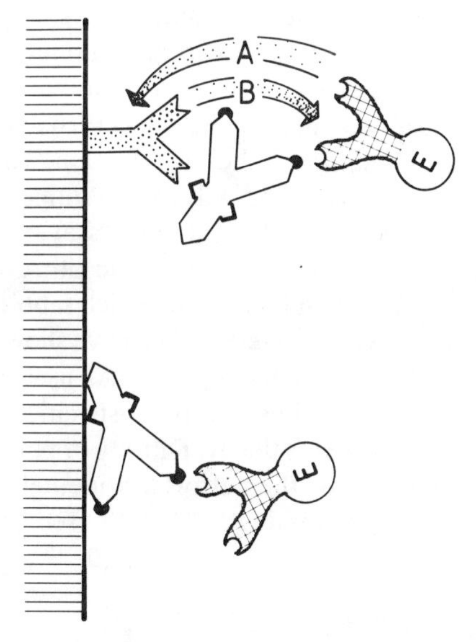

FIGURE 6 Immunochemical short-circuits encountered in sandwich ELISAs. In the model shown, rabbit anti-IgG_1 (stippled) was purified on an IgG_1 affinity column and used to coat the plastic (vertical bars) as a CAb (*Top*). Arrows indicate the various short-circuits. The unshaded immunoglobulin is IgG_1. A = asymmetrical enzyme–conjugate (anti-Fab; chequered) recognizes the CAb. B = the CAb recognizes the antibody in the enzyme–antibody detection system. The diagram below illustrates how leached antigen (e.g. IgG_1) present in the affinity purified anti-IgG_1 CAb, can cause unacceptable background problems. The enzyme–antibody conjugate recognizes the leached ligand (IgG_1) which also becomes adsorbed on the solid phase

that has leached from the affinity column. This can increase background noise through short-circuitry (Figure 6). Hence, selection of PoAbs of high antibody abundance, or monoclonals with at least moderate AgCC and resistance to denaturation, is desirable to avoid the work of affinity purification as well as the negative consequences of ligand leakage.

The need for CAbs of high Ab abundance is a requirement that can be fulfilled by MoAbs, particularly those raised in certain *in vitro* systems. Ascites fluid can be processed by a combination of neutral salt fractionation and gel permeation and hydroxyl apatite or Protein-A chromatography to yield preparations of >50 per cent functional MoAb. Such preparations function as well as affinity-purified CAb. While MoAb may solve the problem of low Ab abundance and circumvent the problems of affinity purification, their functional affinity (i.e. AgCC) is often low (Butler *et al.*, 1986) and this is apparently further decreased by immobilization through adsorption on microtitre wells (Suter and Butler, 1986).

Successful quantitation of Igs by sandwich ELISAs also depends on the specificity of the CAb. Many commercial PoAbs are not specific for the isotype claimed (Butler, Peterman, and Koertge, 1985; see also this volume Chapter 5). Unfortunately, immunological reagents differ from pharmaceutical reagents in that *the proof of specificity and efficacy resides with the user not with the supplier*. Accepting this situation as the current *status quo* means that one must test his/her commercial reagents for their specificity. In general, MoAbs are specific through the very nature of their production. The specificity of anti-Igs can be conveniently tested using the Adsorbed Antigen Activity Assay system (formerly called EADA; Butler, 1981). In this assay, purified Igs are adsorbed directly on plastic over a range of 3–200 ng/microtitre well. The specificity of this assay system is illustrated in Figure 7 in which rabbit antibodies specific for two different bovine allotypes are tested. We have shown that specificity at this level ensures the specificity of such reagents when used as capture antibodies in sandwich assays or as antiglobulins in antigen-specific ELISA. Demonstration of specificity by this test requires the availability of purified Ig, which for some investigators, might complicate their efforts. Nonetheless, such purified Igs are also required to standardize sandwich ELISAs (see below).

Affinity purification of CAbs

The decision to affinity purify a CAb will depend on whether the poor performance of the neat reagents is a consequence of low AgCC or low Ab abundance. Affinity purification can only resolve the latter. The affinity purification of anti-protein antibodies, like those specific for Igs, is more difficult than affinity purification of, e.g. anti-hapten antibodies. In the case of the latter, various competitive elution schemes using a ligand of one affinity and competitors of another at neutral pH, can be employed. The performance

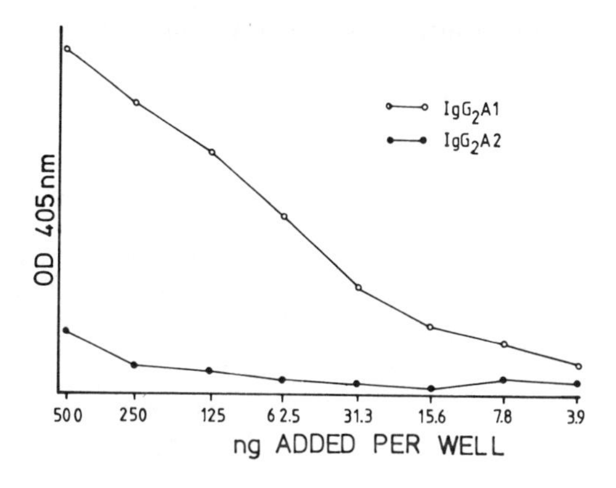

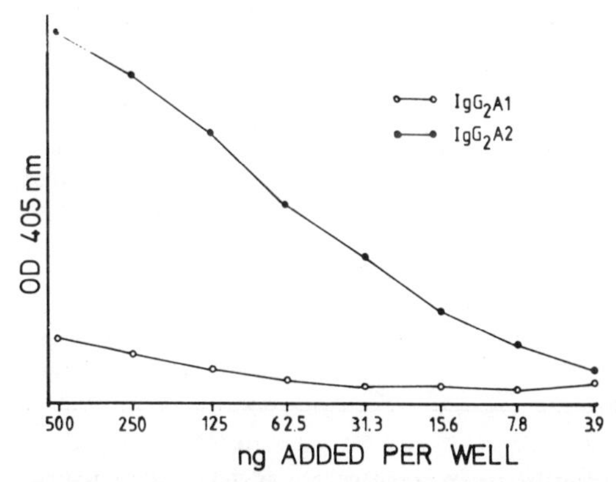

FIGURE 7 The specificity of two rabbit anti-bovine IgG_{2a} anti-allotypic reagents when tested in the A_4 system. Both IgG_{2a} proteins were absorbed on plastic over the concentration range indicated. *Top*: reactions when either anti-IgG_{2a} (A1) or anti-IgG_{2a} (A2) were tested on wells containing IgG_{2a}(A1). *Bottom*: result obtained when either anti-IgG_{2a}(A1) or anti-IgG_{2a}(A2) were tested on wells containing IgG_{2a}(A2)

of competitive elution at neutral pH and without large changes in ionic strength, reduces the problem of ligand leaching and the elution of non-specifically adsorbed proteins which can contaminate the affinity-purified CAbs. Table 1 lists criteria which contribute to the successful affinity purification of CAbs.

A variety of commercially available solid phase matrices are available. Most common are agarose and acrylamide. The latter suffers from slow flow rates and solvent accessibility as well as a propensity for hydrophobic interaction. Cross-linked agarose, available from several suppliers, fulfils most of the requirements (Cuatrecasus, 1970). Its mild hydrophilic character is not a serious drawback and its ease of derivatization is highly advantageous.

TABLE 1 Factors important in the preparation of affinity columns for the purification of CAbs

Criterion	Desirable feature
1. Chemistry of the solid phase support	(a) Easy to manipulate physically (b) Resistance to a wide variety of pH and ionic strength (c) Microstructure easily accessible to macromolecules (d) Readily modifiable for substitution of active conjugation groups (e) Minimal hydrophilic or hydrophobic properties (f) Resistance to proteolytic and microbial attack
2. Derivation or activation procedure	(a) Straightforward and reproducible (b) High degree of substitution (c) Economical (d) Generation of a single reaction group
3. Nature of ligand–solid phase linkage	(a) Covalent and stable to changes in pH and ionic strength (b) Ligand readily accessible for combination with Ab (c) Minimal non-specific hydrophilic or hydrophobic properties (d) Resistance to proteolysis and microbial damage

Trisacyl® has recently been employed and is more inert than cross-linked agarose.

Oxidation of the hydroxyl groups of agarose to ethoxide groups at high pH has been widely used to allow activation with cyanogen bromide (CNBr; Axen and Vretblad, 1971). The cyanate ester produced is in relatively low yield (15 per cent) compared to carbamates (60 per cent) and immunocarbamates (25 per cent), but the absolute quantity of cyanate ester groups nevertheless allows agarose activated in this manner to have a high capacity to bind non-ionized amino groups of proteins; >90 per cent of the protein added at a concentration of 5–10 mg/ml of agarose is covalently coupled by this method. Unfortunately, the unstable linkages produced by the carbamates, as well as the isourea linkage to the cyanate ester, contribute to the problem of ligand leakage. Newer and more expensive cyanylating reagents are available which significantly increase the relative yield of cyanate esters and to some extent lessen this

Trisacyl is the registered trademark of Réactifs IBF, France.

problem (Köhn and Wilchek, 1984) although the covalent attachment is still the isourea linkage.

The negative performance of CNBr-activated agarose can be somewhat lessened by extensive pre- and post-treatment with various buffers, e.g. alternative washing with low and high pH buffers of various high ionic strength. Furthermore, such columns can be treated with ethylene glycol and high ionic strength buffer *after* the desired antibody has bound and prior to pH- or chaotrophic agent-dependent elution. If elution is done competitively, e.g. purification of anti-hapten antibodies, ligand leakage is typically reduced because pH changes do not occur. Nevertheless such procedures are not 'cure-alls' for CNBr-activated agarose and it is preferable to choose affinity matrices in which antigen is bonded by more stable linkages.

Despite problems with affinity purification of antibodies on CNBr-activated agarose, the high capacity of this type of affinity column is advantageous when used to render antisera specific. When used in such a capacity, small amounts of ligand leaked into the unbound fraction are usually of little consequence.

Among the alternatives to CNBr-activated agarose are: (a) Affigels® (Bio Rad, Richmond, CA) in which the ligand is linked via a succinimide ester; (b) agarose activated with oxiranes or epichlorohydrins (Sundberg and Porath, 1974); (c) Trisacyl G-2000® activated with carbodiimadazole (CDI; Bethell *et al.*,) 1979; Hearn, 1983); (d) agarose activated with 2-fluoro-1-methyl-pyridinium toluene-4-sulphonate (FMP; Ngo, 1986); or (e) agarose substituted with primary amines (i.e. Actigel A) and coupled to proteins through an amino–methyl linkage during reduction with cyanoborohydride (Murphy *et al.*, 1977; Frost *et al.*, 1981). Although rigorous quantitative experiments remain to be performed, our experience indicates that all gels in which ligand is bound through carbamate and isouve A linkages leak ligand although leakage from oxiranes, epichlorohydrin, and CDI-activated gels is less than from CNBr-agarose. FMP-activated gels utilize a theoretically stable amino–methyl bond and appear to leak little ligand. Stability has also been claimed for Actigel A.

Much has been published about the value of so-called 'spacer linkages' in affinity chromatography (Cuatrecasus and Anfinsen, 1971). While these no doubt increase ligand accessibility, most procedures which produce spacers also produce activated matrices with lower ligand capacity. Furthermore, long 'spacer' hydrocarbon chains encourage hydrophobic interactions. Because spacer linkages reduce antigen concentration, their value can be questioned as effective ligand–receptor (antibody) binding depends on ligand concentration. This is a fundamental principle in all antigen–antibody interaction of which affinity chromatography is not excepted. The practical value of spacers for the affinity purifications of CAbs is best established empirically for the ligand–Ab system employed. The use of porous glass, while more difficult to derivatize

Affigel is the registered trademark of the BioRad Corporation.

(Rauterberg Schieck, and Hansch, 1979), has been recommended because of its inert characteristics. The complicated activation process, low capacity, and expense of this procedure have heretofore prevented its practical implementation.

This section on affinity chromatography, while seeming independent of the theme of this chapter, has been included because: (a) in some cases PoAb for use as CAb must be affinity purified; and (b) the practical use of affinity chromatography for purification of CAbs is far more complicated than implied by the sales brochures supplied by commercial manufacturers of affinity chromatography products. The consequence of ligand leakage was briefly mentioned above and is illustrated in Figure 6; this topic will be further discussed below. The most important message to be gained is that: (a) alternative solutions which avoid affinity purification of CAb are most desirable; and (b) affinity purification of CAbs with low AgCC will have only a modest effect on their performance in sandwich ELISA and, because of ligand leakage, may function less well after affinity purification!

Detection systems for sandwich ELISAs

The types of detection systems most often used in sandwich ELISAs are reviewed in Figure 1. All four systems have been used. The indirect system (Figure 1; lower right) requires an additional step and permits a corresponding increase in the number of potential short-circuit reactions. The 'reverse asymmetrical' configuration shown at the lower left of Figure 1 is dangerous immunochemically and may only be valid in systems where the Ig to be measured is the predominant Ig in the test sample. Even in such systems, data should be viewed with caution because the proportion of a given isotype bound in the test sample may not be the same as in the Rstd, especially if the latter is purified Ig.

An advantage of both the 'forward' and 'reverse' asymmetrical sandwich assay is that anti-L-chain or anti-Fab reagents are often available as very potent antisera. For example, up to 75 per cent of the Ab made by rabbits to heterologous Igs is directed toward the Fab region (Pringnitz and Butler, 1981). Hence the globulin fraction of such a reagent will be one with higher Ab abundance; this is advantageous either for adsorption on plastic or for the preparation of a potent antibody–enzyme conjugate for use as the detection system.

The symmetrical configuration depicted in Figure 1 (upper left) is commonly used but suffers from the potential of steric hindrance. The principle of all sandwich ELISAs is that both the capture and detector antibody recognize the Ig to be measured. If the determinants are located on the symmetrical heavy chains of an IgG protein, steric hindrance can prevent the detection system (which may be a large antibody–enzyme aggregate) from binding the antigen

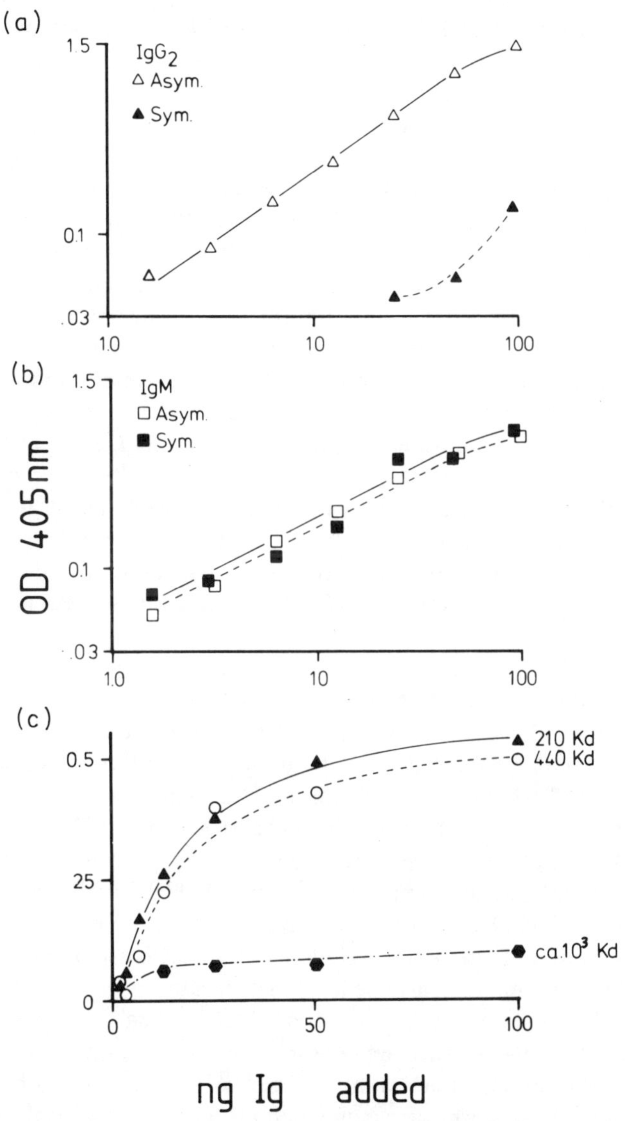

FIGURE 8 The role of steric hindrance in the effectiveness of detection systems for sandwich ELISAs. The comparative ability of asymmetrical and symmetrical one-step glutaraldehyde conjugates to detect IgM and IgG_2 in sandwich ELISAs is shown in (a) and (b). (a) Detection of captured IgG_2 using an asymmetrical (△), and a symmetrical (▲) anti-IgG_2 detection reagents. (b) Detection of captured IgM using asymmetrical (□) and symmetrical (■) anti-μ reagents. (c) The influence of two different-sized HRP-anti-IgG_1 conjugates (⬢ = 10^3 kilodaltons ○ = 200 kilodaltons) on the detection of captured IgG_1 in a symmetrical, isotype-specific capture configuration (see Figure 1) versus detection of captured IgG_1 using an asymmetrical detection configuration (▲; see also Figure 1). (From Butler *et al.*, 1986)

being measured. This theoretical prediction is borne out experimentally; anti-IgG subclass-specific conjugates are sterically hindered from recognizing their respective subclasses, when first captured by a subclass-specific capture antibody (Figure 8a). This steric hindrance is not seen when the much larger IgM is the Ig being measured (Figure 8b). Further studies using horseradish peroxidase (HRP)–antibody conjugates of different sizes, suggest that conjugate size is responsible for this effect (Figure 8c).

In our laboratory, asymmetrical configurations using anti-Fab as the detection reagent have proven to be the most reliable and sensitive configurations for sandwich ELISA for the quantitation of Igs. The use of an asymmetrical configuration means that the same antibody–enzyme conjugate can be used for the quantitation of all isotypes of Igs in a particular animal species. This is the system promoted in this chapter for routine use (see below).

The source of the anti-Fab or anti-L-chain reagent is unimportant. If the reagent is truly an anti-L-chain reagent, both κ- and λ-specificities must be present for the final assay to be valid. The species origin of the anti-Fab or L-chain must be compatible with the capture antibody, i.e. it must not short-circuit to the immobilized CAb (Figure 6). Rabbit anti-human Fab may require absorption with mouse IgG when mouse MoAbs are used as capture antibodies.

The method of preparation of the antibody–enzyme conjugate is perhaps dictated more by convenience and familiarity to the investigator than by overwhelming biochemical or immunochemical principles. If a conjugate of modest molecular size is desired to reduce the effect of steric hindrance, preparation of horseradish peroxidase (HRP)–antibody conjugates using a modification of the method of Nakane and Kawaoi (1974; as described by Suter, Pike, and Nossal, 1985), can be used. This procedure produces conjugates which are mostly composed of 1 mol of HRP and 1 mol of Ab. This method also appears to minimize enzyme alteration compared to glutaraldehyde and carbodiimide methods (Williams, 1984). The use of a variety of newer cross-linking reagents can also be used to generate conjugates of small size and of controllable stoichiometry (see Pierce Chemical Company Catalogue, Rockford, IL). One- or two-step glutaraldehyde conjugates (Avrameas, 1969; Engvall, Jonsson, and Perlmann, 1971) of Ab and alkaline phosphatase are widely used and perform very well when used in asymmetrical configurations were steric hindrance is not a problem (Figure 1). Two-step glutaraldehyde conjugates are more complicated to produce but more controllable in size. In our experience, conjugates of alkaline phosphatase appear more stable than those with HRP.

In addition to these two common types of conjugates, the use of a biotinylated detection antibody, followed by enzyme-labelled avidin, or streptavidin, can also be used. However, this indirect system offers very little advantage over the more conventional HRP and alkaline phosphatase conjugates, and requires an additional reaction step. An experimental application of the biotin–avidin system to evaluate the effect of surface adsorption on CAbs was described above. In addition, Abs can be readily biotinylated in microgram amounts and in a short time (1 hr) using biotin hydrazide (Guesdon, Ternynck, and Avrameas, 1979); the excess biotin can be quickly removed on a BioGel P-10 column. Such biotinylated Abs can then be studied for their specificity and feasibility for use in eventual sandwich ELISAs. By contrast, HRP-periodate and especially glutaraldehyde alkaline phosphate conjugates, require larger amounts of Ab and a longer time to prepare. The major difficulty with biotinylation is its effect on Ab activity which may vary with the Ab in question (Gretch, Suter, and Stinski, 1987).

Antibody–enzyme conjugates used in our laboratory are prepared and stored as originally described by others (Engvall, Jonsson, and Perlmann, 1971) or at −20 °C in 50 per cent glycerol.

MEASUREMENT OF IMMUNOGLOBULINS USING ASYMMETRICAL SANDWICH ELISAS

Preparation of the solid phase CAb

The globulin fraction or an affinity purified form of the desired Ab, is first allowed to adsorb directly on plastic. The chemical basis of protein–plastic bonding is largely hydrophobic (see this volume Chapter 5). This process is known to result in some protein alteration, i.e. conformational change (Bull, 1956; Burghardt and Axelrod, 1983) so that solid phase CAb prepared in this manner may not be as active, i.e. have the same AgCC as in free solution or when it is immobilized by an alternative non-denaturing method (see Figure 5 and Suter and Butler, 1986). In our laboratory, we permit adsorption of CAbs to occur at room temperature overnight at a concentration of 5 μg/ml of protein.

As discussed earlier, the two-stage PABC system significantly improves the AgCC of MoAb to both fluorescein and Igs but does not substantially improve the performance of PoAb CAbs with high AgCC and Ab abundance (Figure 5c; Suter, Butler, and Peterman, 1988). One negative aspect of the PABC system is that significantly less CAb can be immobilized. Unpublished data from manufacturers producing solid phase surfaces with active groups for covalent attachment of CAb indicate that either this method does not greatly increase CAb concentration or when it does, the CAbs do not show a proportional improvement in AgCC beyond that seen by simple adsorption. Furthermore, modification of plastic surfaces for covalent attachment seriously impairs the optical qualities of the plastic so they do not function adequately in standard plate readers (Butler, unpublished). Hence there is a demand for a practical and reproducible method for the immobilization of CAbs at high concentration and with retention of their AgCC. The PABC system in its present form is but an experimental tool which raises the awareness of investigators to the loss in AgCC which CAbs suffer during surface adsorption.

In summary, the selection of a method for CAb immobilization depends on the situation. In general, when a PoAb with high AgCC is available, adsorption of the globulin fraction of this PoAb directly on plastic is simple and performance will be adequate. When the Ab abundance of a PoAb with good AgCC is low, affinity purification is likely to solve the problem. When MoAb and perhaps even some PoAbs lose their activity when adsorbed to plastic, use of the PABC system is a possible alternative. It should again be emphasized that for good performance of sandwich ELISAs there is no substitute for a CAb with high Ab abundance and high AgCC.

Application of standard Rstd and samples

The preparation of a solid phase CAb is then followed by addition of the samples to be tested as dilutions in PBS–T[a]. The use of a serum or, in special circumstances, a secretion, as an Rstd for routine sandwich ELISAs is highly recommended. Serum treated with protease inhibitors (ε-amino caproic acid and 10 mM benzamidine-HCl) and azide, are stable for years when frozen at −20 °C. Such sera should be aliquoted (0.5–1.0 ml) into small vials with rubber gasket seals and removed as needed from the freezer. In special cases, such as for secretory IgA (SIgA), an Ig-rich fraction of colostrum or parotid saliva can be stored in aliquots at −70 °C in 50 per cent glycerol containing azide and protease inhibitors. Short-time studies on Ig stability, i.e. 1–2 yr, indicate no significant loss of Ig antigenicity when stored as serum but a gradual loss of activity when stored as exocrine body fluid (Butler, unpublished).

The Rstd must sooner or later be standardized against a purified preparation of the Ig in question. The use of purified Igs as standards is not recommended because of their lack of long-term stability. For many investigators, obtaining such preparations may also be difficult and it is therefore best to purchase or obtain the Rstd from an another investigator rather than to prepare one's own.

Dilutions of test samples and the Rstd are made in separate 100 × 13 mm glass test tubes in the amounts and number of dilutions as is required for the standard microtitre plate format. This format is the same as used for antigen-specific ELISA (see this volume Chapter 5 and Figure 5) and is designed for data acquisition and analysis by ELISANALYSIS I. Briefly, the dilutions of the Rstd are prepared in independent duplicate: a total of six serial dilutions spanning the range of 3.12–100 ng/ml or 1.56–50 ng/ml, are prepared in a minimum volume of 200 μl/dilution. These are applied to CAb-containing wells in columns 6 and 7, rows A–F of a 96-well microtitre plate. Wells 6G, H and 7G, H receive no sample, only PBS–T.

Test samples are prepared such that four dilutions of each of 20 test samples are tested per microtitre plate. Their approximate concentrations can be estimated if the investigator is familiar with the samples being tested. Hence, dilutions are prepared such that they contain the concentrations of the Ig to be measured which fall within the concentration range of the Rstd. Dilution sequences may be twofold, threefold or, if nothing at all is known about the samples, log (tenfold) dilutions can be prepared. In the case of the latter, final values cannot be obtained because of limitation in the data analysis program (see below). Hence, the log dilution sequences are only used as a method of assessing the correct range over which twofold or threefold dilutions should be prepared for final analysis.

The dilutions of the test samples are then transferred to the appropriate wells of the CAb-coated microtitre plate (see this volume Chapter 5, Figure 5).

The reaction of the CAb and test samples and Rstd is allowed to proceed for 4 hr at RT on a Minishaker (Dynatech, Alexandria, VA), the plates emptied[g], washed[d], and the appropriate asymmetrical antibody–enzyme detection system (rabbit or goat anti-Fab depending on the species of the CAb) added at proper dilution (usually 1 : 500 to 1 : 4000 in PBS–T). The reaction of the detection system is allowed to proceed for 2–3 hr on a Minishaker, the plates emptied, washed, and the appropriate substrate for the enzyme is added. The dilution of the antibody–enzyme detection system should be such that the highest concentration of the Rstd develops its optimal absorbancy (highest concentration = 1.2 at OD_{405} nm for alkaline phosphate or HRP) in about 1 hr. In this period, values for the control well (6G, H and 7G, H) should remain low (<0.2). If not, the conjugate is not compatible with the capture antibody, i.e. a short-circuit has been formed (Figure 6). This can occur because: (a) immobilized CAb contains too much contaminating ligand which has also been immobilized (a consequence of purification by affinity chromatography: see above); (b) the conjugate antibody is from a different species than the CAb and light chain cross-reactivity is occurring; or (c) the conjugate is 'sticky'. Sticky conjugates are sometimes obtained when excess reactive groups such as the carbonyl groups of glutaraldehyde are not completely blocked with excess amine.

Most modern plate readers can be programmed to monitor one or more critical wells. When the standard format for ELISANALYSIS I is used, the BioTek EL310 is programmed to monitor wells A6 or A7 (the highest concentration of the Rstd). When one of these wells reaches the desired OD_{405} of 1.2, the entire plate is read and the data transferred to a diskette through an interfaced personal computer. When *preliminary assays* are performed with test samples for which little is known, it may be advisable to read the results of the entire plate at various times both earlier and later than when well A6 or A7 reaches OD 1.2. This makes it easier to estimate the proper dilution range needed for final analyses of the samples.

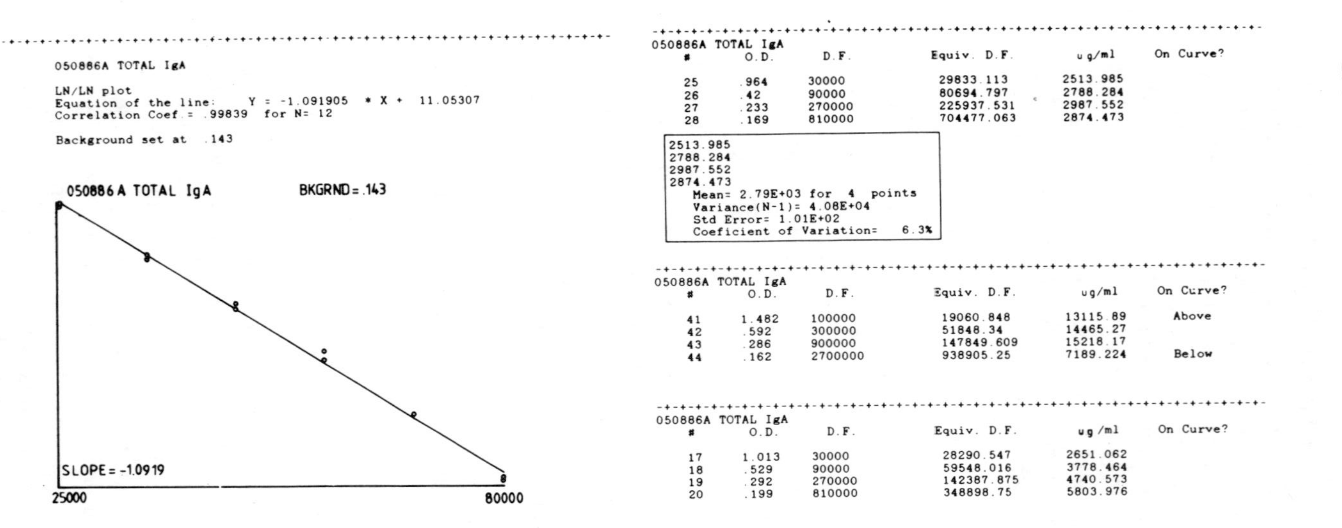

050886A TOTAL IgA

LN/LN plot
Equation of the line: Y = -1.091905 * X + 11.05307
Correlation Coef. = .99839 for N= 12

Background set at .143

050886A TOTAL IgA

#	O.D.	D.F.	Equiv. D.F.	u.g/ml	On Curve?
1	1.081	25000	26406.375	2366.853	
2	1.107	25000	25753.383	2426.866	
3	.579	50000	53262.391	2346.872	
4	.513	50000	49722.629	2513.946	
5	.363	100000	99650.898	2508.758	
6	.379	100000	93445.344	2675.361	
7	.251	200000	191192.734	2615.162	
8	.263	200000	173606.375	2880.079	
9	.193	400000	387056.688	2583.601	
10	.193	400000	387056.688	2583.601	
11	.164	800000	856672.125	2334.616	
12	.163	800000	895818.063	2232.596	

050886A TOTAL IgA

#	O.D.	D.F.	Equiv. D.F.	u g/ml	On Curve?
25	.964	30000	29833.113	2513.985	
26	.42	90000	80694.797	2788.284	
27	.233	270000	225937.531	2987.552	
28	.169	810000	704477.063	2874.473	

2513.985
2788.284
2987.552
2874.473
Mean= 2.79E+03 for 4 points
Variance(N-1)= 4.08E+04
Std Error= 1.01E+02
Coeficient of Variation= 6.3%

050886A TOTAL IgA

#	O.D.	D.F.	Equiv. D.F.	ug/ml	On Curve?
41	1.482	100000	19060.848	13115.89	Above
42	.592	300000	51848.34	14465.27	
43	.286	900000	147849.609	15218.17	
44	.162	2700000	938905.25	7189.224	Below

050886A TOTAL IgA

#	O.D.	D.F.	Equiv. D.F.	ug/ml	On Curve?
17	1.013	30000	28290.547	2651.062	
18	.529	90000	59548.016	3778.464	
19	.292	270000	142387.875	4740.573	
20	.199	810000	348898.75	5803.976	

FIGURE 9 Actual computer printout of the data analysis obtained with ELISANALYSIS I for the quantitation of human IgA. The graph in the upper quadrant is that generated using the Rstd serum. The equation of the log–log regression is shown. Below the graph the left-most column of the table gives the actual OD values generated at each dilution of the Rstd tested and at the extreme right, the concentration of IgA calculated for each of these dilutions. To the right of both graph and table, data are presented for four dilutions of three different test samples. Data are presented as for the dilution of the Rstd shown at the lower left. Note that only the top test sample titrates with the same slope as the Rstd as indicated by the fact that the calculated concentration, when corrected for dilution by ELISANALYSIS I, is independent of the dilution tested. For sample titrating properly, the operator can ask ELISANALYSIS I to calculate a mean value (enclosed in box). For additional information and interpretation, refer to corresponding text

Data acquisition and analysis

Data acquisition and analyses are done using ELISANALYSIS I. Plate reader values are transferred by an IBM-PCjr interfaced to an EL310 plate reader, multiskan plus plate reader or a Dynatech MR600 plate reader. The data, stored on diskettes, can then be analysed using ELISANALYSIS I. As the workings of this program have already been described in Chapter 5 and published elsewhere (Butler, Peterman, and Koertge, 1985; Peterman and Butler, 1988) they will not be repeated here. The only difference between the ELISANALYSIS I used for antigen-specific ELISAs (Chapter 5) and that used for sandwich ELISA is that for the former, the Rstd is assigned a value of 100 ELISA Units/ml of undiluted Rstd. When analysing data for sandwich ELISAs, the Rstd is assigned its actual value in μg Ig/ml, e.g. 2500 μg IgA/ml (Figure 9). Hence ELISANALYSIS I calculates the values for each dilution of the Rstd and test samples in μg/ml (Figure 9). ELISANALYSIS I will calculate the mean values and their corresponding statistics for all values of a test sample which fall within the range of the Rstd plot. In the same manner as in antigen-specific ELISAs, values for test samples, which indicate that the test sample is titrating with a different slope than the Rstd, can be immediately recognized. More often in sandwich ELISAs, failure of a test sample to titrate similarly to the Rstd reflect dilution errors rather than incorrect assessment of background values.

One aspect of ELISANALYSIS which can be usefully employed for analysis of data from sandwich ELISAs is the log–logit transformation. This capability is not available in ELISANALYSIS I; ELISANALYSIS II must be used (Peterman and Butler, 1988). As already described, the curved nature of

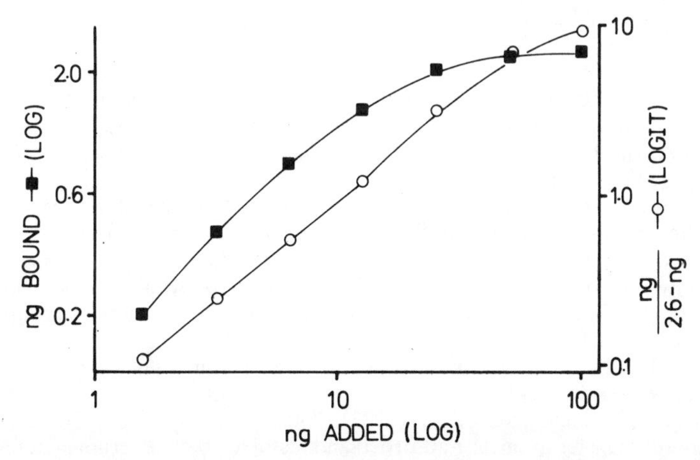

FIGURE 10 The demonstration of the logit-log capacility of ELISANALYSIS II in lengthening the usable linear region of the Rstd titration plot. ■ = ng bound (log); ○ = ng/(ng−2.6) (logit)

sandwich ELISA titration plots (when it occurs) is a reproducible effect resulting from Mass Law consideration for the binding of Ag and CAb when the latter becomes limiting. Hence, such curved plots can be linearized to allow further analyses to be performed by linear regression. Figure 10 illustrates how use of the logit–log transformation converts the curved sandwich ELISA plot into a straight line. ELISANALYSIS II contains an algorithm for estimating Y_{max} and automatically converting the data to a logit–log plot.

NOTES

(a) PBS–T = 0.01 phosphate buffer saline, pH 7.1, containing 0.05 per cent Tween 20 and 0.02 per cent sodium azide.
(b) The terms isotype, allotype, and idiotype define antigenic determinants which, respectively, denote those which are class or subclass specific (isotypic), the result of genetic variants (allotypic) and those which are individually specific and located in the variable region of the Ig (idiotypic).
(c) AgCC is a functional property reflecting both antibody affinity and its concentration. It is best equated to the 'extent of reaction' (α) where ($\alpha = Kc/(1+Kc)$, $K = Ka$ and c = free antigen concentration.
(d) The nearly universal method for protein immobilization on plastic microtitre plates is surface adsorption which results in <25 per cent of the surface being covered with protein (Butler *et al.*, 1986).
(e) Antibody abundance is the application of a radiochemical expression which, in the situation described here, refers to the ratio of specific Ab to total globular protein.
(f) Functional affinity refers to the actual behaviour of an antibody in the circumstances under discussion. For example, MoAbs reacting with solid phase antigen behave with different (typically lower) affinities than in free solution (Peterman, 1986).
(g) Plates are emptied by inverting and simultaneously 'flicking' over a sink. Wells are washed with a sixteen-well gravity feed system developed in our laboratory in collaboration with master machinist K. Breier (Butler, Peterman, and Koertge, 1985).

REFERENCES

Avrameas, S. (1969). Coupling of enzymes to protein with glutaraldehyde. Use of the conjugates for the detection of antigens and antibodies. *Immunochemistry*, **6,** 43–52.

Axen, R., and Vretblad, P. (1971). Chemical fixation of proteins to water-soluble carriers. In: *Protides of the Biological Fluids* (ed. H. Peeters). Pergamon Press, Oxford. Vol. 18, pp. 383–9.

Berson, S. A., and Yalow, R. S. (1959). Quantitative aspects of the reaction between insulin and insulin-binding antibody. *J. Clin. Invest.*, **38,** 1996–2016.

Bethell, G. S., Ayers, J. S., Hancock, W. S., and Hearn, M. T. (1979). A novel method of activation of cross-linked agaroses with 1,1′-carbonyldiimidazole which gives a matrix for affinity chromatography devoid of additional charged groups. *J. Biol. Chem.*, **254,** 2572–4.

Bull, H. B. (1956). Adsorption of bovine serum albumin on glass. *Biochim. Biophys. Acta*, **19,** 464–71.

Burghardt, T. P., and Axelrod, D. (1983). Total internal reflection fluorescence study of energy transfer in surface-adsorbed and dissolved bovine serum albumin. *Biochemistry*, **22,** 979–85.

Butler, J. E. (1981). The amplified ELISA: Principles of and applications for the comparative quantitation of class and subclass antibodies and the distribution of antibodies and antigen in biochemical separates. In: *Methods in Enzymology*, Vol.

73, *Immunochemical Methods* (eds. H.J. Vunakis and J.J. Lagone, eds.) pp. 482–523.
Butler, J. E., Peterman, J. H., and Koertge, T. E. (1985). The amplified enzyme-linked immunosorbent assay (a-ELISA). In: *Enzyme-Mediated Immunoassay* (eds. T. T. Ngo and H. M. Lenoff). Plenum Press, NY, pp. 241–76.
Butler, J. E., Spradling, J. E., Suter, M., Dierks, S. E., Heyermann, H., and Peterman, J. H. (1986). The immunochemistry of sandwich ELISAs. I. The binding characteristics of immunoglobulins to monoclonal and polyclonal antibodies adsorbed on plastic and their detection by symmetrical and asymmetrical antibody–enzyme conjugates. *Molec. Immunol.*, **23,** 971–82.
Butler, J. E., Peterman, J. H., Suter, M., and Dierks, S. E. (1987). The immunochemistry of solid-phase sandwich enzyme-linked immunosorbent assays. *Fed. Proc.* **46,** 2548–56.
Catt, K., Niall, H. D., and Tregear, G. W. (1966). Solid-phase radioimmunoassay of human growth hormone. *Biochem. J.*, **100,** 31c–33c.
Cuatrecasus, P. (1970). Agarose derivatives for purification of protein by affinity chromatography. *Nature (Lond.)*, **228,** 1327–8.
Cuatrecasus, P., and Anfinsen, C. B. (1971). Affinity chromatography. *Methods in Enzymology*, **22,** 345–85.
Engvall, E., Jonsson, K., and Perlmann, P. (1971). Enzyme-linked immunosorbent assay. II. Quantitative assay of protein antigen, immunoglobulin G, by means of enzyme labelled antigen and antibody coated tubes. *Biochim. Biophys. Acta*, **251,** 427–34.
Frost, R. G., Monthony, J. F., Engelhorn, S. C., and Siebert, J. C. (1981). Covalent immobilization of N-hydroxysuccinimide ester derivatives of agarose. *Biochim. Biophys. Acta*, **670,** 163–9.
Gretch, D. R., Suter, M., and Stinski, M. F. (1987). The use of biotinylated monoclonal antibodies and streptavidin affinity chromatography to isolate herpesvirus hydrophobic proteins or glycoproteins. *Analyt. Biochem.* **163**, 270–277.
Guesdon, J. L., Ternynck, T., and Avrameas, S. (1979). The use of avidin–biotin interactions in immunoenzymatic techniques. *J. Histochem. Cytochem.*, **8,** 1131–9.
Hearn, T. W. (1983). Preparative and analytical applications of CDI-mediated affinity chromatography. In *Affinity Chromatography and Biological Recognition* (eds. I.M. Chaiken, M. Wilchek, and I. Parikh). Academic Press, NY pp. 191–6.
Koertge, T. E., and Butler, J. E. (1985). The relationship between the binding of primary antibody to solid-phase antigen in microtiter plates and its detection by the ELISA. *J. Immunol. Meth.*, **83,** 283–99.
Köhn, J., and Wilchek, M. (1984). The use of cyanogen bromide and other novel cyanylating agents for the activation of polysaccharide resins. *Appl. Biochem. Biotech.*, **9,** 285–305.
Murphy, R. F., Conlon, J. M., Inman, A., and Kelly, G. J. C. (1977). Comparison on non-biospecific effects in immunoaffinity chromatography using cyanogen bromide and bifunctional oxirane as immobilizing agents. *J. Chromatography*, **135,** 427–33.
Nakane, P. K., and Kawaoi, A. (1974). Peroxidase-labeled antibody: A new method of conjugation. *J. Histochem. Cytochem.*, **22,** 1084–91.
Ngo, T. T. (1986). Facile activation of Sepharose hydroxyl groups by 2-fluoro-1-methylpyvidinium toluene-4-sulfonate: preparation of affinity and covalent chromatographic matrices. *Bio/Technology*, **4,** 134–37.
Peterman, J. H. (1986). Factors which influence the binding of antibody to solid phase antigens: Theoretical and experimental investigations. Ph.D. Thesis, University of Iowa.
Peterman, J. H., and Butler, J. E. (1988). ELISANALYSIS: A data acquisition and analysis system for antigen-specific and sandwich ELISAs which considers their immunochemistry. *J. Imm. Methods* (submitted).

Pringnitz, D. J., and Butler, J. E. (1981). The response of rabbits to the subclass-specific epitopes of bovine IgG1 and IgG2. *Vet. Immunol. Immunopath.*, **2,** 353–66.
Rauterberg, E. W., Schieck, C., and Hansch, G. (1979). Isolation of late complement components by affinity chromatography: I. Purification of the human complement component C9 and production of a C9-defective human serum. *Z. Immun. Forsch.*, **155,** 365–77.
Sundberg, L., and Porath, J. (1974). Preparation of adsorbents for biospecific affinity chromatography. I. Attachment of group-containing ligands to insoluble polymers by means of bifunctional oxiranes. *J. Chromatography*, **90,** 87–98.
Suter, M., and Butler, J. E. (1986). The immunochemistry of sandwich ELISAs. II. A novel system prevents the denaturation of capture antibodies. *Immunol. Lett.*, **13,** 313–16.
Suter, M., Butler, J. E., and Peterman, J. H. (1988). The immunochemistry of sandwich ELISAs. IV. The stoichiometry and efficacy of the protein–avidin–biotin capturc (PABC) system. *Molec. Immunol.* (submitted).
Suter, M., Pike, B.L., and Nossal, G.J.V. (1985). An ELISA assay efficiently detects cloned antibody formation single hapten-specific B lymphocytes. *J. Imm. Methods* **84**, 327–41.
Williams, D. G. (1984). Comparison of three conjugation procedures for the formation of tracers for use in enzyme immunoassays. *J. Imm. Methods*, **72,** 261–8.

ELISA and Other Solid Phase Immunoassays
Edited by D.M. Kemeny and S.J. Challacombe

CHAPTER 8

The Use of ELISA in the Characterization of Protein Antigen Structure and Immune Response

Amadeo J. Pesce and **J. Gabriel Michael**

University of Cincinnati Medical Center, Cincinnati, Ohio, USA

CONTENTS

INTRODUCTION

ELISA can be used to: (a) accurately measure polyclonal and monoclonal antibody to the same antigen; (b) measure antibodies to both the whole protein molecule and to its components; (c) distinguish between strongly and weakly reacting determinants of the same antigen; (d) discriminate between determinants on native and denatured structures; (e) detect the effects of charge on antigen–antibody reactions; and (f) rapidly evaluate idiotypic anti-idiotypic interactions. Thus, the ELISA technique is useful in studying immunochemical properties of antigens and antibodies. However, ELISA exhibits certain characteristics which distinguish it from other procedures. These include: (a) steric limitations of antibody–antigen reactivity; (b) affinity differences in inhibition assays; (c) potential for denaturation of the antigen on plastic surfaces; and (d) difficulties in utilization of the charged antigen. Results of our experiments presented here should provide the reader with examples of how to recognize such problems and resolve them when using this assay.

In our studies bovine serum albumin served as a model antigen. The antibody response to the native and modified protein was measured by ELISA. The modifications consisted of enzymatic fragmentation, denaturation, and alteration of the charge. To study immunochemical properties of these modified antigenic structures, a comprehensive antigen and antibody measuring procedure was needed. We report here our experiences with ELISA in characterizing immunochemical properties of modified BSA and its fragments, as well as the reactivity between these antigens and their antibodies.

Selection and handling of reagents

In our hands all reagents used in ELISA assay, except for the antigen, were quite stable. They did not degrade readily at room temperature and could be kept for weeks in the cold. In our experimental system, plate coating albumin solution could be re-used several times and was not discarded, but rather mixed with fresh solution. Care had to be taken to eliminate bacterial growth and therefore filtering and the addition of azide (0.01 per cent final concentration) were done to maintain the stability of antigen solutions. When utilizing albumin as the test antigen, it was necessary to use an additional protein to stabilize and block non-specific absorption. In our hands, gelatine served this function very well.

Antibodies used for coating of ELISA plates should be affinity purified, or if mouse ascites containing monoclonal antibody is used, it must be of high titre. Partially purified low titre antibody decreased the sensitivity of the assay to the point where it gave very little or no discrimination.

Antigenic reagents

Crystalline BSA was purchased from Miles Laboratories, Elkhart, IN. Methods of preparation of BSA fragments have been described by Peters, Feldhoff, and Reed (1977). Fragments used were $T_{377-582}$, $T_{115-184}$, $P_{505-582}$, P_{1-306}, $P_{307-582}$, $P_{307-385}$, and $CNBr_{505-546}$. The letter indicates the method of preparation of the fragments: T for trypsin digestion, P for pepsin digestion, and CNBr for cyanogen bromide cleavage. The subscripts indicate the amino acid positions. Human serum albumin and its CNBr fragments A, B, and D were also utilized. Each fragment has been shown to maintain its native structure following preparation. Reduced carboxyl methylated (RCM) BSA was prepared by urea denaturation followed by reduction with cysteine and carboxylation with iodoacetate (Apple *et al.*, 1984). Cationized BSA was prepared using ethyl carbodiimide to couple ethylene diamine to the carboxyl groups (Pesce *et al.*, 1986).

Antibody standard

The antibody standard was prepared by affinity chromatography. The usual procedure is to couple antigen to CNBr-activated Sepharose (Marsh, Parikh, and Cuatrecasas, 1974). Antibody was adsorbed to the column and eluted with either pH 3.0 or pH 1.0 glycine buffer. The solution was neutralized immediately after elution and stored in small aliquots until used. The protein content was determined from the absorption at 280 nm or by a biuret assay.

Antibody assay

For determination of antibody titres, a modified ELISA (Engvall and Perlmann, 1972) was used. Antibody concentrations were derived from standard curves generated by the use of anti-BSA polyclonal antisera of known antibody concentration. Polystyrene microtitre plates were coated with 50 μg/ml antigen in 0.1 M $NaHCO_3/Na_2CO_3$ buffer, pH 9.5. Before use, the plates were washed three times with wash buffer (0.01 M K_2HPO_4/KH_2PO_4, pH 7.4, 0.85 per cent NaCl, 0.1 per cent Tween 20). Serum or hybridoma supernatant samples diluted in wash buffer with 0.5 per cent gelatine were added. After incubation for 1 hr at 37 °C, the plates were washed and an alkaline phosphatase rabbit anti-mouse IgG conjugate (Zymed, Burlingame, CA) then added. Following incubation at 37 °C for 1 hr, the plates were washed. *p*-Nitrophenyl phosphate (1 mg/ml in 1 M diethanolamine HCl, 0.5 mM $MgCl_2$, pH 9.8) was added and the colour reaction stopped at the appropriate time with 1 M NaOH.

USE OF ANTIGENIC FRAGMENTS FOR MEASUREMENT OF THE ANTIBODY RESPONSE TO DEFINED EPITOPES

Antibody concentrations in mouse sera were determined by a modified enzyme-linked immunosorbent assay (ELISA). Polystyrene microtitre plates (Immulon II, Dynatech Corp., Alexandria, VA) were coated with 50 μg/ml antigen in 0.1 M $NaHCO_3/Na_2CO_3$ buffer, pH 9.5. Following addition of antigen solution, plates were sealed with plastic sealing tape and stored at 4 °C until used. Immediately before use, plates were washed three times with 0.01 M phosphate buffer (K_2HPO_4/KH_2PO_4) containing 0.1 per cent Tween 20 (PBS/Tween). Antisera were then diluted in PBS/Tween containing 0.5 per cent gelatine, and 0.2 ml of each dilution was added to the antigen-coated wells.

The remainder of the assay procedure was performed as described above under antibody assay. With every determination, an affinity-purified mouse anti-BSA standard was run and values from unknown sera were compared with a standard curve. Values are reported as micrograms of anti-BSA antibody per millilitre of mouse sera.

USE OF ANTIGENIC FRAGMENTS TO CHARACTERIZE EPITOPE ANTIBODY CROSS-REACTIVITY

Labelled fragment method

Proteins were labelled with alkaline phosphatase by the method of Avrameas (1969). Human albumin fragments and alkaline phosphatase were dialysed against 0.1 M potassium phosphate buffer, pH 8.5. To 0.5 mg of human albumin, or fragments in 0.3 ml of buffer, 80 μl of a solution containing 2 mg of alkaline phosphatase and 10 μl of 1 per cent solution of glutaraldehyde were added. After incubation at room temperature for 3 hr, 50 μl of 1 M lysine, pH 7, was added to stop the reaction. The solution was dialysed overnight at 4°C against 0.01 M phosphate-buffered saline, pH 7.4. Insoluble material was removed by centrifugation at 20 000 rpm. The supernatant was mixed with an equal volume of glycerol and stored at −20 °C.

The enzyme-labelled antigen was used with antibodies fixed to a solid support. Buffer A contained 0.01 M potassium phosphate pH 7.4, 0.15 M NaCl and 0.1 (v/v) Tween 20. Buffer B contained 0.5 per cent w/v gelatine in addition to these components. Coating of the plates was done as described for the antigen. Purified antibody was used at a concentration of 10 μg/ml. After coating of the plates with antibody, various concentrations of alkaline-phosphatase-labelled antigen, human serum albumin, fragment A, fragment D in 0.2 ml of buffer B were placed in wells and incubated for 2 hr at 37 °C. The plates were washed six times with buffer A. The amount of labelled antigen fixed on the plate was determined by reacting with 0.2 ml of 10^{-3} M *p*-nitrophenyl phosphate in 0.1 M Tris-HCl buffer, pH 8, containing 1.5 M NaCl. The reaction was stopped by addition of 50 μl of 1 M K_2HPO_4 and colour was read at 405 nm. The optimal labelled antigen concentration produced maximum absorbancy.

The inhibition assay using antibody fixed to a solid support was run as follows: plates were coated with purified antibody. Various concentrations of inhibitor (antigen or fragment) in 0.1 ml of buffer were placed in wells and incubated for 1 hr at 37 °C. Then, 0.1 ml of a solution of labelled albumin or fragment was added. The mixture was incubated for 2 hr at 37 °C and the reaction measured as described above.

Labelled antibody method

This assay uses antigen fixed to the solid support. Each well of a Linbro EIA microtitration plate was coated using 0.2 ml of different antigens, for example human albumin (50 μg/ml), fragment D (10 μg/ml), fragment A (100 μg/ml) in 0.1 M sodium carbonate buffer, pH 9.5, by incubation at 37 °C for 1 hr and then overnight at 4 °C. (The concentration required for optimum coating with the different antigens had been established in previous assays.) The coated plates were washed six times with buffer to remove unbound material. The test antigen or standard was added in a volume of 0.1 ml of buffer. Antibody solution was previously titred to be about 70 per cent of the optimal colour yield. The plates were washed six times with buffer. Then, 0.2 ml of 5 μg/ml solution of sheep anti-mouse IgG coupled to peroxidase in buffer was added. This was allowed to react for 2 hr at 37 °C and the plates washed six times. To reveal the reaction, 0.2 ml of a solution containing 25 mg of *ortho*-phenylenediamine and 20 μl of H_2O_2 in 100 ml of phosphate buffer, 0.01 M, pH 6, was added to each well. The reaction was stopped with 50 μl of 6 M HCl and the colour read at 450 nm.

Inhibition studies with denatured BSA

The antibody concentration of sera or hybridoma ascites was determined by ELISA, then adjusted in wash buffer to a concentration of 500 ng/ml. Antigen dilutions were also made in wash buffer and inhibitions were conducted by adding 0.9 ml antibody with 0.1 ml antigen (or wash buffer as a control) and incubating overnight at 5 °C in polypropylene tubes. Antibodies were incubated with both native BSA and RCM-BSA and were tested against native BSA and RCM-BSA coated plates described in the antibody assay method using overnight incubation. ELISA was conducted as described earlier. Per cent inhibition was determined relative to antibodies that had been incubated overnight at 5 °C with wash buffer alone.

ANTIBODY ASSAY FOR DETERMINATION OF ANTIBODY TO CATIONIZED ANTIGEN

Polystyrene microtitre plates were coated with antigen as described above in the antibody assay and washed three times with buffer (0.01 M K_2HPO_4/KH_2PO_4, pH 7.4, 0.85 per cent NaCl, 0.1 per cent Tween 20). A modified wash buffer, containing either 10 mM ethylenediaminetetraacetic acid (EDTA) and 5 USP units sodium heparin/ml or 5 USP units heparin/ml, was then incubated with the plates for 1 hr before addition of test serum. The test serum was diluted in the modified wash buffer with 0.5 per cent gelatine added. After incubation for 1 hr at 37 °C, the wells were washed and filled with 150 μl of an alkaline phosphatase rabbit anti-mouse IgG conjugate (Zymed, Burlingame, CA), diluted in the modified wash buffer with 0.5 per cent gelatine. After incubation for 1 hr the plates were washed and *p*-nitrophenyl phosphate added as described above. The colour reaction was stopped at the appropriate time with NaOH. The absorbancy values were read with a MICROELISA reader (Dynatech).

Inhibition studies

The antibody concentrations of the sera were determined using the modified ELISA technique. Sera were diluted so that antibody concentrations were approximately 500 ng/ml. Antigens to be tested as inhibitors were dissolved in the modified wash buffer with gelatine. Mixtures of antibody and inhibiting antigen were prepared by adding 0.5 ml of diluted antibody with 0.5 ml of antigen at various concentrations. Modified wash buffer was added to antibody as a control. Inhibition assays were conducted by overnight incubation of antibody and antigen mixtures in 55 × 12 mm polypropylene tubes (Sarstedt, West Germany) at 4 °C. ELISA was then conducted as described above, using the modified wash buffer. Per cent inhibition was determined relative to antibodies that had been incubated overnight at 4 °C with modified buffer alone.

Production of anti-idiotypic antibodies

BDF_1 mice were injected with 100 μg of affinity-purified monoclonal antibodies in CFA i.p. on day zero. On days 30 and 45 mice received additional injections of 50 μg of monoclonal antibodies in Freund's incomplete adjuvant. Prior to the second and third immunizations, on days 29 and 44, mice were bled. Their sera were pooled and tested for the presence of anti-idiotypic antibody. Anti-idiotypic antibody was demonstrated by the ability of the pooled sera to inhibit the monoclonal antibody from binding to $P_{505-582}$-coated plates in an ELISA assay.

Determination of anti-idiotypic antibodies to 17.5.5E idiotype

The presence of anti-idiotypic antibody was determined by an inhibition assay. The inhibition assays were performed by first diluting affinity-purified 17.5.5E monoclonal antibodies to a concentration of 500 ng antibody/ml. Sera to be assayed were serially diluted, beginning at a dilution of 1 : 5. Five hundred microlitres of both monoclonal antibodies and sera were then incubated for

1 hr in glass tubes at 37 °C. Following incubation, 200 μl of each mixture were placed in $P_{505-582}$-coated wells for assay in the ELISA system for measurement of antibody as described above.

ANTIBODY RESPONSE

Intact protein antigen

One of the most common ways of measuring antibody response by ELISA is to coat a plastic surface, usually a polystyrene microtitre plate, with antigen, allow it to react with the test antibody, and then react it with a second antibody coupled with enzyme. This procedure is very effective and measures antibody concentration in the nanogram range. The most straightforward method of quantifying the reaction is to correlate the absorbancy of the ELISA reaction of the test antibody with an affinity-purified standard of known antibody concentration. However, it is important to note that the antigen–antibody reaction on the plate quantitatively differs from that in solution or in gels such as agarose. For example, as shown in Figure 1, the colour yield (absorbance 405 nm) is the same for both polyclonal and monoclonal antibody (Pesce, Krieger, and Michaol, 1983). This implies that the same number of antibody molecules has reacted on the plate. Therefore, it appears that in ELISA assay due to steric or other factors one molecule of antibody reacts only with one molecule or antigen.

Antigenic fragments

BSA, a globular protein, can be degraded by proteolytic enzymes such as

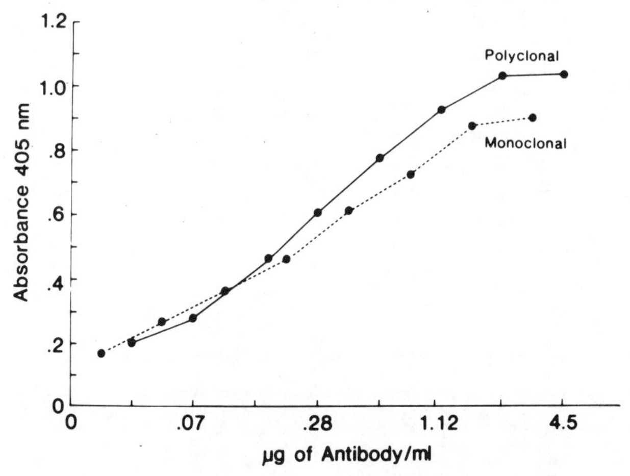

FIGURE 1 Titration curves of polyclonal and monoclonal antibodies to bovine albumin

TABLE 1 Binding of BSA fragments to antifragment antibody anti-BSA antibody response

	Fragment bound to solid phase†					
Immunogen*	BSA	P_{1-306}	$P_{307-582}$	$T_{115-184}$	$T_{377-582}$	$P_{505-582}$
BSA	1638‡ (100)§	348 (21)	1048 (64)	50 (3.1)	1015 (62)	78 (4.8)
P_{1-306}	441 (100)	312 (100)	3.2 (1)	20 (6.4)	2.0 (0.6)	2.2 (0.7)
$P_{307-582}$	2028 (100)	81.5 (4.7)	1740 (100)	7.9 (0.5)	1825 (100)	108 (6.2)
$T_{115-184}$	220 (100)	201 (100)	2.5 (3.9)	189 (100)	2.3 (2.6)	1.5 (0.7)
$T_{377-582}$	2418 (100)	13 (0.6)	2480 (100)	5.9 (0.3)	2121 (100)	191 (9.1)
$P_{505-582}$	492 (100)	0 (0)	509 (100)	0 (0)	484 (100)	476 (100)

* Groups of BDF_1 mice were immunized with 100 μg of the fragment or BSA (Day 0) in CFA, followed 17 days later by booster immunization with 100 μg of the same antigen in IFA. Sera were obtained 20 days after the second immunization (Day 37).

† Polystyrene plates were coated as described earlier under 'Antibody assay' with each fragment.

‡ Amount of anti-BSA bound by each fragment, μg/ml.

§ Number in parentheses is per cent binding based on the μg/ml antifragment bound to the fragment used to obtain the antisera. Binding of other antisera to that fragment is a percentage thereof.

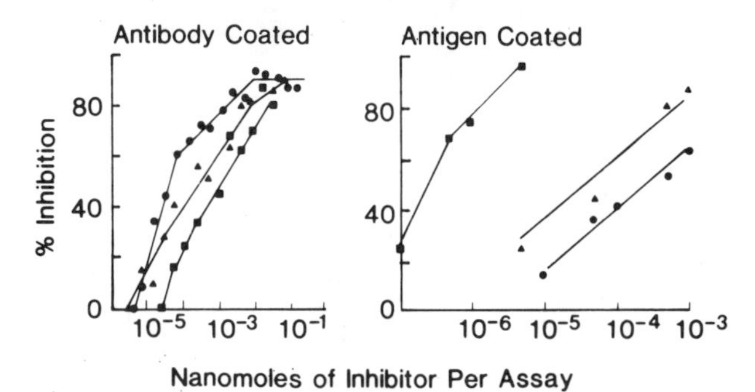

FIGURE 2 Affinity differences between ELISA inhibition assays. Polystyrene plates were coated with either antibody specific for human albumin fragment B_{1-123} (antibody coated) or human albumin (antigen coated). For antibody-coated plates, inhibitor was added followed by the homologous fragment labelled with alkaline phosphatase. For antigen-coated plates, the inhibitor was added to the coated plate followed by the anti-fragment antibody. A second antibody, goat anti-rabbit IgG coupled to alkaline phosphatase, was used as the developer reagent. Symbols: ■–■ albumin; ▲–▲ fragment D_{1-298}; ●–● fragment B_{1-123}

pepsin and trypsin to yield fragments with defined sequences and structure. Measurement of the antibody response to these fragments can be accomplished by binding the fragments to polystyrene following the procedure described above. Table 1 illustrates the specificity of the antibody to the whole molecule as well as the specificity of the antibody to each of the antigenic fragments (Ferguson *et al.*, 1983). Quantitation of this reaction is accomplished by comparing the reaction to an affinity-purified BSA standard. It is important to note that in the case of BSA fragments, as few as 70 residues bind to the polystyrene surface. Reading across the data of Table 1, line 1, we show that the polyclonal response is greater for the C terminal portion of the molecule as compared to the N terminal. Antibody produced against the entire BSA molecule reacted with individual antigenic fragments providing the opportunity to identify major antigenic determinants.

Cross-reactivity between antigen determinants

Human serum albumin contains 585 residues and is comprised of three domains of about 190 residues which evolved from gene duplication. There is a homology of 18 per cent between the domains. The problem we addressed was whether or not there was cross-reactivity between the domains. One of the first observations made was concerned with the differences in affinity, when the inhibition ELISA system was tested using antigen-coated plates versus antibody-coated plates (Figure 2) (Pesce, Krieger, and Michael, 1983). In order to study weakly cross-reacting antigens, the best system appeared to be an antibody-coated plate and antigen labelled with enzyme. Using this system

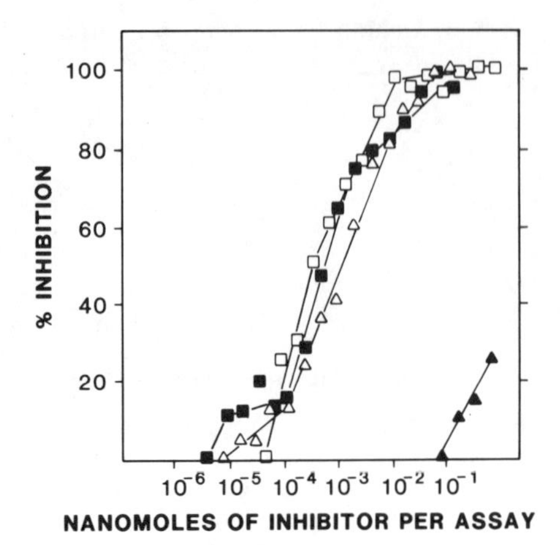

FIGURE 3 Enzyme immunoassay inhibition curves. Polystyrene wells were coated with (a) HA-1 and (b) HA-2 monoclonal antibodies. Inhibitor was added followed by human serum albumin labelled with alkaline phosphatase. The per cent inhibition was calculated as follows for each inhibitor concentration:

$$\text{Per cent inhibition} = \frac{I_O - I_C}{I_O} \times 100$$

where I_O is the absorbancy in the absence of inhibitor and I_C is the absorbancy in the presence of inhibitor. Inhibitors: ■–■, albumin; ▲–▲, D; △–△, A; □–□, F1

and monoclonal antibodies, we were able to show that fragment D_{1-298} cross-reacted weakly (approximately 1/1000) with a monoclonal antibody specific for fragment F-1 which is near the C terminus of the molecule (Figure 3) (Doyen, Pesce, and La Presle, 1981).

Native versus denatured determinants

The role of protein determinants in immunoregulation remains controversial. This controversy is related to the role of native versus denatured determinants. We used ELISA to study the immune response and to characterize the antigenic determinants important to the response to native and denatured BSA. The denatured BSA (RCM-BSA) was urea denatured, reduced, and carboxymethylated, yielding a protein with essentially no disulphide cross-linking. A series of antibodies was tested: included were early antibody (9 days after second immunization) and late antibody (28 days after second immunization). Antigens, native BSA, and RCM-BSA were tested for inhibition on both native BSA and RCM-BSA-coated plates. The results of such experiments are shown in Figure 4 (Apple *et al.*, 1984).

Early anti-native BSA antibodies bound to both native BSA and RCM-BSA-

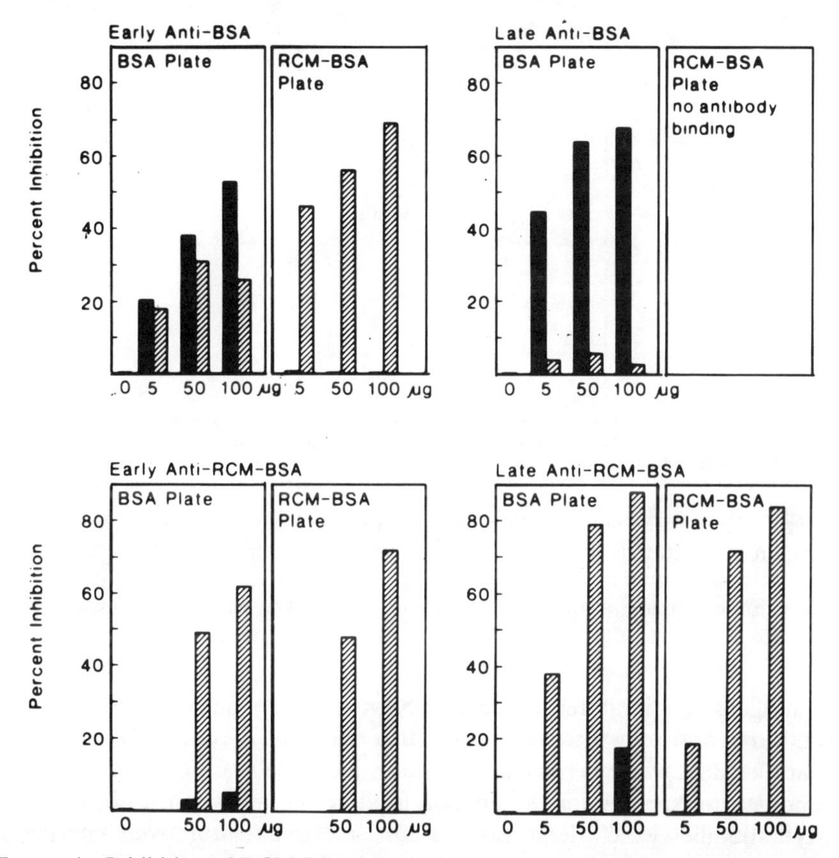

FIGURE 4 Inhibition of RCM-BSA MaAbs by native and denatured BSA. Inhibition conducted with BSA (■) and RCM-BSA (□). Early anti-BSA, early anti-RCM BSA, later anti-BSA, and late anti-RCM-BSA antisera were prepared as described in the text. Antisera were diluted to 500 ng/ml and inhibited by varying concentrations of native BSA or RCM-BSA overnight at 5 °C. ELISA was then conducted as described earlier and per cent inhibition was determined relative to antibody incubated overnight at 5 °C without antigen

coated plates. Binding of these antibodies to native BSA was inhibited by both native BSA and RCM-BSA. However, early anti-native BSA antibodies that bound to RCM-BSA were inhibited only by RCM-BSA. These data indicate that early response to BSA was directed to determinants found on both native and denatured BSA. Late anti-native BSA antibodies from hyperimmunized animals showed a greater specificity for the homologous antigen, since only native BSA inhibited the binding on native BSA plates. No binding of these antibodies occurred on RCM-BSA plates (Figure 4).

Antibodies elicited as a response to RCM-BSA administration bound to both native BSA and RCM-BSA-coated plates. Binding of such antibodies to

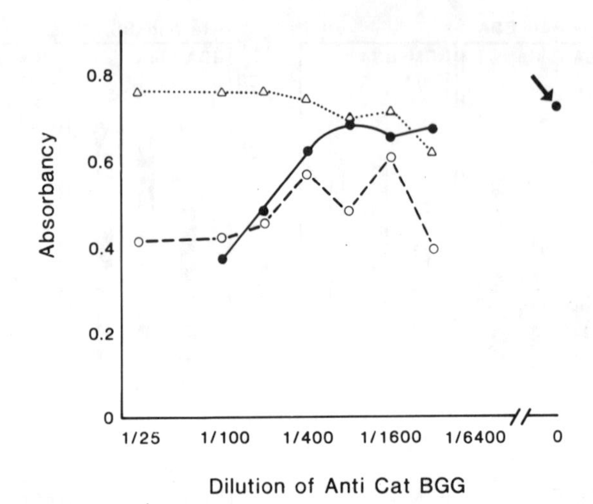

FIGURE 5 Example of non-specific binding is cationized antigen (BGG)-coated plates. Absorbancy is plotted versus dilution of mouse anti-cationized BGG. △ · · · △ Affinity-purified antibody to cat BGG; ○ · · · ○ antisera to cat BGG; ● · · · ● mouse serum control. Arrow indicates control with no antisera added

the antigens was inhibited by RCM-BSA but not by native antigen. These results are consistent with the concept that native BSA-coated plates contain denatured BSA which is recognized by anti-RCM-BSA antibodies. Thus, late antibodies are specific for the antigen used in immunization, as opposed to early antibodies which demonstrated cross-reactivity. Moreover, coating of native BSA on polystyrene plates resulted in its partial denaturation.

Cationic determinants

Chemically modified proteins have been used for many years as a tool to understand specificity of the immune reaction. We chemically modified albumin bovine serum (cat BSA) and bovine gammaglobulin (cat BGG) by coupling ethylenediamine to the carboxyl groups using a carbodiimide (Pesce *et al.*, 1986). The resulting protein was polycationic with an isoelectric point greater than 9.5. When we attempted to measure the antibody against cationized BGG, we observed a non-specific binding of immunoglobulins to charged antigen (Figure 5). This non-specific binding was minimized by the addition of heparin to the diluent buffer (Figure 6). Likewise, measurement of antibody to cat BSA was improved by the addition of polyanions such as heparin and dextran sulphate (Figure 7). The effect of polyanions on a system involving positively charged antigen indicate that positive charges may block antigenic sites on the surface of polystyrene plates.

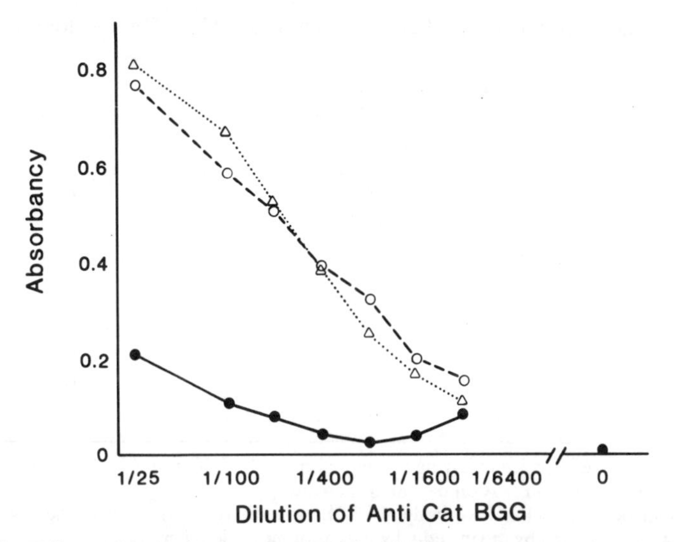

FIGURE 6 Effect of heparin on non-specific binding to cationized antigen (BGG)-coated plates. Symbols same as for Figure 5

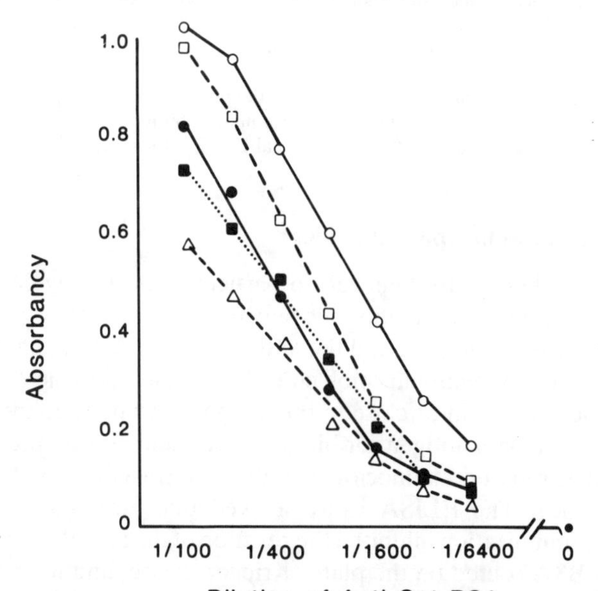

FIGURE 7 Effect of polyanions on colour yield of cationized BSA-coated plates. Absorbancy is plotted versus dilution of mouse antibody to cat-BSA in various gelatine wash buffer solution. ● · · · ● wash buffer only; ○—○ wash buffer–gelatine + heparin; △ — △ wash buffer–gelatine + 1.0 mg/ml dextran sulphate; □—□ wash buffer–gelatine + 100 μg/ml dextran sulphate; ■ · · · ■ wash buffer–gelatine + 10 μg/ml dextran sulphate

TABLE 2 Demonstration of anti-idiotypic antibodies to 17.5.5E monoclonal antibodies

Serum dilution*	Amount of 17.5.5E bound to fragment $P_{505-582}$† (ng/ml)	Per cent inhibition‡
1/5	34.1	86
1/10	46.5	81
1/20	57.3	77
1/40	175.0	30
1/80	225.0	10
1/160	249.0	0.4
1/320	251.0	0
Control§	250.1	0
Buffer	251.8	—

* Pooled sera from eight mice receiving two injections of 17.5.5E monoclonal antibody; 100 μg in CFA on day 0 and 50 μg in IFA on day 30. Sera were obtained 44 days after the initial injection.
† Affinity-purified monoclonal antibody, 17.5.5E, was diluted to a concentration of 500 ng/ml. The assay was performed by mixing equal volumes of monoclonal antibody with various dilutions of anti-idiotypic serum or buffer and pre-incubating the mixture for 1 hr at 37 °C. These mixtures were then incubated in BSA fragment $P_{505-582}$-coated wells for 1 hr. Binding of the monoclonal antibody to $P_{505-582}$ was detected using alkaline-phosphatase-conjugated rabbit anti-mouse Ig as described earlier in 'Antibody assay for determination of antibody to cationized antigen'.
‡ The per cent inhibition of binding of the monoclonal antibody to $P_{505-582}$ was calculated according to the equation:

$$\% \text{ Inhibition} = \left(1 - \frac{\text{ng of 17.5.5E bound in the presence of serum}}{\text{ng of 17.5.5.E bound in buffer}}\right) \times 100$$

§ Control consisted of a 1/5 dilution of pooled normal $B6D2F_1$/J serum.

Measurement of anti-idiotypic antibodies

Anti-idiotypic antibodies are believed to participate in the regulation of the immune response, through the idiotypic network. Since immunoglobulins are proteins, they are immunogenic and the antibody produced can be directed to any part of the immunoglobulin molecule. When the idiotypic determinant (idiotope) is located within or close to the antigen binding site, anti-idiotypic antibodies compete with antigen for binding. This competition prevents antigen binding. We prepared a monoclonal antibody to BSA (17.5.5E) and used it to immunize mice. The ELISA assay showed that anti-idiotype antibody formed in this immunization inhibited the reaction of the anti-BSA monoclonal antibody with BSA coated on the plate (Krieger, Pesce, and Michael, 1984). The results of a typical experiment shown in Table 2 indicate that this anti-idiotypic antibody binds to idiotopes located within or close to the antigen binding site. However, this system of detection is not useful if the idioptopes are located at some distance from the antigen binding site. In this instance a different type of assay must be used which involves binding between idiotope bearing immunoglobulin and anti-idiotypic antibodies.

ACKNOWLEDGEMENTS

We wish to thank Dr Victor E. Pollak and Dr Roger D. Smith for their encouragement, and Dr Annette Muckherheide for initiating this project. This work was supported in part by a grant from Dialysis Clinics, Inc., and USPHS Grants AM 17196 and AI 15520. We also wish to thank Ms Mary Grannen for preparation of the manuscript and the biomedical illustrations group at the University of Cincinnati Medical Center for preparation of the drawings.

REFERENCES

Apple, R., Knauper, B., Pesce, A., and Michael, G. (1984). Shared determinants of native and denatured bovine serum albumin are recognised by both B- and T-cells. *Molec. Immunol.* **21,** 901.

Avrameas, S. (1969). Coupling of enzymes to proteins with glutaraldehyde. Use of the conjugates for the detection of antigens and antibodies. *Immunochemistry*, **6,** 43.

Doyen, N., Pesce, A.J., and LaPresle, C. (1981). Specificity of two anti-human albumin monoclonal antibodies. *Immunol. Lett.*, **3,** 365.

Engvall, E., and Perlmann, P. (1972). Enzyme-linked immunosorbent assay (ELISA) III: Quantitation of specific antibodies by enzyme-labelled anti-immunoglobulin in antigen coated-tubes. *J. Immunol.*, **109,** 29.

Ferguson, T.A., Peters, T., Jr., Reed, R., Pesce, A.J., and Michael, J.G. (1983). Immunoregulatory properties of antigenic fragments from bovine serum albumin. *Cellular Immunol.*, **78,** 1.

Krieger, N.J., Pesce, A.J., and Michael, J.G. (1984). Induction of multispecific antibodies to bovine serum albumin after production of anti-idiotype antibodies to an albumin-specific monoclonal antibody. *Annals of the New York Academy of Sciences*, **418,** 305.

Marsh, S.C., Parikh, I., and Cuatrecasas, P. (1974). A simplified method for Cyanogen Bromide activation of Agarose for affinity chromatography. *Anal. Biochem.*, **60,** 149.

Pesce, A.J., Krieger, N.J., and Michael, J.G. (1983). *Immunoenzymatic Techniques* (eds S. Avrameas *et al.*). Elsevier, Amsterdam, p. 127.

Pesce, A.J., Apple, R., Sawtell, N., and Michael, J.G. (1986). Cationic antigens. Problems associated with measurement by ELISA. *J. Imm. Methods*, **87,** 21.

Peters, T., Jr., Feldhoff, R.C., and Reed, R.G. (1977). Immunochemical studies of fragments of bovine serum albumin. *J. Biol. Chem.*, **252,** 8464.

ACKNOWLEDGEMENTS

ELISA and Other Solid Phase Immunoassays
Edited by D.M. Kemeny and S.J. Challacombe

CHAPTER 9

The Modified Sandwich ELISA (SELISA) for the Detection of IgE and Other Antibody Isotypes

D.M. Kemeny

UMDS, Guy's Hospital, London

CONTENTS

INTRODUCTION

Despite their widespread popularity in ELISA, microtitre plates have a lower capacity for protein than cyanogen bromide-activated cellulose or agarose. This makes them unsuitable for the measurement of IgE antibodies where ELISA performs less well than established immunoradiometric assays (IRMA) such as the radioallergosorbent test (RAST) (Eriksson and Ahlstedt, 1977). These IRMAs use high capacity matrices such as Sephadex (Wide, Bennich, and Johnasson, 1967), cellulose filter paper (Ceska, Eriksson, and Varga, 1972), and nitrocellulose strips (Derer *et al.*, 1984) or discs (Walsh, Wrigley, and Baldo, 1984). There does not appear to be anything inherent in ELISA that prevents it from being used to measure IgE and a two-site assay for total IgE developed some thirteen years ago (Hoffman, 1973) works just as well as a similar radioisotopic assay. Furthermore, when enzyme-labelled anti-IgE is used with the same allergen-coated discs that are normally used for RAST, there is close agreement between the methods (Negrini, Troise, and Voltolini, 1985). Thus the limiting factor in using ELISA to measure allergen-specific IgE antibodies appears to be the type of solid phase rather than the method of detection.

It is well recognized that when mixtures of proteins are used to coat microtitre plates there can be competition between them for binding sites on the surface of the plastic (Cantarero, Butler, and Osborne, 1980). Allergen extracts often contain large amounts of irrelevant material and, even where this has been removed, the complex mixture of proteins present may result in competition between them for binding to the plate. Furthermore, not all purified allergens bind well to plastic and for this reason we have investigated alternative methods of linking allergens to microtitre plates.

BACKGROUND

The assays described here were developed as a result of experiments aimed at reducing the background in solid phase immunoassays for allergen-specific IgG antibodies. Assays for IgE antibodies do not generally suffer from high background binding because the proportion of total IgE that is specific for a given allergen is high, as much as 60 per cent (Gleich and Jacob, 1975; Gleich *et al.*, 1977) but only a small proportion of IgG is specific for a single allergen (Paull *et al.*, 1978). Attempts to use RAST to measure IgG antibodies resulted in a poor signal to noise rate (Kemeny, Lessof, and Trull, 1980; Platts-Mills, 1981) although this could be improved by the addition of excess unlabelled anti-IgG (Shimizu *et al.*, 1978) or trace-labelled protein A (Hamilton and Adkinson, 1980). Preliminary experiments carried out using bee venom phospholipase A_2 (PLA_2)-coated microtitre plates indicated that high background binding was also a problem in ELISA (Urbanek, Kemeny, and Samuel, 1985). Different ways of reducing this were investigated and the modified sandwich ELISA (SELISA) described here is the product of those experiments.

Non-specific binding of ^{125}I-radiolabelled antigen and serum immunoglobulin is low in fluid phase radioimmunoassays. As sensitivity is inversely proportional to the amount of radiolabelled antigen the quantities of ^{125}I antigen used are generally low (1–2 ng) (Figure 1). Even with larger amounts of antigen (10–20 ng), background binding to a negative serum is still relatively low. Bound and free ^{125}I antigen can be separated using antibody directed against a particular class of immunoglobulin but such reagents can be costly (e.g. anti-IgG subclass antibodies). For this reason the possibility of using a rabbit antibody directed against the antigen rather than the immunoglobulin was investigated. By carrying out the antibody–antigen reaction in the fluid phase and subsequently precipitating the immune complexes thus formed on

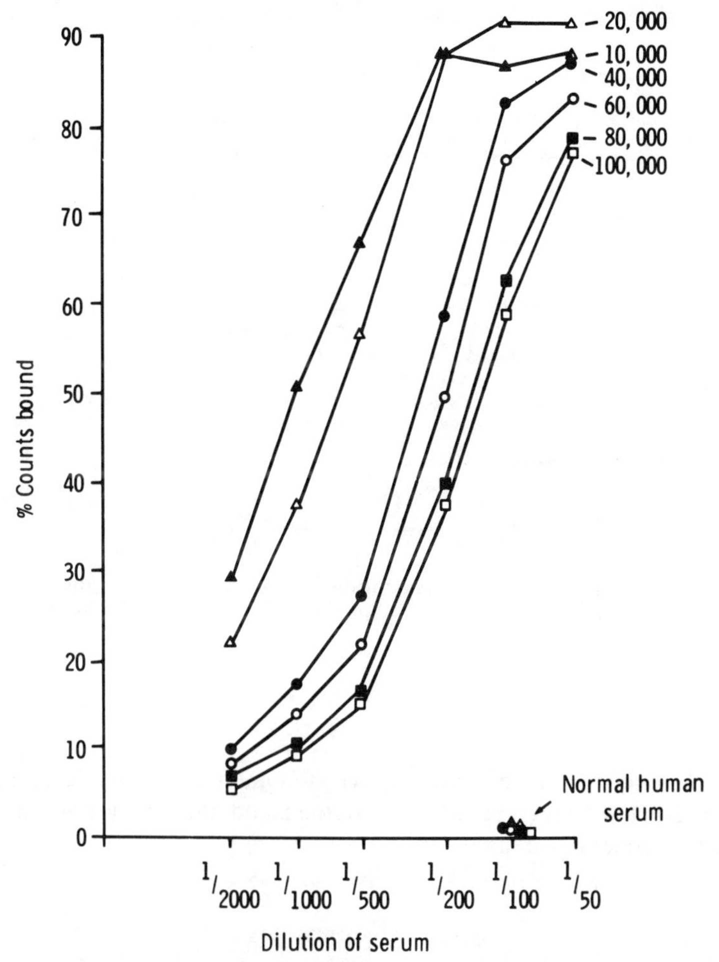

FIGURE 1 Radiommunoassay for IgG antibodies to bee venom PLA_2 using different amounts of ^{125}I PLA_2 from 10 999 to 100 000 counts per minute (1.3–13 ng PLA_2). Sensitivity is inversely proportional to the quantity of antigen added although background binding with normal human serum is similar at 1.3–13 ng ^{125}I PLA_2

an anti-antigen-coated plate it was hoped that the proportion of immunoglobulin that was antigen specific would be increased and that, following incubation with labelled anti-IgG, background binding would be lower. However, this did not happen and, using affinity-purified rabbit anti-PLA_2-coated microtitre plates to bind the antigen and antibody complexes thus formed in the serum, the background was similar to plates coated directly with PLA_2 (see Figure 6, below) although the amount of IgG antibody bound was increased 10–20-fold.

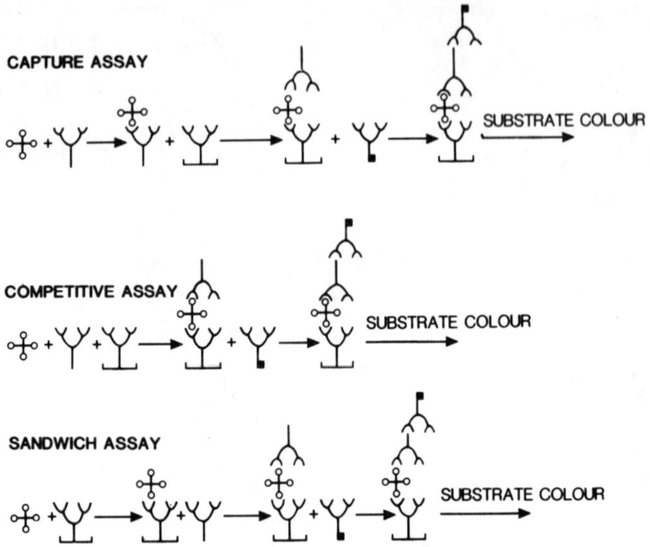

FIGURE 2 The different assay configurations tested. ✛ = antigen. ⅄ = rabbit antibody-coated microtitre plate. Y = human antibody. ⅄ = HRP-anti-IgG or IgE

In this chapter the investigation of this phenomenon and the development of the sandwich ELISA (SELISA) described and the wider implications for ELISA in general are discussed.

ASSAY CONFIGURATION

The increased sensitivity achieved using affinity-purified rabbit anti-PLA_2-coated microtitre plates was investigated further in a number of different assay procedures (Figure 2). The first was the capture assay, referred to above, in which the antigen was incubated with the sample in the fluid phase and then transferred to the antibody-coated plate where the antigen–antibody complexes formed were captured by the anti-PLA_2 on the plate. In the second, the antigen and sample were added to the anti-PLA_2-coated plate together in a competitive or simultaneous procedure. The third configuration was a modified sandwich ELISA (SELISA) in which the PLA_2 was first bound to the anti-PLA_2-coated plate and, after a subsequent incubation and wash step, the sample was added. For each procedure detection of bound IgG antibody was with horseradish-peroxidase-labelled rabbit anti-human IgG (HRP anti-IgG). All three methods gave similar results for detection of IgG antibody (Figure 3)

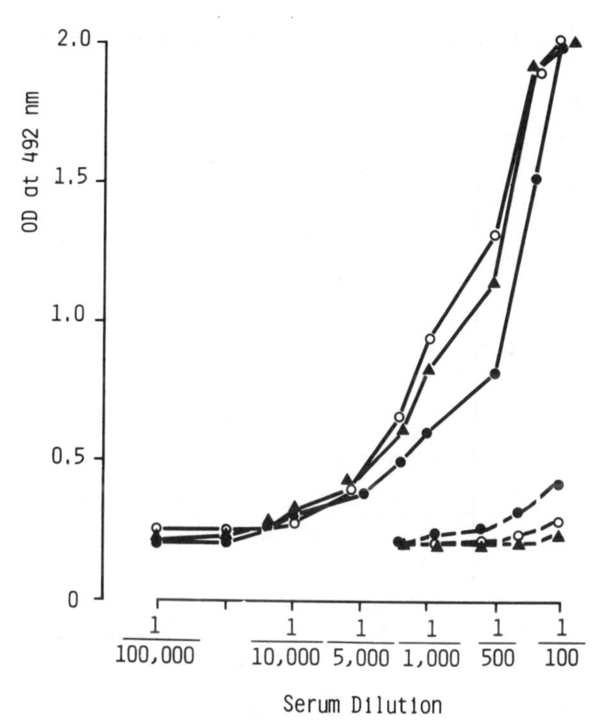

FIGURE 3 Comparison of the competitive (●), capture (▲), and sandwich ELISA (○) for IgG antibodies to PLA_2 (——). Comparable results were obtained with each assay (--- = background binding with a negative control serum). (Kemeny *et al.*, 1985a)

and each was more sensitive than direct coating (Urbanek, Kemeny, and Samuel, 1985; Kemeny *et al.*, 1985a).

Comparison of assay procedures

When these different assay procedures were compared in an IgE antibody ELISA (HRP anti-IgE substituting for the HRP anti-IgG) the outcome was different. At higher serum dilutions than those shown, the sandwich assay was more sensitive than the competitive or capture procedures and at high concentrations of the serum sample, there was reduced binding in the capture and, to a lesser extent, the competitive assays (Figure 4). The serum sample used contained much more IgG than IgE antibody to PLA_2 and it seems likley that the reduced binding of IgE is due to competition with IgG antibody for the small amount of fluid phase antigen (10 ng). This competition did not occur in the sandwich assay—presumably because of the greater capacity of solid phase antigen for antibody. Thus the sandwich procedure would seem to be more appropriate for the measurement of IgE antibodies.

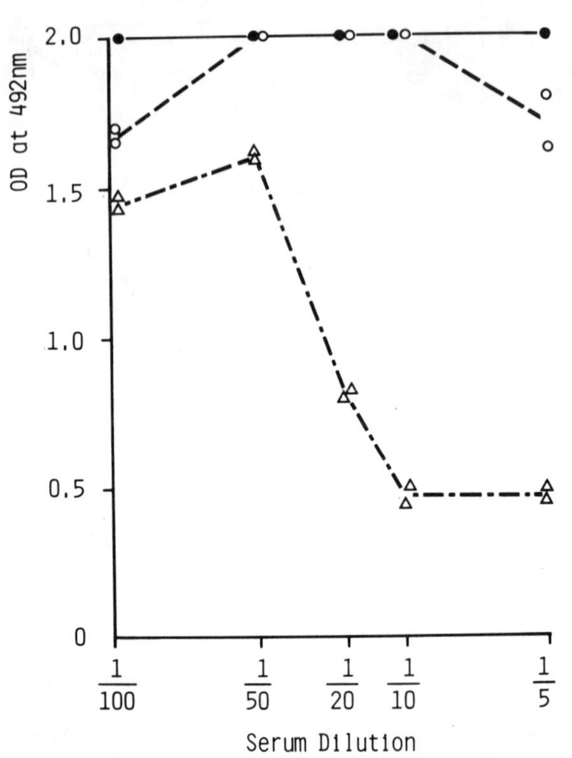

FIGURE 4 Comparison of the sandwich (●), competitive (○), and capture ELISA (△) for IgE antibodies to PLA_2. There was marked inhibition of IgE binding at high serum concentrations in the capture assay

SANDWICH ELISA (SELISA)

Comparison with antigen-coated plates and RAST

The sensitivity of RAST, ELISA, and SELISA for detection of IgE antibodies was compared (Figure 5). Using either low (Figure 5a) or high (Figure 5b) capacity microtitre plates the sensitivity of SELISA was much greater than the RAST or the conventional ELISA. This increased sensitivity is due to the greatest efficiency with which the anti-PLA_2/PLA_2 plates bind IgE antibody and not to other factors, such as the amount of HRP anti-IgE, which were constant between the two types of ELISA. Background binding was tested using a negative control serum diluted 1/10 and was similarly low for all the assays. The sensitivity of the SELISA was ten times greater than the RAST using the low capacity plates and 40-fold greater with the high capacity plates (Table 1).

A similar improvement in sensitivity was found in the IgG antibody SELISA

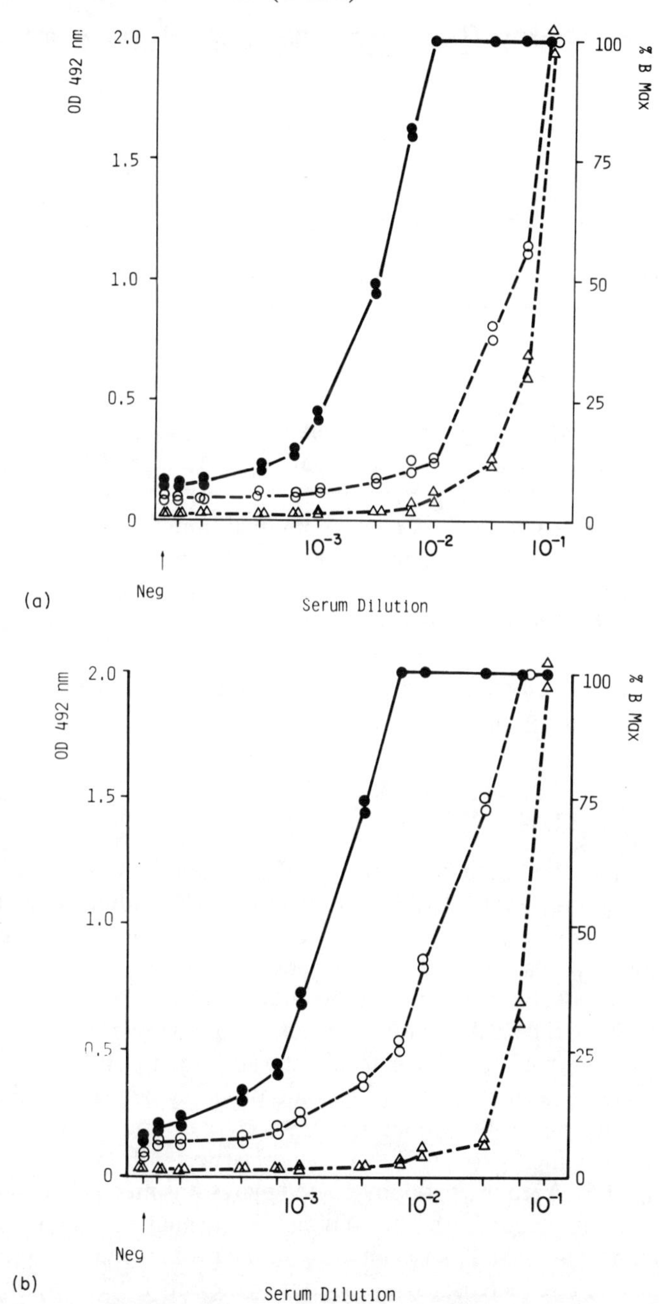

FIGURE 5 A comparison of the RAST (△ - - - △), ELISA (○ - - - ○), and SELISA (●——●) for IgE antibody to bee venom PLA_2 using (a) Dynatech M129A and (b) Nunc Immuno-1 microtitre plates

TABLE 1 The limit of detection (1.5 × background) of the RAST, ELISA, and SELISA for IgE antibodies to PLA_2

Method	Sensitivity (pg of IgE antibody per ml)	
	Dynatech Immulon 1	Nunc F
RAST	387	387
Indirect ELISA	194	97
Sandwich ELISA	39	10

(Figure 6) and again the high capacity plates gave the greatest sensitivity (Figure 6a) although binding of the negative control serum was greater compared with the low capacity plates (Figure 6b). This assay works well with a number of other allergens such as bee venom hyaluronidase (Kemeny *et al.*, 1985b) and purified cow's milk allergens (Figure 7). An additional advantage of SELISA is that the specificity is greater than for direct solid phase coating (Kemeny *et al.*, 1983a,b) as the only antigen that will bind is the one recognized by the antibody on the plate. If monoclonal antibodies are used (Kemeny *et al.*, 1986), this makes it possible to measure antibodies to individual antigens without having to purify them.

Optimization of SELISA

The optimal time for coating the plates with antibody was studied and near-maximal binding, assessed by the capacity to bind ^{125}I radiolabelled PLA_2, was reached in 18 hr at 4 °C (Figure 8a). This agrees with previous studies (Herrmann and Collins, 1976; Ishikawa and Kato, 1978). It is, of course, possible that this could be increased at higher temperatures. Prolonged incubation (e.g. 1 week) of the antibody with the plate resulted in a reduced capacity for PLA_2 which is likely to be due to denaturation of the bound antibody (Pesce, Ford, and Graizutis, 1978). This could probably be prevented by the addition of stabilizing proteins once the antibody had bound or by freeze-drying the coated plates (Voller, Bidwell, and Bartlett, 1979).

The binding of PLA_2 to the antibody-coated plates was much faster and was complete within 1 hr (Figure 8b). The binding of very small amounts of antigen or by small quantities of solid phase antibody would probably take longer. The amount of PLA_2 bound increased in proportion to the anti-PLA_2 concentration used to coat the plates (Table 2) and there was much less desorption of bound antigen (<1 per cent) from antibody-coated plates compared with direct coating (see Chapter 2).

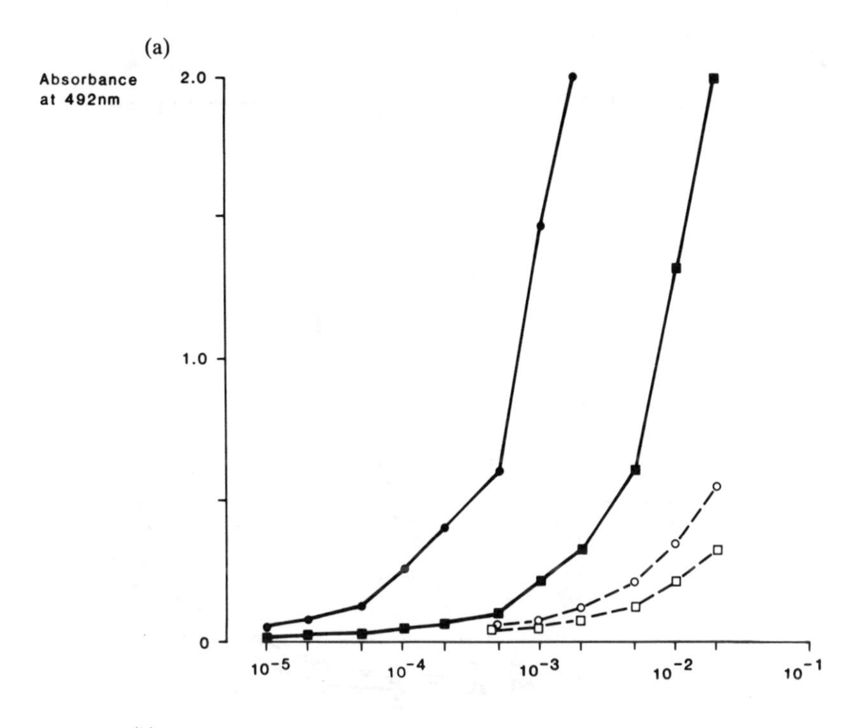

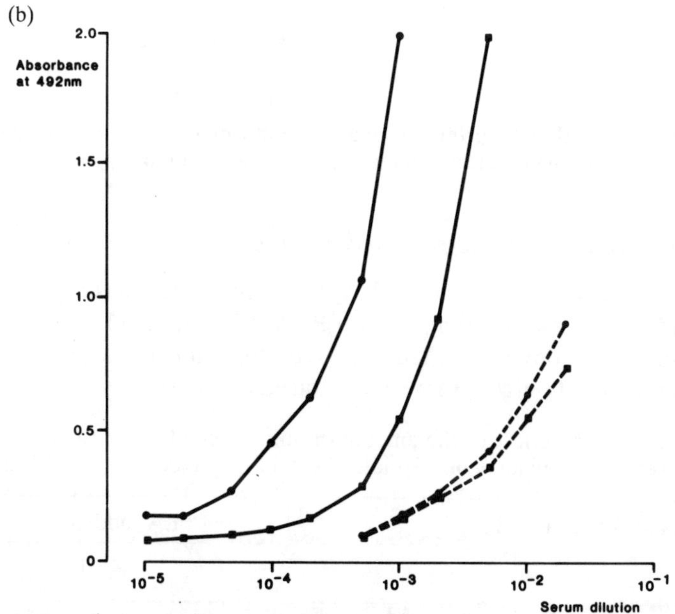

FIGURE 6 A comparison of the ELISA (■–■) and SELISA (●–●) for IgG antibody to PLA_2 using (a) Dynatech M129A and (b) Nunc Immuno-1 microtitre plates (– – – = binding of a negative control serum). (Kemeny *et al.*, 1985a; Urbanek, Kemeny and Samuel, 1985)

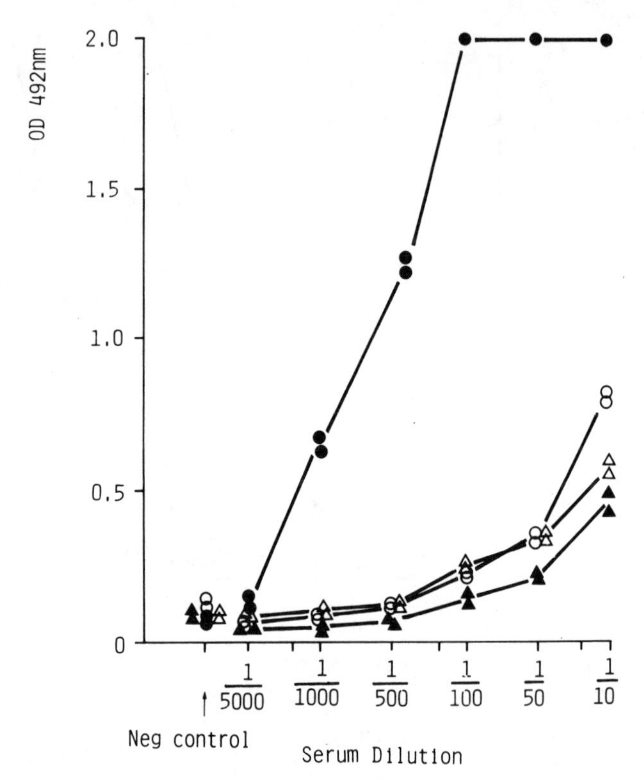

FIGURE 7 SELISA for IgE antibodies to individual cow's milk antigens: (•) Casein, (○) bovine serum albumin, (▲) α-lactalbumin, (△) β-lactoglobulin

Effect of antibody/allergen concentration

The effect of different anti-PLA_2/PLA_2 concentrations on the assay was also studied. Over a wide range of anti-PLA_2/PLA_2 ratios (1 : 1 to 1 : 100) identical results were obtained (Figure 9). At a high anti-PLA_2 coating concentration (Figure 9a, 100 μg/μl) there was reduced binding with smallest amount (10 ng)

TABLE 2 The effect of the amount of anti-PLA_2 added to the microtitre plate (100 μl/well) on the binding of ^{125}I-radiolabelled PLA_2 added at 1, 10, 100 and 1000 ng/well

μg anti-PLA_2 added	ng PLA_2 added			
	1	10	100	1000
0.01	0.013 (1.3%)	0.17 (1.7%)	0.7 (0.7%)	56 (5.6%)
0.1	0.1 (10%)	0.17 (1.7%)	2.3 (2.3%)	130 (13%)
1.0	0.31 (31%)	2.3 (23%)	31 (31%)	230 (23%)
10	0.26 (26%)	3.5 (35%)	60 (60%)	560 (56%)

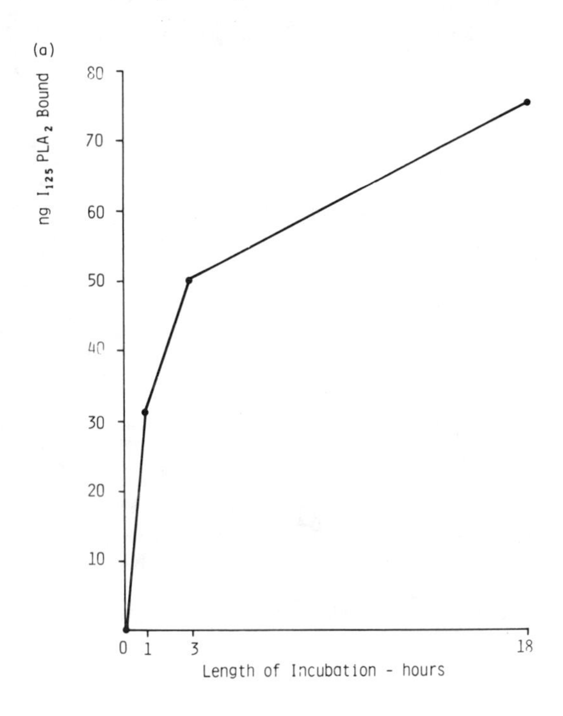

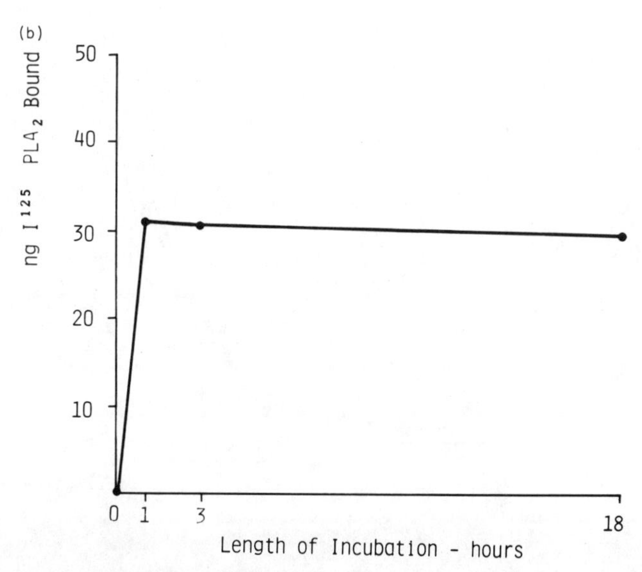

FIGURE8 (a) The affect of anti-PLA_2 coating time on binding of ^{125}I PLA_2. (b) The rate of binding of PLA_2 to an anti-PLA_2-coated microtire plate. (Kemeny *et al.*, 1985a)

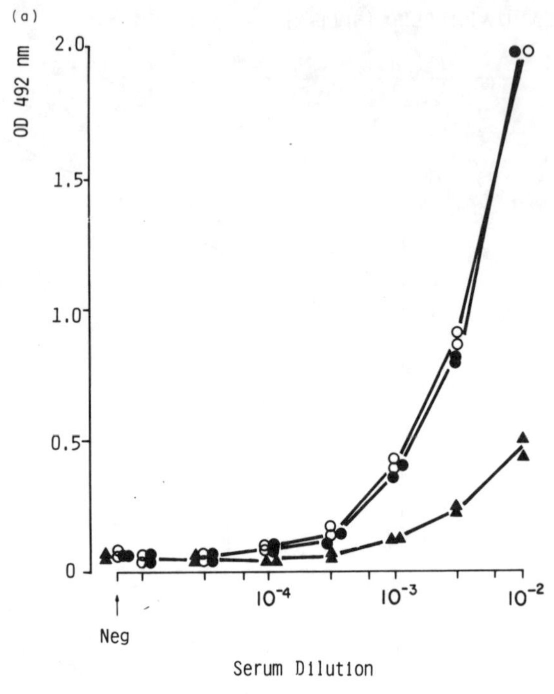

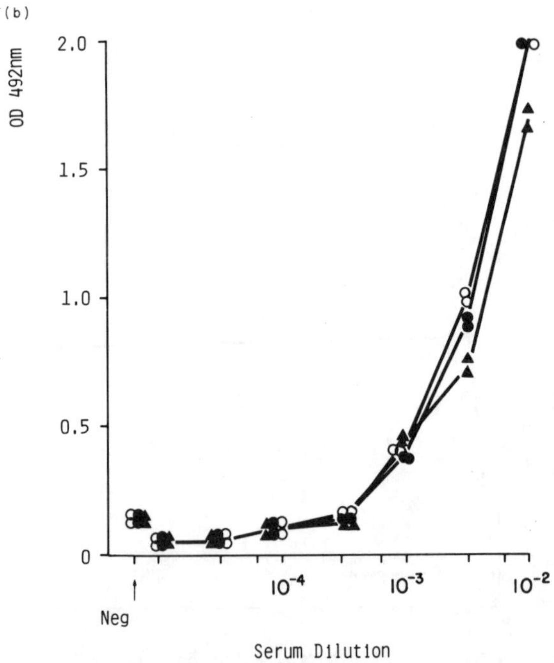

FIGURE 9 The effect of varying anti-PLA_2 (a) = 100 μg/ml (b) = 10 μg/ml on the performance of IgE SELISA using (●) 10 000, (○) 1000, (▲) 100 ng/ml PLA_2.

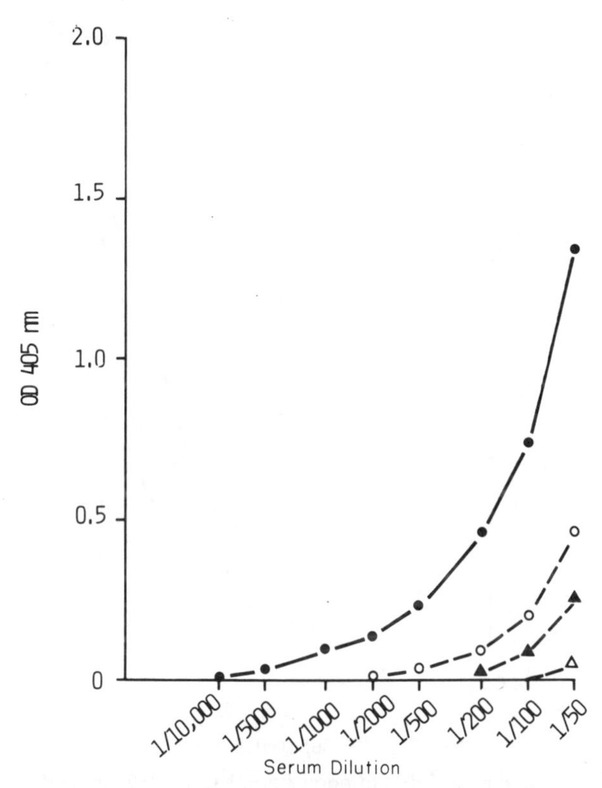

FIGURE 10 A comparison of SELISA for IgE antibodies (●–●) using rabbit anti-*D.pteronyssinus* and *D.pteronyssinus* (0.1 mg/ml) and ELISA with 0.1 (○–○), 0.01 (▲–▲), and 0.001 (△–△), mg/ml *D. pteronyssinus* extract. (Kemeny *et al.*, 1986)

of PLA_2 used but this was not seen when less anti-PLA_2 was used (Figure 9b) or with an excess of PLA_2. It is also interesting to note that optimal results could be obtained with as little as 2.3 ng of PLA_2 bound to the well (Table 2).

Application of SELISA to complex antigen mixtures

Few allergens have been purified and even fewer are available in sufficient quantities for general use in immunoassays. Furthermore, there can be considerable variation in the sensitivity of individual patients to different allergens. It is therefore important that the majority of allergens that comprise the extract are represented on the solid phase. The fact that SELISA worked over such a wide range of antibody–antigen ratios suggested that it might be possible to use it to detect IgE antibodies to mixtures of proteins (Kemeny *et al.*, 1986). Plates can be coated with the isolated serum immunoglobulin fraction of rabbits that have been hyperimmunized with grass pollen or dust mite extracts. These

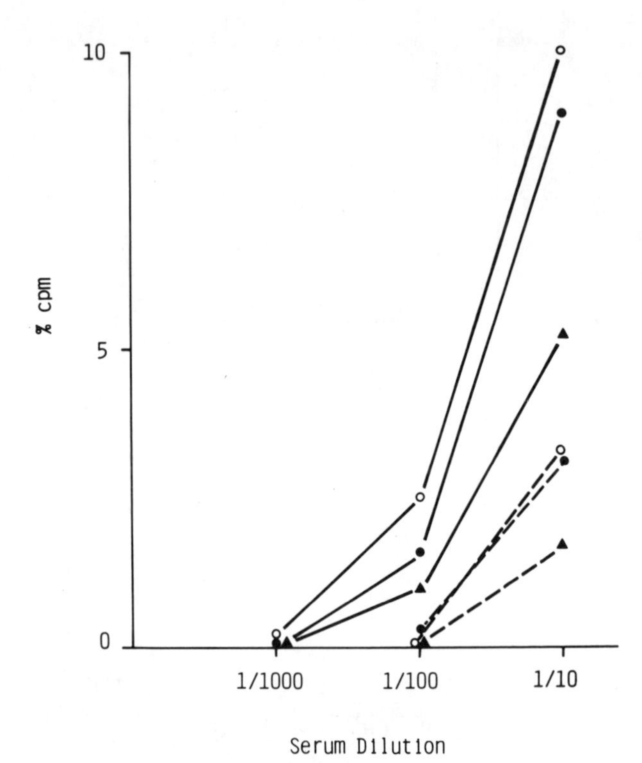

FIGURE 11 Comparison of IgE antibody bound using ^{125}I anti-IgE with Dynatech M129A removawells coated with rabbit anti-*D. pteronyssinus* and *D. pteronyssinus* extract (●- - -● 0.1, ○- - -○ 0.01, ▲- - -▲ 0.001 mg/ml) and agarose coated with anti-rabbit antibody, rabbit anti-*D. pteronyssinus* and *D. pteronyssionus* extract ●—● 0.1, ○—○ 0.01, ▲—▲ 0.001 mg/ml)

antisera are normally used for two-dimensional radio immunoelectrophoresis (crossed radio immunoelectrophoresis, CRIE).

As for single allergens, the sensitivity of SELISA using these rabbit antisera is greater than with direct coating of the allergen extract, here house dust mite (Figure 10) although the sensitivity is similar to RAST (Kemeny *et al.*, 1986). The fact that sensitivity was not much greater than RAST may be explained, at least in part, by the inability of the microtitre plate to bind sufficient of different antibodies. When the same rabbit antibodies and dust mite extract were used in a similar assay but were insolubilized with agarose coated with anti-rabbit antibody and the bound IgE detected with ^{125}I radiolabelled in place of enzyme-labelled anti-IgE, the sensitivity is greater than for SELISA (in this case carried out using the lower capacity Dynatech removawells) (Figure 11). It seems, therefore, that while sufficient purified antibodies may be bound to

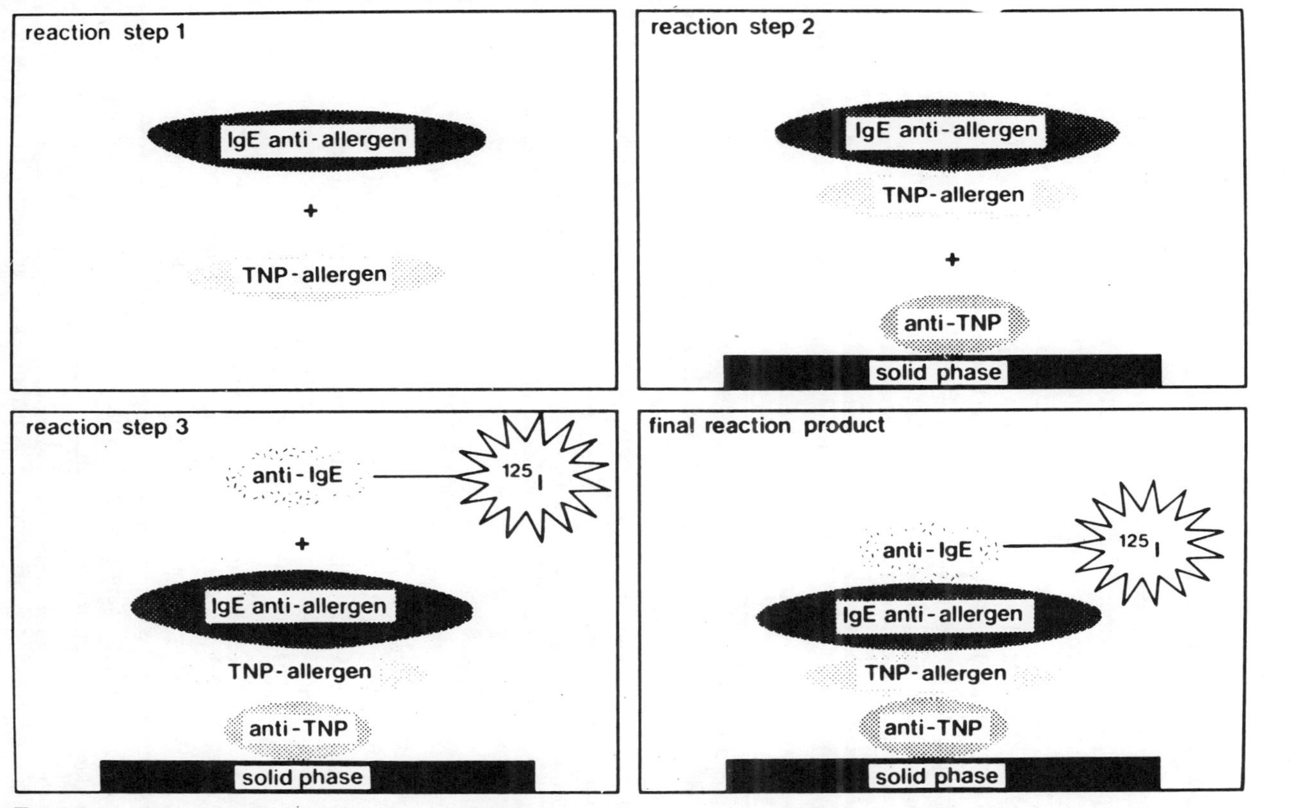

FIGURE 12 The principle of the central laboratory of the Netherlands Blood Transfusion Service (CLB) -RAST. (Reproduced with the kind permission of Dr R.C. Aalberse, Red Cross, Amsterdam; Aalberse, *et al.*, 1985)

microtitre plates for single antigens, for mixtures of antigens a higher capacity matrix is required.

Other allergen-capture assays

While we were developing the SELISA to try to overcome the limitations of solid phase allergen-specific antibody assays, Aalberse and co-workers at the Red Cross, Amsterdam, were trying to bind complex mixtures of proteins to nylon balls for use in their IgE antibody radioimmunometric assay (Aalberse *et al.*, 1986). The system they used is shown diagrammatically in Figure 12 and involves TNP-labelled allergens and anti-TNP nylon balls. Incubation of the test sample with haptenated allergen is carried out in the fluid phase. The complex thus formed is then bound to nylon balls coated with anti-TNP and the IgE subsequently detected with ^{125}I-radiolabelled anti-IgE. Their results show that this provides comparable results to RAST although it is somewhat more sensitive. It does not appear to be subject to the same interference with IgG antibody as our competitive SELISA for bee-venom-PLA_2-specific IgE antibodies.

There are other advantages to this technique. First, only one type of antibody (i.e. anti-TNP) is required in addition to the anti-IgE. Second, the capacity of the nylon balls for anti-TNP is greater than the microtitre plate. We did, however, only show interference in our PLA_2 system and it is recognized that much higher levels of IgG than IgE are found in response to bee venom as compared with inhalant allergens (Kemeny *et al.*, 1982). Many other forms of 'universal' label such as biotin–avidin could be used (Aalberse *et al.*, 1986).

CONCLUSIONS

As emphasized in Chapter 2 many of the problems associated with ELISA stem from the interaction of the antigen or antibody with the microtitre plate. Their advantages in terms of low background binding and convenient handling make them more attractive than many other solid phase supports. In common with all such matrices binding of protein to the plastic may alter the immunological reactivity of the bound material in an unpredictable fashion — with the exception of antibody, which only has one site of action (the combining site). The overall capacity of the microtitre plate can be increased so that even if some of the activity is lost sufficient is retained to bind antibody. Alternatively, the way in which the antigen is presented can be improved through using antibodies, or other binding systems, to link the antigen to the plate as described in this chapter. Harvey and Longbottom (1986) have subsequently used this approach to measure IgG antibodies to antigen 7 of *Aspergillus fumigatus* and virologists have recognized that this may be important when assessing the immune response to vaccines. (see Chapter 14). Affinity-purified rabbit anti-mouse immunoglobulin, for example, can be used to bind monoclonal antibodies which otherwise perform poorly on microtitre plates (Mangili *et al.*, 1987) and this can also reduce the amount of monoclonal antibody used.

The value of using the sort of approach described here goes beyond a simple improvement in assay performance. When screening cell culture supernatants for monoclonal antibodies it is possible that antigen directly bound to the plate may not display some epitopes which may be obscured by the plastic surface, with the result that some antibody-secreting clones would be missed. Furthermore, in human serum samples the presence of free antigen (e.g. foods), may interfere with the binding of antibody. If a sandwich assay, such as that described here, is used, then any immune complexes present in the sample will also be detected. Whether antigens should be bound directly to the microtitre plate, to antibody or to some other linkage reagent will depend on the type of antigen and the needs of the user. However, it is worth noting that the agreement between ELISA and other methods of measuring antibody such as RIA can be poor with antigen directly bound to the plate but that agreement between the SELISA and solid or fluid phase RIA is usually very good.

REFERENCES

Aalberse, R.C., Van Zoonen, M., Chemens, J.G.J., and Winkel, I. (1986). The use of hapten-modified antigens instead of solid-phase coupled antigens in a RAST-type assay. *J. Imm. Methods*, **87,** 51.

Cantarero, L.A., Butler, J.E., and Osborne, J.W. (1980). The absorptive characteristics of proteins for polystyrene and their significance in solid-phase immunoassays. *Anal. Biochem.*, **105,** 375.

Ceska, M., Eriksson, R., and Varga, J.M. (1972). Radioimmunosorbent assay of allergens. *J. All. Clin. Immun.*, **49,** 1.

Derer, M.M., Miescher, S., Johansson, B., Frost, H. (1984). Application of the dot immunobinding assay to allergy diagnosis. *J. All. Clin. Immun.*, **74,** 85.

Eriksson, N.E., and Ahlstedt, S. (1977). Diagnosis of reaginic allergy in house dust, animal dander and pollen allergens in adult patients. A comparison between ELISA, provocation skin test and RAST for diagnosis of reaginic allergy. *Int. Arch. All. Appl. Immun.*, **54,** 88.

Gleich, G.J., and Jacob, G.L. (1975). Immunoglobulin E antibodies to pollen allergens account for high percentages of total immunoglobulin E protein. *Science*, **190,** 1106.

Gleich, G.J., Jacob, G.L., Yunginger, J.W., and Henderson, L.L. (1977). Measurement of the absolute levels of IgE antibodies in patients with ragweed hayfever. Effect of immunotherapy on seasonal changes and the relationship to IgG antibodies. *J. All. Clin. Immun.*, **60,** 188.

Hamilton, R.G., and Adkinson, N.F. (1980). Quantification of antigen-specific IgG in human serum: Standardization by a Staph A solid-phase radioimmunoassay elution technique. *J. Immunol.*, **124,** 1966.

Harvey, C., and Longbottom, J.L. (1986) Development of a sandwich ELISA to detect IgG and IgG sub-class antibodies specific for a major antigen (Ag 7) of *Aspergillus fumigatus*. *Clin. Allergy*, **16,** 323.

Herrmann, J.E., and Collins, M.E. (1976). Quantitation of immunoglobulin adsorption to plastics. *J. Imm. Methods*, **10**, 363.

Hoffman, D.R. (1973). Estimation of serum IgE by an enzyme-linked immunosorbent assay (ELISA). *J. All. Clin. Immun.*, **51,** 303.

Ishikawa, E., and Kato, K. (1978). Ultrasensitive enzyme immunoassay. *Scand. J. Immunol.*, **8** (suppl 7), 43.

Kemeny, D.M., Lessof, M.H., and Trull, A.K. (1980). IgE and IgG antibodies to bee venom measured by a modification of the RAST. *Clin. Allergy*, **10,** 413.

Kemeny, D.M., and Richards, D. (1987). ELISA for the detection of IgE : speed and sensitivity. *Immunological techniques in microbiology,* **24**, 47.

Kemeny, D.M., Miyachi, S., Platts-Mills, T.A.E., Wilkins, S., and Lessof, M.H. (1982). The immune response to bee venom : comparison of the antibody response to phospholipase A2 with the response to inhalant antigens. *Int. Arch. All. Appl. Immun.*, **68,** 268.

Kemeny, D.M., Harries, M.G., Youlten, L.J.F., Mackenzie-Mills, M., and Lessof, M.H. (1983a). Antibodies to purified bee venom proteins and peptides. I. Development of a highly specific RAST for bee venom antigens and its application to bee sting allergy. *J. All. Clin. Immun.*, **71,** 505.

Kemeny, D.M., Mackenzie-Mills, M., Harries, M.G., Youlten, L.J.F., and Lessof, M.H. (1983b). Antibodies to purified bee venom proteins and peptides. II. A detailed study of the changes in IgE and IgG antibodies to individual bee venom antigens. *J. All. Clin. Immun.*, **72,** 376.

Kemeny, D.M., Urbanek, R., Samuel, D., and Richards, D. (1985a). Improved sensitivity and specificity of sandwich, competitive and capture enzyme-linked immunosorbent assays for allergen-specific antibodies. *Int. Arch. All. Appl. Immun.*, **77,** 199.

Kemeny, D.M., Urbanek, R., Samuel, D., and Richards., D. (1985b). Increased sensitivity and specificity of a sandwich ELISA for measurement of IgE antibodies. *J. Imm. Methods.*, **78,** 212.

Kemeny, D.M., Urbanek, R., Samuel, D., Richards, D., and Maasch, H. (1986). The use of monoclonal and polyspecific antisera in the IgE ELISA. *J. Imm. Methods*, **87,** 45.

Mangili, R., Kemeny, D.M., Li, L.K., and Viberti, G.C. (1987). Development of a sensitive enzyme-linked immunosorbent assay (ELISA) for quantification of human IgG subclasses. *J. All. Clin. Immun.* **79**, 222.

Negrini, A.G., Troise, C., and Voltolini, S. (1985). ELISA in the diagnosis of respiratory allergy. *Allergy*, **40,** 238.

Paull, B.R., Jacob, G.L., Yunginger, J.W., and Gleich, G.J. (1978). Comparison of binding of IgE and IgG antibodies to honey bee venom phospholipase A2. *J. Immunol.*, **120,** 1917.

Pesce, A.J., Ford, D.J., and Graizutis, M.A. (1978). Quantitative and qualitative aspects of immunoassays. *Scand. J. Immunol.* **8** (Suppl. 7), 1.

Platts-Mills, T.A.E. (1981). Laboratory techniques in immediate hypersensitivity. In: *Immunological and Clinical Aspects of Allergy* (ed. M.H. Lessof). MTP Press, Lancaster, pp. 85–99.

Schimizu, M., Wicher, K., Reisman, R.E., and Arbesman, C.E. (1978). A solid-phase radioimmunoassay for detection of human antibodies. I. Measurement of IgG antibody to bee venom antigens. *J. Imm. Methods*, **19,** 317.

Urbanek, R., Kemeny, D.M., and Samuel, D. (1985). Use of the enzyme-linked immunosorbent assay for measurement of allergen-specific antibodies. *J. Imm. Methods*, **79,** 123.

Voller, A., Bidwell, D.E., and Bartlett, A. (1979). *The Enzyme-Linked Immunosorbent Assay (ELISA)*. Dynatech Europe, Guernsey.

Walsh, B.J., Wrigley, C.W., and Baldo, B.A. (1984). Simultaneous detection of IgE binding to several allergens using a nitrocellulose 'polydisc'. *J. Imm. Methods*, **66,** 99.
Wide, L., Bennich, H., and Johansson, S.G.O. (1967). Diagnosis of allergy by an *in vitro* test for allergen antibodies. *Lancet.* **ii,** 1105.

ELISA and Other Solid Phase Immunoassays
Edited by D.M. Kemeny and S.J. Challacombe

CHAPTER 10

The Solid Phase Enzyme-Linked Immunospot Assay (ELISPOT) for Enumerating Antibody-Secreting Cells: Methodology and Applications

Cecil Czerkinsky, Lars-Åke Nilsson, Andrej Tarkowski, William J. Koopman, Jiri Mestecky and **Örjan Ouchterlony**

Universities of Göteborg, Sweden and Birmingham, Alabama, USA

CONTENTS

INTRODUCTION

Measurement of antibodies and antibody-secreting cells

The presence of specific antibodies (Ab) in body fluids may have considerable biological and pathological significance and can be documented by a variety of techniques. These include solid phase immunoassays (SPIA) which have gained increasing importance in the armamentarium of serologists.

The quantification of Ab accumulated in various fluids has, however, limited value for analysing the dynamic aspects of humoral immune responses. For instance, the precision of these analyses is influenced by the half-life and the catabolism of Ab molecules which vary among the different classes and subclasses of immunoglobulins (Ig). In addition, the suitability of these approaches for quantitative estimations of autoantibody repertoires is particularly questionable since *in vivo* absorption of self-reactive Ab by circulating or tissue-bound autoantigens is likely to interfere with their detection. Perhaps the most obvious limitation of these analyses is in their inability to yield information concerning the precise anatomical location(s) of Ab formation.

The analysis of Ab formation at the cellular level appears considerably more informative in this regard. Immunohistochemical methods based on cytoplasmic staining of Ab-containing cells, using antigens (Ag) labelled with either a fluorochrome (Coons, Leduc, and Conolly, 1955), an enzyme (Bosman, Feldman, and Pick, 1969) or an isotope (autoradiography) (Pick and Feldman, 1967) are often employed. However, such approaches are cumbersome and, more importantly, make it difficult to distinguish Ab-secreting cells (ASC) from cells that are synthesizing Ab without secreting it, or even from cells that might have internalized Ab and hence labelled Ag.

Haemolytic plaque assay

For almost three decades, the haemolytic plaque-forming cell (PFC) assay (Jerne and Nordin, 1963) has been the most widely used indicator system to study Ab secretion and to analyse the regulatory mechanisms that control the expansion of ASC populations. The original technique and its numerous modifications (reviewed by Jerne *et al.*, 1974) have been fundamental tools for the clonal analysis of Ab formation. The method offers the unique possibility of detecting individual ASC even in situations where only a few clones are activated, such as during the early period of a primary immune response or in attempts to induce Ab production *in vitro*. However, the applicability of the procedure has met with technical and theoretical problems mainly inherent in the use of the haemolysis reaction as the indicator system of localized Ag–Ab reactions.

The difficulties encountered in coupling many antigens to erythrocytes in order to provide lysable targets with an appropriate density of antigenic epitopes are well documented. Moreover, the reliability of PFC assays for isotype-, allotype-, or idiotype-specific determinations is largely based on theoretical predictions (Jerne *et al.*, 1974; Sterzl and Riha, 1965; Goidl *et al.*, 1979) and has been questioned in certain circumstances (Plotz, Talal, and Asofsky, 1968; Pasanen and Makela, 1969; Baker and Stashak, 1969; Shepart,

Mayers, and Bankert, 1985). Additionally, the suitability of PFC assays for estimating Ab affinity at the single cell level (Anderson, 1974), has also been questioned on the basis of theoretical considerations (De Lisi, 1976) and also on experimental evidence (Bankert, Mazzaferro, and Mayers, 1981). Further, the method does not permit quantification of Ab molecules secreted. Such information might be particularly useful when estimating the metabolic state of the cell population under study.

Alternative assays

The limitations of traditional PFC assays have prompted a number of investigators to develop alternative approaches for studying Ab secretion at the cellular level. One such approach consists of quantifying Ab produced and accumulated in supernatants of short-term cultures of lymphoid cells. The popularity of this approach largely owes to the availability of simple and relatively sensitive methods such as solid phase radio-, enzyme-, or fluoro-immunoassays. Being most pertinent to the theme of this volume, one should be cognizant of the pitfalls of each application of SPIA in the study of Ab formation at the cellular level.

Hybridoma technology has now made possible critical evaluation and careful interpretation of SPIA data, obtained on complex biological fluids such as tissue culture supernatants. For instance, the accuracy of these assays for class- and subclass-specific determinations is critically influenced by phenomena of competition between Ab populations belonging to distinct isotypes and/or expressing different affinities for binding to solid phase Ag. In addition, the threshold of sensitivity of these assays, although rather low, limits their use in situations where relatively small numbers of cells are producing Ab. Most importantly, these procedures do not enable an estimation of the frequency of the corresponding ASC, and thus do not permit clonal analyses of Ab production.

Enzyme assays

Recently, an alternative strategy based also on SPIA technology has been developed for enumerating specific ASC. The method, termed enzyme-linked immunospot (ELISPOT) assay (Czerkinsky *et al.*, 1983) or ELISA-plaque assay (Sedgwick and Holt, 1983a; see also this volume Chapter 11), utilizes the principle of the diffusion-in-gel ELISA (DIG-ELISA) (Elwing and Nygren, 1979). This modification of the classical ELISA (Engvall and Perlmann, 1972) allows localized Ag–Ab reactions to be visualized. In principle (Figure 1), a lymphoid cell suspension containing putative ASC is incubated for a few hours in plates that have previously been coated with the pertinent Ag. Following removal of the cells, zones of secreted Ab bound to the solid phase are

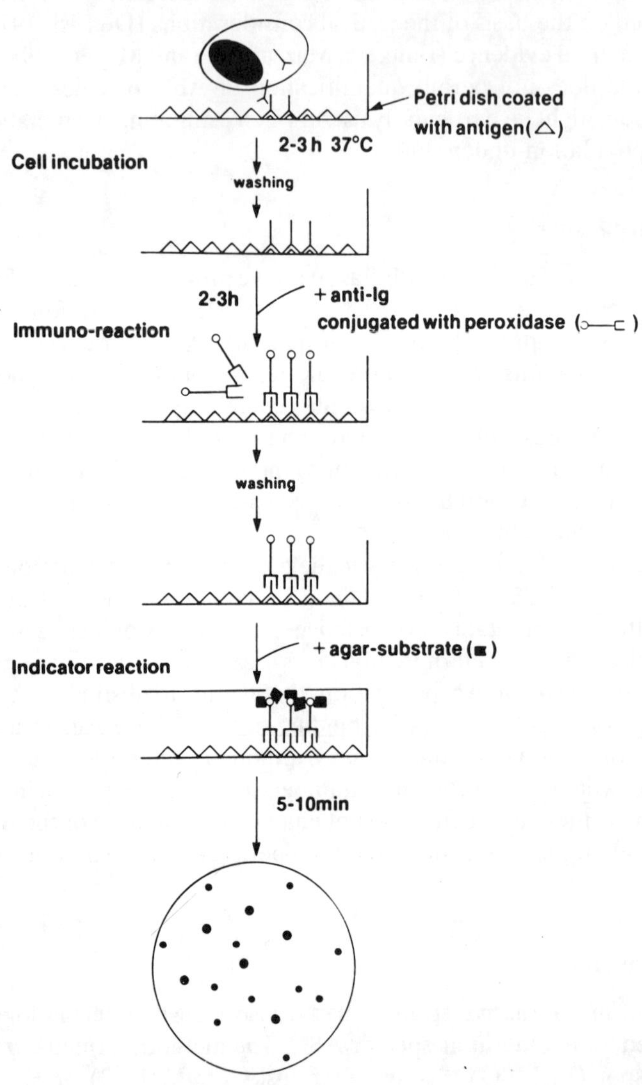

FIGURE 1 Principle of the ELISPOT assay for enumerating specific antibody-secreting cells

demonstrated by stepwise addition of enzyme-labelled antiglobulin conjugate and agar containing enzyme substrate. Coloured foci or spots appear at the former location of specific ASC and can be enumerated with the naked eye or under low magnification (Figure 2). The indicator enzymes used in the assay are horseradish peroxidase (HRP) (Czerkinsky *et al.*, 1983) or alkaline

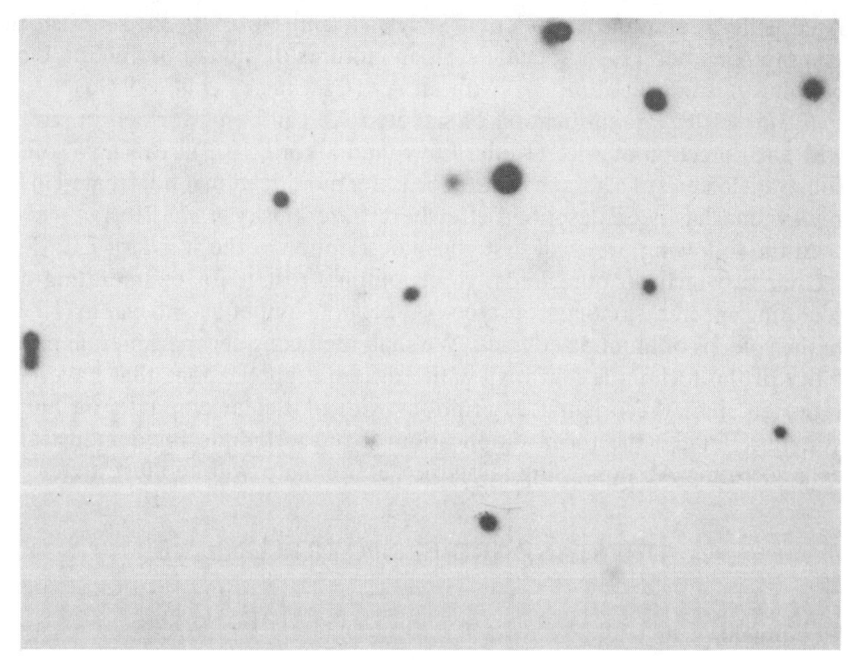

FIGURE 2 Typical appearance of spots generated by specific ASC (×6)

phosphatase (Sedgwick and Holt, 1983a) which can be conjugated with Ab. The conjugated Abs are used in direct or indirect techniques that are identical to those described for conventional solid phase ELISA. The indicator reactions depends on the interaction of numerous molecules of substrate, with a limited number of enzyme molecules bound to the solid phase. The enzyme–substrate complexes formed react with a chromogenic electron donor to yield a visible coloured product that precipitates into an agar matrix. We have employed HRP and *p*-phenylenediamine–H_2O_2 as the indicator system. Diaminobenzidine, aminoethyl carbazole, and 4-chloro-1-naphthol can also be used with HRP conjugates (Czerkinsky, unpublished). Others (Sedgwick and Holt, 1983a) have employed alkaline phosphatase together with the substrate 5-bromo-4-chloro-3-indolyl phosphate (BCIP).

Advantages of enzyme assays for ASC

Using the appropriate enzyme-labelled antiglobulin conjugate, the assay theoretically allows for detection of cells secreting Ab of a given isotype, allotype, or even idiotype. The sensitivity of this method is at least equal to that of conventional PFC assays (Sedgwick and Holt, 1983a; Czerkinsky *et al.*, 1983) and permits the detection of extremely low numbers of ASC such as IgE

ASC cells in immunized rodents (Sedgwick and Holt, 1983b; see also this volume, Chapter 11) or specific ASC in cultures of human peripheral blood lymphocytes educated *in vitro* with an Ag (Czerkinsky *et al.*, 1984a).

In situ ELISA quantification of secreted Ab can be performed in parallel with the detection of ASC (Kelly, Levy, and Sikora, 1979) by using a soluble substrate instead of a gel substrate. The latter modification is not treated in this review and has been described elsewhere (Czerkinsky *et al.*, 1983).

In the following, we shall first give a description of the standard ELISPOT technique which, in our hands, yields optimal results for enumerating cells secreting specific Ab to a prototype Ag, keyhole limpet haemocyanin (KLH), in the spleens of immunized mice. We shall then consider the general applicability of the technique and review the different applications that have been reported since the original descriptions with particular emphasis on human systems. Finally, the recent application of the technique for demonstrating cells secreting Ab possessing idiotypic determinants will be illustrated.

THE STANDARD ELISPOT TECHNIQUE

Equipment

- Humidified incubator set at 37 °C; if a carbonate-buffered medium is used the atmosphere should contain CO_2 (5–10 per cent).
- Level platform.
- Polystyrene Petri dishes, 40 to 50 mm diameter, bacteriological grade, e.g. from Nunc, Roskilde, Denmark, or from Falcon (type 1006), Oxnard, CA. Smaller dishes are not recommended.
- Water bath set at 46–48 °C.
- Magnifying device equipped or connected with a light source, e.g. colony counter, stereomicroscope.

Reagents and solutions

Chemicals should be of analytical grade.

(1) *Assay culture medium*: our standard medium consists of Iscove's medium (Gibco Europe, Glasgow, UK) supplemented with Hepes, 25 mM; $NaHCO_3$, 36 mM; foetal bovine serum (FBS) 10 per cent; pH 7.43. Prepare under sterile conditions and store at 4 °C. Other commonly used tissue culture media are also suitable.

(2) *Phosphate-buffered saline (PBS)*: phosphates, 10 mM, NaCl, 0.15 M; pH 7.2.

(3) *Coating solution*: dissolve 1 mg of KLH, e.g. from Calbiochem, San Diego, CA, in 10 ml of PBS (2); aliquot and store at −20 °C until use.

(4) *Blocking solution*: add 10 ml FBS to 90 ml PBS (2); freshly prepared.

(5) *PBS–Tween*: add 5 ml Tween 80 to 10 1 PBS (2).
(6) *PBS–EDTA*: dissolve 18.61 g EDTA–$Na_2H_2.2H_2O$ in 1 l PBS (2); adjust pH to 7.2 if necessary with NaOH, 0.1 M.
(7) *PBS–agar*: dissolve 1 g Noble Agar (Difco, Detroit, MI) in 100 ml PBS (2). Boil for 15–20 min until the agar is completely dissolved. Allow the molten agar solution to cool to 48 °C in a water bath. If the solution contains aggregates, filter it through a clean hot sintered glass filter (No. 2 or 3).
(8) *Gel–substrate solution*: dissolve 50 mg 1,4-*p*-phenylenediamine free base (Aldrich Chemicals, Milwaukee, WI, or Sigma Chemical Company, St Louis, MO) in a minimum volume (1–2 ml) of methanol. Agitate in water bath (48 °C) until completely dissolved (30 sec to 1 min). Add 50 μl of 30 per cent H_2O_2. Mix with 100 ml of 46–48 °C molten agar solution (7) and use immediately. The pH of this solution must be comprised between 7.0 and 7.2.
(9) *Enzyme-labelled reagents*
  (a) HRP-conjugated anti-mouse Ig can be purchased commercially, e.g. from Dakopatts, Copenhagen, Denmark; usual working dilution 1 : 100 to 1 : 500.
  (b) Biotinylated anti-mouse Ig can be prepared, or can be purchased commercially, e.g. from Tago, Inc., Burlingame, CA or from Calbiochem; usual working dilutions (1 : 500 to 1 : 2000; 0.5–2 μg/ml).
  (c) HRP-conjugated avidin can be obtained from, e.g. Vector Laboratories, Burlingame, CA or from Sigma.
  (d) Avidin–biotin–HRP complexes can be purchased from, e.g. Dakopatts or Vector.
  Dilute all reagents before use in PBS–Tween (5) to a predetermined concentration.

Preparation of cell suspensions

- Prime mice, 10–20 weeks' old, with an intraperitoneal injection of KLH (100 μg) emulsified in Freund's complete adjuvant; challenge the animals 2 to 3 weeks later with a similar injection and sacrifice 4–5 days later.
- Remove the spleens and place them in Petri dishes containing chilled medium (1) (without FBS); dissociate the organs by pressing them with a smooth instrument through a fine wire grid. Filter the suspensions through a layer of nylon cloth to remove aggregates.
- Pellet the suspension by centrifugation at 200 × *g* for 10 min. The erythrocytes may be lysed by brief exposure (2–5 min) to 0.85 per cent (w/v) Tris buffered NH_4Cl solution.
- Wash the cells twice by centrifugation in medium (1).
- Count the number of nucleated cells and estimate their viability using a Trypan blue dye exclusion test. Approximately 90 to 95 per cent of the cells

should be found viable. Keep the cell suspensions on ice until use.

Assay procedure

The assay consists of five stages: first the preparation of a solid phase immunoadsorbent, then the cell incubation stage, followed by the immunoperoxidase stage. The fourth stage consists of adding a gelling substrate which will yield a coloured product. The corresponding spots are then enumerated in the final stage.

Preparation of assay dishes

- Fill polystyrene Petri dishes with 1.5 ml of coating solution (3) and let stand in humidified chambers for 2 hr at 37 °C, 4 hr at room temperature or overnight at 4 °C whichever is most convenient. Wash the plates three times with PBS (2). Store at 4 °C (filled with buffer) or at −20 °C (decanted of buffer) until use.
- In order to saturate remaining binding sites on the plastic surface, expose the dishes for 2 hr at 37 °C to 3 ml of blocking solution (4) before use.
- Appropriately treated, the inner surface of the dish should be hydrophilic (a very small volume of PBS, i.e. 300 µl, should spread evenly on to the coated solid phase of a decanted dish).

Notes:

- In order to minimize the effects of cell edging, one can apply a thin vaseline film along the edge of the coated dish using a narrow cotton tip.
- Other supports can be used as solid phase in the assay. Nitrocellulose membranes are particularly well suited for this purpose (Möller and Borrebaeck, 1985), their binding capacity being superior to that of polystyrene surfaces. The membranes are coated, washed, and blocked as above. Following drying, they can be cut into rectangles (3 × 2 cm) and used as 'cover-slips' by placing them on to small volumes (100 µl) of cell suspension dispensed dropwise on the inner surface of a large Petri dish (uncoated) (Czerkinsky *et al.*, in preparation).
- Alternatively, individual rings from 24 wells tissue culture plates can be coated and employed as assay culture vessels.

Cell incubation

- Add 300 µl aliquots of assay culture medium (1) containing appropriate numbers of cells to coated plates immediately after they have been decanted of blocking solution; care must be taken to remove any residual solution prior to addition of the cells (plates are emptied by flicking over a sink and remaining solution is removed by aspiration).

- Incubate the dishes for 3–4 hr at 37 °C undisturbed on a levelled surface in a 10 per cent CO_2-humidified atmosphere.
- At the completion of the culture period, recover the cells for other purposes or discard them by rinsing the plates twice with PBS (2).
- Elute remaining plastic-adherent cells from the solid phase by incubating the plates with 3 ml chilled PBS–EDTA solution (6) for 15 min.
- Wash the plates three additional times with PBS–Tween (5).
- Keep the plates filled with this solution to avoid drying.

Notes:

- The optimal cell density should be determined by experimentation and adjusted so that the cells form a non-confluent monolayer; we find it convenient to assay most lymphoid suspensions at three different densities (serial fourfold dilutions above and below the expected optimum density) to provide a margin of safety if there are unexpectedly high or low numbers of ASCs.
- Culture periods of 3–4 hr are suitable for most mouse and human lymphoid cells. Prolonging the incubation time does not substantially affect the number of detectable ASCs.

Immunoperoxidase stage

- Pipette 2.5 ml of HRP-anti-mouse Ig conjugate (9) diluted in PBS–Tween (5) into each dish.
- Incubate for 2 hr at room temperature or overnight at 4 °C; wash plates four times with PBS (2) (without Tween) to remove unbound HRP antiglobulins; keep the plates filled with washing buffer (2).
- Carefully decant each dish and develop immediately (see below).

Enzymatic indicator reaction (development)

- Add 3–5 ml gel–substrate solution (8).
- Remove excess agar immediately by shaking it off the dish. The inner surface of the plate is covered with a thin film of gelling substrate.
- Leave the plates undisturbed in a horizontal position until gelling is completed (2 min); after 5 min, brownish spots recognizable by the naked eye begin to appear. Their number and their intensity are maximal by approximately 30 min. They can then be enumerated directly by eye or under low power magnification.

Notes

- The use of a thin layer of agar–substrate prevents spontaneous darkening of the background (which would occur progressively with larger volumes of

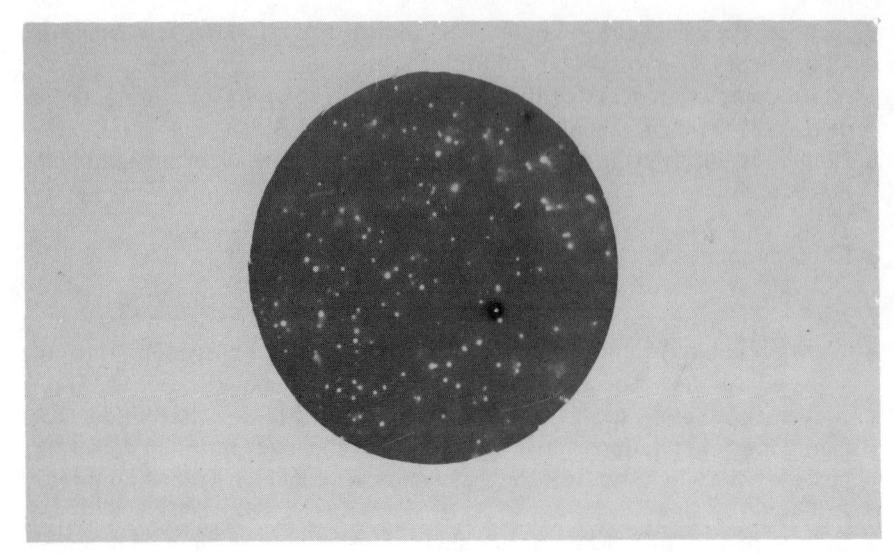

FIGURE 3 Contact photograph of a developed dish

gel–substrate solution). We have kept developed dishes for several years at room temperature without apparent loss of definition. The use of inhibitors of enzyme–substrate reactions is thus unnecessary when using a thin agar–substrate overlay.

- Other HRP substrates can be used, e.g. diaminobenzidine, aminoethyl carbazole, 4-chloro-1-naphthol, but we favour the use of *p*-phenylenediamine which appears to yield more intense reactions (unpublished observations). As recently reported by Sedgwick and Holt (1986) for the alkaline phosphatase–BCIP system, the use of agar is not necessary also for several HRP substrates but the definition of the spots is improved.
- Contact photographs can also be made, allowing the count to be performed at leisure on the print and permitting convenient record of results (Figure 3): place developed dishes on a suitable high contrast photographic paper (e.g. Agfa Brovira BHH1) and trans-illuminate with a white light source; develop the print according to standard photographic procedures; count the reactions directly on the print with the aid of a magnifying device if necessary.

Enumeration

- Count spots first with the naked eye and indicate their position on the back of the dish with a marker pen.
- Then confirm the positivity of the reactions accurately under low magnification (×6 to ×20) on a white background under direct illumination. Bacter-

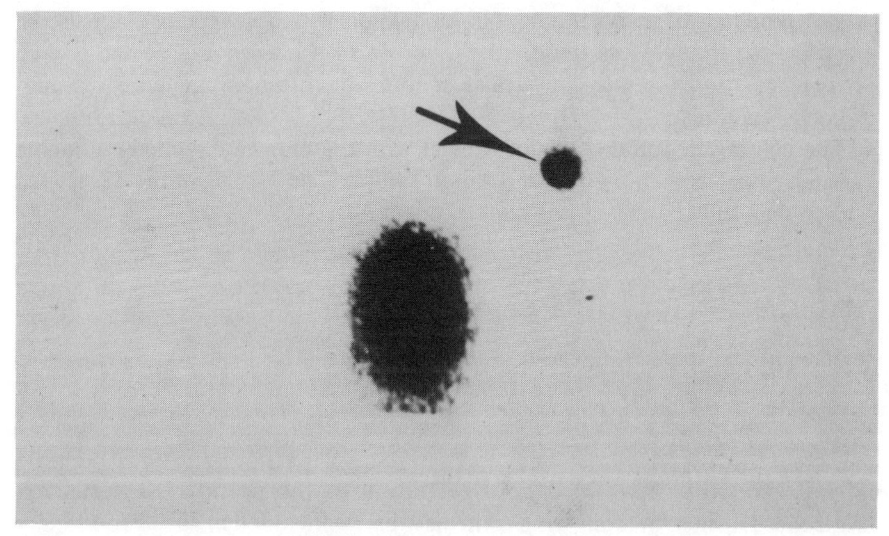

FIGURE 4 Typical granular appearance of a spot at high magnification (×60). Note small non-granular artefact (arrow)

iological colony counters or stereomicroscopes equipped with a white light source are ideal for this purpose.

Appearance of spots

- Macroscopically, spots are defined as brown, circular, and well-individualized foci. The colour intensity of these reactions varies from light brown to almost black. Their diameter also varies, ranging from 0.1 mm to 0.5 mm, rarely larger. There is no relationship between a given isotype of Ig secreted and a particular pattern of spots. However, differences in the appearance of spots generated may reflect heterogeneity in the secretion rate of individual cells and/or differences in the affinity of monoclonal Ab populations secreted (Holt *et al.*, 1984a; Nygren, Czerkinsky, and Stenberg, 1985).
- Under low magnification, spots exhibit a typical granular appearance (Figure 4). These dark granules consist of precipitates of enzyme–substrate reaction products which are homogeneously distributed within each spot. The latter feature is the most useful criterion to distinguish true spots from falsely positive reactions that might be occasionally observed.

Source of error

- Artefacts due to the presence of cell debris on the solid phase and/or of aggregates in the gel–substrate layer can be observed occasionally. Often, these reactions appear as small non-granular dots (Figure 4), sometimes surrounded by a granular halo. Granular spots with a clear centre can also be observed, though less frequently. The use of a magnifying device is then required to distinguish these false spots from true homogeneously granulated reactions.

- The coating material should not react with the enzyme-labelled conjugate used. Such an eventuality will result in complete darkening of the developed dish and will require appropriate absorptions.

VALIDATION OF THE ASSAY

Specificity

Antigen specificity

Since secreted Ig may bind non-specifically to the polystyrene surface of Ag-coated dishes, the following controls must be included in each experiment:

- Dishes coated with the immunizing Ag but exposed to a cell suspension prepared from either non-immunized or sham-immunized animals.
- 'Uncoated' dishes, i.e. blocked only, or dishes coated with an irrelevant Ag preparation and exposed to cell suspensions prepared from the test animal(s) and control animal(s).

Note: In certain situations, cells of animals that have not been actively immunized can generate spots when assayed in dishes coated with a variety of Ag. In general, the frequency of these reactions is negligible and the corresponding cells are thought to arise spontaneously or following previous exposure to a cross-reacting Ag. However, in certain situations, the frequency of these reactions can be considerably higher, such as with certain strains of animals whose immune system undergoes polyclonal B cell hyperactivation resulting in the expansion of numerous clones of Ab-secreting cells. In these situations, spot formation might be specific as far as the Ag-binding property of the corresponding Ig secreted is concerned, but this can only be documented by inhibition and/or elution experiments.

Inhibition

Spot formation can be inhibited during the cell incubation period by addition of soluble Ag of the same specificity as that of the Ag used for coating. In general, the degree of inhibition increases with increasing concentration of free Ag. However, experimental systems using monoclonal Ab and hapten-conjugated proteins, with different hapten substitution ratios, have indicated that the degree of antibody binding to solid phase Ag is influenced by the affinity of the different Ab populations and by the density of the corresponding antigenic epitope immobilized on the solid phase (Herzenberg *et al.*, 1980). Thus, a low epitope density preferentially allows monovalent Ab binding which is mainly due to Ab of relatively high affinity. In that situation, and by analogy, addition of free Ag during the cell incubation period inhibits spot formation in a dose-dependent manner and the lower the amount of free Ag required to inhibit spot formation to an arbitrary degree, the higher is the average affinity of the Ab populations secreted (Nygren, Czerkinsky, and Stenberg, 1985). At high epitope density, which allows binding of both high and low affinity Ab, comparatively larger amounts of free Ag might be required to inhibit spot formation. Thus, spot formation due to cells secreting Ab of relatively low affinity is more difficult to inhibit than spot formation due to cells secreting Ab of high affinity. Elution studies can be performed for this purpose.

Elution

Advantage is taken of the reversible nature of Ag–Ab reactions and of the relatively weak association between solid phase Ag and secreted Ab of low affinity. Addition of free soluble Ag immediately after the cell incubation period, and for a few hours prior to the enzymatic reaction, can dissociate secreted Ab from solid phase Ag and inhibit subsequent development of the corresponding spots. The lower the amount of free antigen required to inhibit spot formation to an arbitrary level, the lower is the average affinity of the Ab secreted and eluted from the solid phase (Nygren, Czerkinsky, and Stenberg, 1985).

Specificity for Ig markers

The suitability of the ELISPOT test for detection of cells secreting Ab of a given isotype, allotype, or idiotype relies entirely on the specificity of the anti-immunoglobulin preparations used during the immunoenzymatic stage. The specificity of such reagents can be evaluated by block titrations against purified Ig preparations, preferably myeloma proteins. Conventional solid phase ELISAs are adequate for this purpose. The specificity of the anti-immunoglobulin preparations should be further documented in the ELISPOT test by inhibition experiments. For example, a preparation which is intended for use as developing antiglobulin for a given Ig isotype should inhibit spot formation due to cells secreting Ab of the corresponding isotype when added during the cell incubation period. In contrast, spot formation due to cells secreting Ab of other isotypes should not be affected. Where necessary, the preparations can be rendered monospecific by absorption against the appropriate insolubilized proteins.

Assessment for *de novo* antibody synthesis

Antibody secretion is a dynamic process and the method whenever applied to the detection of ASC must provide evidence of active synthesis of Ab by a cell population rather than merely reveal its association with a cell. Indeed, certain Ab, termed cytophilic Ab, bind to cells which are incapable of secreting Ab. It is, therefore, essential to confirm that what is being detected as a spot is not merely the result of release of pre-formed Ab, but that Ab synthesis by spot-forming cells (SFC) is taking place during the *in vitro* incubation period. This can be accomplished in various ways, such as the use of an inhibitor of protein synthesis: dissolve cycloheximide, e.g. from Sigma, in assay culture medium (1) at a final concentration of 100 μg/ml. Resuspend cells to be tested in the above medium at a density of 2×10^6 cells/ml and dispense 1 ml aliquots in plastic tissue culture tubes. Incubate the tubes for 3 to 4 hr at 37 °C. Wash cells three times with cycloheximide-containing medium by centrifugation at 200 $\times g$ for 10 min. Then resuspend the peletted cells in the above medium at the appropriate densities and assay for SFC numbers by the ELISPOT test. Spot formation should be inhibited by 70–90 per cent after exposure of the cells to cycloheximide as compared to untreated cells, providing evidence for active metabolism of spot-forming cells.

Precision and sensitivity

The percentage of variation between replicate dishes generally does not exceed 20 per cent. However, considerably higher variations can be observed between replicate dishes that have been exposed to suspensions containing small numbers of ASC (less than 20 detectable SFC per dish).

Under optimal conditions the ELISPOT test described here is as sensitive as the PFC assay for enumeration of splenocytes secreting anti-ovalbumin Ab (Sedgwick and Holt, 1983a; Czerkinsky et al., 1983). However, the sensitivity of the ELISPOT assay can be substantially increased by using relatively simple immunoenzymatic amplication procedures. A complete description of these techniques for use in the ELISPOT assay is beyond the scope of this review and can be found elsewhere (Czerkinsky, 1986). We shall only mention the use of the avidin–biotin techniques which, in our hands, have been the most sensitive immunoenzyme techniques used in the ELISPOT test. Advantage here is taken of the considerable affinity of avidin for the vitamin biotin. Ab can easily be conjugated with biotin. The conjugation procedure is mild and yields conjugates that have retained virtually all their original activity (Heitzmann and Richards, 1974). Biotinylated antiglobulins can then be used both in direct and indirect ELISPOT tests in combination with HRP–avidin. Alternatively, a complex made of unconjugated avidin and HRP–biotin conjugate can be used

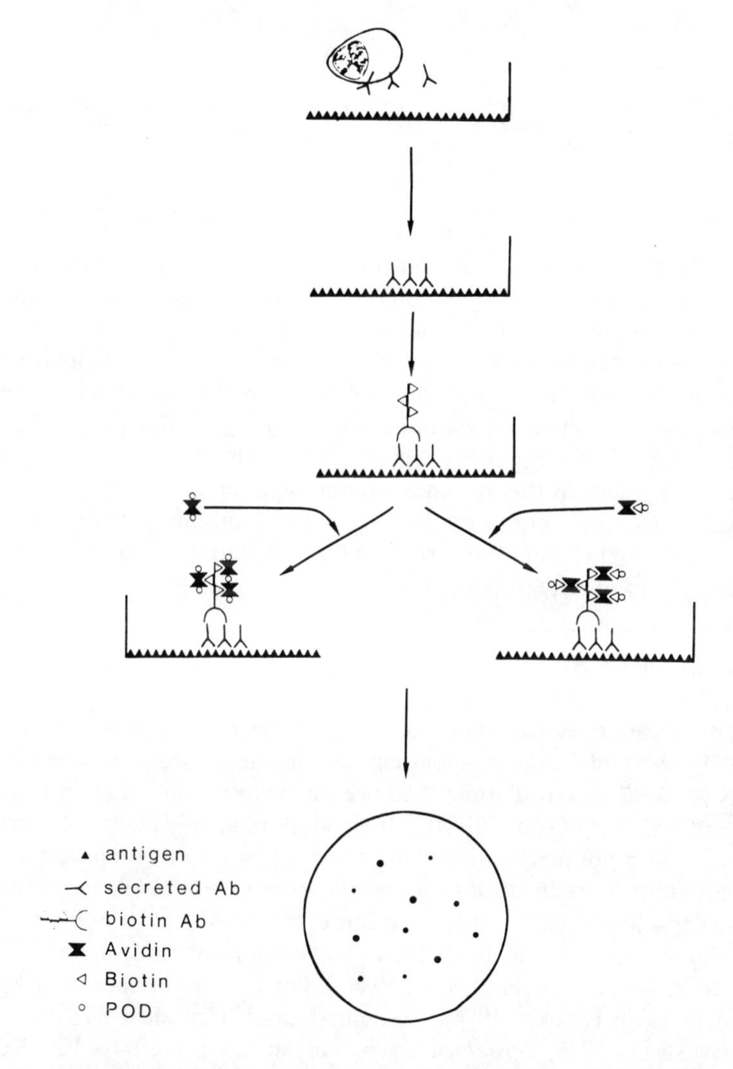

FIGURE 5 Diagrammatic reaction scheme of the ELISPOT assay using aviding–biotin techniques. Direct avidin–biotin technique (left) and avidin–biotin complex (ABC) technique (right)

for the same purpose. Since one molecule of avidin is capable of binding several molecules of biotin, the use of avidin–HRP–biotin complexes increases the surface concentration of HRP molecules giving the method even greater sensitivity (Figure 5).

APPLICATIONS

Since its original description, the ELISPOT assay has been used to detect both animal and human ASC against a variety of antigens.

Antigens

The ELISPOT technique is, in principle, applicable to the detection of cells secreting Ab against any Ag, soluble or particulate, that can be adsorbed (directly or indirectly) on to a solid surface. The adsorption of most macromolecules in solution on to solid supports involves random hydrophobic interactions between solid phase and solute. Thus, the statistical representation of a given antigenic epitope in the vicinity of a secreting cell can limit the number of specific binding sites geographically accessible to the corresponding secreted Ab. Bearing this in mind, it will be clear that Ag concentrations required to achieve appropriate coating of the solid phase in the ELISPOT technique are higher than those required in conventional solid phase ELISA and must be accurately determined.

Soluble antigens

Proteins: A variety of soluble protein Ag have been successfully used in the ELISPOT test to detect corresponding ASC in animal systems. These include prototype Ag such as ovalbumin (Sedgwick and Holt, 1983a; Czerkinsky *et al.*, 1983) and BSA (Holt *et al.*, 1984a). In general, relatively high concentrations are required for optimal coating. This drawback has been minimized by using relatively simple modifications such as the use of glutaraldehyde to increase binding capacity of plastic surfaces (Holt *et al.*, 1984b) or to polymerize low molecular weight Ag which, in their native conformation, appear to bind poorly to these surfaces (Holt *et al.*, 1984a). Purified bacterial protein Ag such as cholera toxin (Lycke, 1986a), tetanus toxoid (Czerkinsky *et al.*, 1984a, Tarkowski *et al.*, 1985), *Streptococcus mutans* surface protein Ag I/II (Russell, Czerkinsky, and Moldoveanu, 1986; Czerkinsky et al., 1987) have been used to enumerate corresponding ASC in rodents and humans. The technique is particularly valuable for the study of auto-Ab repertoires and has been used to evaluate the size and organ distribution of cell populations secreting auto-Ab against a variety of soluble protein auto-Ag such as IgG (for detection of rheumatoid-factor-secreting cells) (Tarkowski, Czerkinsky, and Nilsson, 1984, 1985), thyroglobulin (Holt *et al.*, 1984c), or collagen type II (Tarkowski *et al.*, 1986).

Polysaccharides: Several carbohydrate Ag have been used as solid phase Ag in the ELISPOT assay. In general, these Ag are poorly hydrophobic and high concentrations are required to achieve appropriate coating of the solid phase. This drawback can be minimized by chemical modification of the Ag. For example, *Streptococcus mutans* serotype *c* polysaccharide can be esterified and the derivative is readily adsorbed on to methylated BSA-coated plates which can then be used for the demonstration of corresponding ASC (Russell, Czerkinsky, and Moldoveanu, 1986; Czerkinsky *et al.*, 1987). Dextran- and lipopolysaccharide (LPS)-specific ASC have also been detected (Franci *et al.*, 1986).

Nucleic acids: Specific ASC to single stranded or double stranded DNA can be detected in high numbers in the spleens of certain autoimmune mouse strains by the ELISPOT method (Ando, Ebling, and Hahn, 1986). Negatively charged nucleotides bind poorly to plastic surfaces and best results are obtained on plates previously coated with poly-L-lysine or methylated BSA which provide layers of positive charges.

Haptens: Chemically defined haptens require covalent coupling to a carrier, e.g. albumin, for use in the ELISPOT assay. The degree of substitution of the carrier conjugate is critical (Herzenberg *et al.*, 1980) as illustrated by experiments using TNP–albumin conjugates with different hapten substitution ratios (Holt *et al.*, 1984a; Nygren, Czerkinsky, and Stenberg, 1985).

Complex antigenic preparations: Soluble extracts consisting of multiple antigenic compounds have been used in the ELISPOT technique. Thus, crude parasite extracts from *Ascaris suum* (Holt *et al.*, 1984a), *Schistosoma mansoni* adult worms and eggs (Czerkinsky 1987), bacterial cell walls (Mestecky *et al.*, 1986; Czerkinsky *et al.*, 1987), or whole lymphocytic choriomeningitis virus (Moskophidis and Lehmann-Grube, 1984), have been employed as coats for enumerating corresponding SFC in rodents and in humans. However, it should be emphasized that the interpretation of ELISPOT assay data in systems using heterogeneous mixtures is subject to caution. Indeed, plastic adsorbable molecules might compete with each other for binding to the plastic support, possibly decreasing the representation of relatively less hydrophobic molecules or even preventing their binding without necessarily being more relevant biologically.

Particulate antigens

Interesting developments have recently been reported for adapting the ELISPOT method to enumerate cells secreting specific Ab to particulate Ag. The use of hydrophilic glass surfaces in combination with polyaldehyde fixatives and/or poly-L-lysine has been described to adsorb either sheep erythrocytes or

Escherichia coli whole cells providing suitable solid phase for detection of corresponding ASC (Franci *et al.*, 1986). Others (Kantele, Arvilommi, and Jokinen, 1986) have reported successful detection of human peripheral blood ASC using direct adsorption of *Salmonell typhi* bacteria to polystyrene wells.

The above examples attest to the versatility of the method. It should be mentioned that 'reverse' modifications of the ELISPOT technique using surfaces coated with either Ab (Holt *et al.*, 1984a; Czerkinsky *et al.*, 1984b) or other ligands can be used to detect virtually any cell, eukariotic or not, secreting a given metabolite provided the product is (a) antigenic, (b) secreted in sufficient quantities, and (c) the corresponding Ab are available. Examples of antigen-secreting cells detected by this modification include lymphocytes secreting Ig regardless of antigenic specificity (Holt *et al.*, 1984; Czerkinsky *et al.*, 1984b) in anti-Ig coated plates; fibronectin-secreting cells in gelatin-coated dishes (Czerkinsky *et al.*, 1984c); enterotoxigenic *Escherichia coli* colonies secreting thermolabile toxin in Petri plates coated with the toxin receptor, ganglioside GM1 (Czerkinsky and Svennerholm, 1983).

Cells

The ELISPOT assay can be employed to detect ASC in cell suspensions prepared from a variety of animal species and from different lymphoid compartments. The cells can be tested either directly following collection (spontaneous ASC) or after *in vitro* exposure to a mitogen or to an Ag (cultured cells).

Spontaneous ASC

The ELISPOT technique has been employed to detect specific ASC not only in the spleens of actively immunized rodents but also in suspensions prepared from mesenteric and respiratory lymph nodes as well as from bone marrow aspirates of parenterally immunized or aerosolized animals (Sedgwick and Holt, 1983b; Holt *et al.*, 1984c; Sedgwick and Holt, 1985). The technique has recently been adapted to enumerate specific ASC in dissociated intestinal lamina propria specimens obtained from orally immunized mice (Lycke, 1986). Specific ASC can also be detected in isolated liver granuloma lesions of mice experimentally infected with *Schistosoma mansoni* (Czerkinsky, 1987). Recently, the technique has been applied to detect antiviral ASC in parenchymatous organs (brain, kidney) of infected mice (Moskophidis, Löhler and Lehmann-Grube, 1987).

In humans, the technique has been particularly useful for detecting spontaneous ASC that appear transiently in the peripheral circulation of normal individuals immunized parenterally (Tarkowski, Czerkinsky and Nilsson, 1985) or perorally (Lycke, Lindholm, and Holmgren, 1985; Kantele, Arvilommi, and Jokinen, 1986; Mestecky *et al.*, 1986; Czerkinsky *et al.*, 1987). The method is currently used to detect gliadin-specific ASC in suspensions prepared from intestinal biopsies of patients with gluten-sensitive enteropathies (Lycke *et al.*, in preparation) and also to detect auto-ASC in inflammatory synovial fluid (Tarkowski, Czerkinsky and Nilsson, 1985) or dissociated synoval tissue (Koopman *et al.*, in preparation) of patients with rheumatoid arthritis.

Cultured cells

The ELISPOT technique has also been used to enumerate individual ASCs generated after *in vitro* exposure of human and mouse mononuclear cells to soluble Ag (Czerkinsky *et al.*, 1984a; Lycke, 1986; Möller and Borrebaeck, 1985). This illustrates the potential of the technique for clonal analyses of Ab formation in situations where very few cells are activated by an Ag to proliferate and to undergo final differentiation into ASC.

Isotype-specific determinations

As already indicated, the suitability of the ELISPOT assay for the detection of cells secreting specific Ab of a given class or subclass, relies entirely on the immunological specificity of the antiglobulin preparations used as enzyme-labelled indicator reagents. The method has been used to detect cells secreting Ab belonging to the major Ig classes and subclasses and even to detect specific IgE-secreting cells (Sedgwick and Holt, 1983b; Holt *et al.*, 1984b). In the latter instance, modifications of the assay conditions have been described for optimal detection of IgE-secreting cells (see Sedgwick and Holt, 1983b; and this volume chapter 11).

Enumeration of idiotype-positive antibody-secreting cells

Recently, we have found the ELISPOT assay to be particularly promising for the detection of cells producing Ab possessing idiotypic determinants recognized by either polyclonal or monoclonal anti-idiotypic reagents. Particularly informative in this regard have been studies performed with peripheral blood lymphocytes from a patient with a B cell lymphoma associated with production of a monoclonal IgM,k rheumatoid factor (RF) (Koopman *et al.*, 1983). In these studies, a monoclonal IgG,k anti-idiotype (Id) Ab directed against a determinant in or near the binding site of the patient's monoclonal RF was prepared and used to enumerate circulating Id-secreting cells in the patient's blood. As shown in Table 1 there was good agreement between the assay for Id-positive IgM-secreting B cells and parallel assay of total IgM-secreting B

TABLE 1 Detection of circulating idiotype-positive B cells in a patient with B cell lymphoma

Developing antibody	SFC numbers/10^6 cells*		
	Expt 1	Expt 2	Expt 3
Anti-human IgM†	151	99	164
Anti-mouse IgG‡			
● following mouse IgG,k anti-Id§	192	71	132
● following control mouse IgG,k«	0	ND	3

* Numbers represent the mean number of spot-forming cells (SFCs) detected with the developing Ab. Plates were coated with 30 μg/ml of anti-human IgM. The patient was studied on three separate occasions.

† Plates were developed with a 1:750 dilution of biotinylated goat anti-human IgM (Tago Inc.) and HRP-avidin.

‡ Plates were developed with biotinylated goat anti-mouse IgG (Tago, 1:500).

§ Following removal of patient cells from the plates, the plates were then incubated for 2 hr at 20 °C with 50 μg/ml of monoclonal IgG1,k anti-Id directed against the patient's IgM RF prior to the addition of the developing Ab (anti-mouse IgG).

« Following removal of patient cells from the plates, the plates were then incubated for 2 hr at 20 °C with control mouse IgG_1k/50 μg/ml, prior to the addition of the developing Ab (anti-mouse IgG).

cells in the patient's blood. The capacity of human IgG to suppress the number of spots detected with the anti-Id Ab (Table 2) provided corroborative evidence that the assay was indeed capable of detecting Id-positive RF-secreting cells in the patient's blood. These results indicate that the ELISPOT assay can be used to quantitate cells secreting Id-positive Ab of a defined antigenic binding specificity.

TABLE 2 Specificity of the ELISPOT assay for detection of idiotype-positive rheumatoid factor secreting B cells

Developing antibody	Inhibitor*	SFC numbers/10^6 cells
Anti-human IgM	−	164
Anti-human IgM	+	135
Anti-mouse IgG		
● following mouse IgG_1,k anti-Id	−	132
● following mouse IgG_1,k anti-Id	+	29

* Cells were incubated on anti-human IgM-coated plates in the presence (+) or absence (−) of human IgG (30 μg/ml) in the medium. After a 2 hr incubation the plates were developed with either biotinylated goat anti-human IgM (Tago, 1:750) or incubated with monoclonal mouse IgG_1,k anti-Id or control at 50 μg/ml followed by biotinylated anti-mouse IgG (Tago 1:500), and HRP-avidin.

REFERENCES

Anderson, B. (1974). Studies on antibody affinity at the cellular level. *J. Exp. Med.*, **135,** 312.

Ando, D.G., Ebling, F.M., and Hahn, B.H. (1987). Detection of native and denatured DNA antibody forming cells by the ELISPOT assay: a clinical study of NZB/W F1 mice. *Arthritis Rheum.* **29,** 1139.

Baker, P.J., and Stashak, P.W. (1969). Quantitative and qualitative studies on the primary antibody response to pneumococcal polysaccharides at the cellular level. *J. Immunol.*, **103,** 1342.

Bankert, R.B., Mazzaferro, D., and Mayers, G.L. (1981). Hybridomas producing hemolytic plaques used to study the relationship between monoclonal antibody affinity and the efficiency of plaque inhibition with increasing concentrations of antigens. *Hybridoma*, **1,** 47.

Bosman, C., Feldman, J.D., and Pick, E. (1969). Heterogeneity of antibody-forming cells. An electron microscopic analysis. *J. Exp. Med.*, **129,** 1029.

Coons, A.H., Leduc, E.H., and Conolly, J.M. (1955). Studies on antibody production. I. A method for the histochemical demonstration of specific antibody and its application to a study of hyperimmune rabbits. *J. Exp. Med.*, **102,** 49.

Czerkinsky, C. (1986). Antibody-secreting cells, In: *Methods of Enzymatic Analysis* (ed. H.U. Bergmeyer). VCH Verlag, Weinheim, p. 11

Czerkinsky, C., and Svennerholm, A.-M. (1983). Ganglioside GM1 enzyme-linked immunospot (GM1-ELISPOT) assay for simple identification of heat-labile enterotoxin-producing *Escherichia coli. J. Clin. Microbiol.*, **17,** 965.

Czerkinsky, C., Nilsson, L.-Å., Nygren, H., Ouchterlony, Ö., and Tarkowski, A. (1983). A solid-phase enzyme-linked immunospot (ELISPOT) assay for enumeration of specific antibody-secreting cells. *J. Imm. Methods*, **65,** 109.

Czerkinsky, C., Nilsson, L.-Å., Ouchterlony, Ö., Tarkowski, A., and Gretzer, C. (1984a). Detection of single antibody-secreting cells generated after *in vitro* antigen-induced stimulation of human peripheral blood lymphocytes. *Scand. J. Immunol.*, **19,** 575.

Czerkinsky, C., Tarkowski, A., Nilsson, L.-Å., Ouchterlony, Ö., Nygren, H., and Gretzer, C. (1984b). Reverse enzyme-linked immunospot (RELISPOT) assay for detection of cells secreting immunoreactive substances. *J. Imm. Methods*, 72, 489.

Czerkinsky, C., Nilsson, L.-Å., Tarkowski, A., Ouchterlony, Ö., Jeansson, S., and Gretzer, C. (1984c). An immunoenzyme procedure for enumerating fibronectin-secreting cells. *J. Immunoassay*, **5,** 291.

Czerkinsky, C. (1987). *Ph.D. Thesis*, Göteborg University.

Czerkinsky, C., Prince, S.J., Michalek, S.M., Jackson, S., Russell, M.W., McGhee, J.R., and Mestecky, J. (1987). IgA antibody-producing cells in peripheral blood after antigen ingestion: evidence for a common mucosal immune system in humans. *Proc. Natl. Acad. Sci. USA,* **84,** 2449.

De Lisi C. (1976). Hemolytic plaque inhibition: physical chemical limits on use as affinity assay. *J. Immunol.*, **117,** 2249.

Elwing, H., and Nygren, H. (1979). Diffusion-in-gel enzyme-linked immunosorbent assay (DIG-ELISA): a simple method for quantitation of class-specific antibodies. *J. Imm. Methods*, **31,** 101.

Engvall, E., and Perlmann, P. (1972). Enzyme-linked immunosorbent assay, ELISA III. Quantification of specific antibodies by enzyme-labelled anti-immunoglobulin in antigen-coated tubes. *J. Immunol.*, **109,** 29.

Franci, C., Ingles, J., Castro, R., and Vidal, J. (1986). Further studies on the

ELISA-spot technique. Its application to particulate antigens and potential improvement in sensitivity. *J. Imm. Methods*, **88,** 225.
Goidl, E.A., Schrater, A.F., Siskind, G., and Thorbeck, G.J. (1979). Production of auto-antiidiotypic antibody during normal immunue response to TNP-Ficoll. II. Hapten-reversible inhibition of anti-TNP plaque-forming cells by immune serum as assay for auto-anti-idiotypic antibody. *J. Exp. Med.*, **150,** 154.
Heitzman, H., and Richards, R.M. (1974). Use of avidin–biotin complex for staining of biological membranes in electron microscopy. *Proc. Natl. Acad. Sci. USA*, **71,** 3537.
Herzenberg, L.A., Black, S.J., Tokuhisa, T., and Herzenberg, L.A. (1980). Memory B cells at successive stages of differentiation: affinity maturation and the role of IgD receptors. *J. Exp. Med.*, **151,** 1071.
Holt, P.G., Sedgwick, J.D., Stewart, G.A., O'Leary, C., and Krska, K. (1984a). ELISA plaque assay for the detection of antibody secreting cells: observations on the nature of the solid-phase and on variations in plaque diameter. *J. Imm. Methods*, **74,** 1.
Holt, P.G., Cameron, K.J., Stewart, G.A., Sedgwick, J.D., and Turner, K.J. (1984b). Enumeration of human immunoglobulin secreting cells by the ELISA-plaque method: IgE and IgG isotypes. *Clin. Immunol. Immunopthol.*, **30,** 159.
Holt, P.G., Sedgwick, J.D., O'Leary, C., Krska, K. and Leivers, S. (1984c). Long-lived IgE- and IgG-secreting cells in rodents manifesting persistent antibody responses. *Cell. Immunol.*, **89,** 281.
Jerne, K., and Nordin, A.A. (1963). Plaque formation in agar by single antibody-producing cells. *Science*, **140,** 405.
Jerne, K., Henry, C., Nordin, A.A., Fuji, A., Koros, A.M.C., and Lefkovits, I. (1974). Plaque forming cells: methodology and theory. *Transplant. Rev.*, **18,** 130.
Kantele, A., Arvilommi, H., and Jokinen, I. (1986). Specific immunoglobulin-secreting human blood cells after peroral vaccination against *Salmonella typhyi. J. Infect. Dis.*, **153,** 1126.
Kelly, B.S., Levy, J.G., and Sikora, L. (1979). The use of the enzyme-linked immunosorbent assay (ELISA) for the detection and quantification of specific antibody from cell cultures. *Immunology*, **37,** 45.
Koopman, W.J., Schrohenloher, R.E., Barton, J.C., and Greenleaf, E.C. (1983). Suppression of *in vitro* monoclonal rheumatoid-factor synthesis by antiidiotypic antibody. *J. Clin. Invest.*, **72,** 1410.
Lycke, N., Lindholm, L., and Holmgren, J. (1985). Cholera antibody production *in vitro* by peripheral blood lymphocytes following oral immunisation of humans and mice. *Clin. Exp. Immuno.*, **62**, 39.
Lycke, N. (1986). A sensitive method for the detection of specific antibody production in different isotypes from single lamina propria plasma cells. *Scand. J. Immunol.* **24**, 393.
Lycke, N., and Holmgren, H. (1986). Intestinal mucosal memory cells in lamina propria and Peyper's patches of mice 2 years after oral immunization with cholera toxin. *Scand. J. Immunol.*, **23,** 611.
Lycke, N. (1986b) *Ph.D. Thesis*, Göteborg University.
Mestecky, J., Czerkinsky, C., Brown, T.A., Prince, S.J., Michalek, S.M., Russell, M.W., Jackson, S., Scholler, M., and McGhee, J.R. (1986). Human immune responses to *Streptococcus mutans* In: *Molecular Microbiology and Immunobiology of* Streptococcus mutans (eds S. Hamada, *et al.*). Elsevier, Amsterdam, p. 297.
Moskophidis, D., and Lehmann-Grube, F. (1984). The immune response of the mouse to lymphocytic choriomeningitis virus. IV. Enumeration of antibody-producing cells in spleens during acute and persistent infection. *J. Immunol.*, **133**, 3366.

Moskophidis, D., Löhler, J., and Lehmann-Grube, F. (1987). Antiviral antibody-producing cells in parenchymatous organs during persistant virus infection. *J. Exp. Med.*, **165**, 705.

Möller, S.A., and Borrebaeck, C.A.K. (1985). A filter immuno-plaque assay for the detection of antibody-secreting cells *in vitro*. *J. Imm. Methods*, **79**, 195.

Nygren, H., Czerkinsky, C., and Stenberg, M. (1985). Dissociation of antibodies bound to surface-immobilized antigen. *J. Imm. Methods*, **85**, 87.

Pasanen, V.J., and Makela, O. (1969). Effect of the number of haptens coupled to each erythrocyte on haemolytic plaque formation. *Immunology*, **16**, 399.

Pick, E., and Feldman, J.D. (1967). Autoradiographic plaques for the detection of antibody formation to soluble proteins by single cells. *Science*, **156**, 964.

Plotz, P.H., Talal, N., and Asofsky, R. (1968). Assignment of direct and facilitated haemolytic plaques in mice to specific immunoglobulin classes. *J. Immunol.*, **100**, 744.

Russell, M.W., Czerkinsky, C., and Moldoveanu, Z. (1987). Detection and specificity of antibodies secreted by spleen cells in mice immunized with *Streptococcus mutans*. *Infect, Immun.*, **53**, 317.

Sedgwick, J.D., and Holt, P.G. (1983a). A solid-phase immunoenzymatic technique for the enumeration of specific antibody-secreting cells. *J. Imm. Methods*, **57**, 301.

Sedgwick, J.D., and Holt, P.G. (1983b). Kinetics and distribution of antigen-specific IgE-secreting cells during the primary antibody response in the rat. *J. Exp. Med.*, **157**, 2178.

Sedgwick, J.D., and Holt, P.G. (1985). Induction of IgE-secreting cells and IgE-isotype specific suppressor T cells in the respiratory lymph nodes of rats in response to antigen inhalation. *Cell. Immunol.*, **94**, 182.

Sedgwick, J.D., and Holt, P.G. (1986). The ELISA–plaque assay for the detection and enumeration of antibody-secreting cells: an overview. *J. Imm. Methods*, **87**, 37.

Shepart, B.S., Mayers, G.L., and Bankert, R.B. (1985). Failure of PFC inhibition assays to distinguish idiotypically between clonotypes that are distinguishable by RIA analysis. *J. Immunol.*, **135**, 1683.

Sterzl, J., and Riha, I. (1985). Detection of cells producing 7S antibodies by the plaque technique. *Nature* (Lond.), **208**, 857.

Tarkowski, A., Czerkinsky, C., and Nilsson, L.-Å. (1984). Detection of IgG-rheumatoid factor-secreting cells in autoimmune MRL/1pr mice: a kinetic study. *Clin. Exp. Immunol.*, **58**, 7.

Tarkowski, A., Czerkinsky, C., and Nilsson, L.-Å. (1985). Simultaneous induction of rheumatoid factor – and antigen-specific antibody-secreting cells during the secondary immune response in man, *Clin. Exp. Immunol.*, **61**, 379.

Tarkowski, A., Czerkinsky, C., Nilsson, L.-Å., Nygren, H., and Ouchterlony, Ö. (1984). Solid-phase enzyme-linked immunospot assay (ELISPOT) for enumeration of IgG-rheumatoid factor secreting cells. *J. Imm. Methods*, **72**, 451.

Tarkowski, A., Holmdahl, R., Rubin, K., Klareskog, L., Nilsson, L.-Å., and Gunnarsson, K. (1986). Patterns of autoreactivity to collagen type II in autoimmune MRL/1 mice. *Clin Exp. Immunol.*, **63**, 441.

ELISA and Other Solid Phase Immunoassays
Edited by D.M. Kemeny and S.J. Challacombe

CHAPTER 11

ELISA – Plaque Assay for the Detection of Single Antibody-Secreting Cells

Jonathon D. Sedgwick[1] and **Patrick G. Holt**[2]

MRC Cellular Immunology Unit, University of Oxford[1] *and Princess Margaret Hospital, Perth, Australia*[2]

CONTENTS

INTRODUCTION

The determination of antibody levels in serum and secretions is without doubt the most fundamental and important measure of immune status. The requirement in both clinical and experimental laboratories for fast, accurate, and sensitive methods for this purpose has resulted in the development of a range of isotopic-, enzymatic-, and fluorescent-based techniques which are now routinely used to measure antibody levels within the nanogram range.

While such an approach has proved invaluable in furthering our general understanding of the processes which regulate immunoglobulin (Ig) produc-

[1]To whom all correspondence should be addressed.

tion *in vivo*, analysis of the cellular interactions which are ultimately responsible for antibody production has required somewhat more refined and specific methodologies. The advent of the haemolytic plaque-forming cell (PFC) assay (reviewed in Jerne *et al.*, 1974) proved timely in this respect in providing a technique which allowed the direct examination of Ig secretion at the single cell level and, employing this method, many important aspects of Ig production and regulation have been investigated.

The haemolytic PFC assay nevertheless has a number of inherent diadvantages, many of which are related to the obligatory use of lysable erythrocyte targets as the indicator for antibody secretion (Jerne *et al.*, 1974). Limitations include the variation in susceptibility of target erythrocytes to lysis, and the difficulties associated with conjugating antigens to red blood cells. Additionally, the nature of the antibody isotype(s) which contribute to direct plaque formation cannot always be assumed (Wortis, Dresser and Anderson, 1969) and furthermore, these may complicate the enumeration of developed or indirect plaques particularly where relatively few of the latter are present.

Following a number of unsuccessful attempts employing haemolytic PFC assays to study the anatomical location of IgE immunoglobulin-secreting cells (ISC) in hyperimmune mice and rats and to detect IgE-ISC in cultured human peripheral blood leukocytes, it became apparent to us that a technique was required for the detection of ISC, that was more discriminating and sensitive. Employing the well-established principles of the enzyme-linked immunosorbent assay (ELISA) we subsequently devised a technique — the ELISA-plaque (EP) assay — which proved to be highly suited to the detection of IgE-ISC in both rodent and human systems, as well as a variety of other applications which included the detection of antigen-specific and total ISC of isotypes other than IgE.

In this chapter we examine the EP assay in detail. All relevant methodology is included together with a discussion of important technical considerations which are necessary to optimize the system. Some applications of the technique are also presented although this aspect is considered in greater detail elsewhere (Czerkinsky *et al.*, this volume Chapter 10; Sedgwick and Holt, 1986).

BASIC PRINCIPLES AND METHODOLOGY

In many respects, the EP assay (see Figure 1) is little different to the standard liquid ELISA. Thus, where a test serum of unknown titre is added in the latter assay, a suspension of cells is substituted in the former. Following these procedures, both assays rely on the addition of specific antisera coupled to an enzyme/substrate system to reveal the antibody now bound to the solid phase. In the standard ELISA, however, a substrate which produces a soluble product after enzymatic cleavage is employed, while in the EP technique, an insoluble

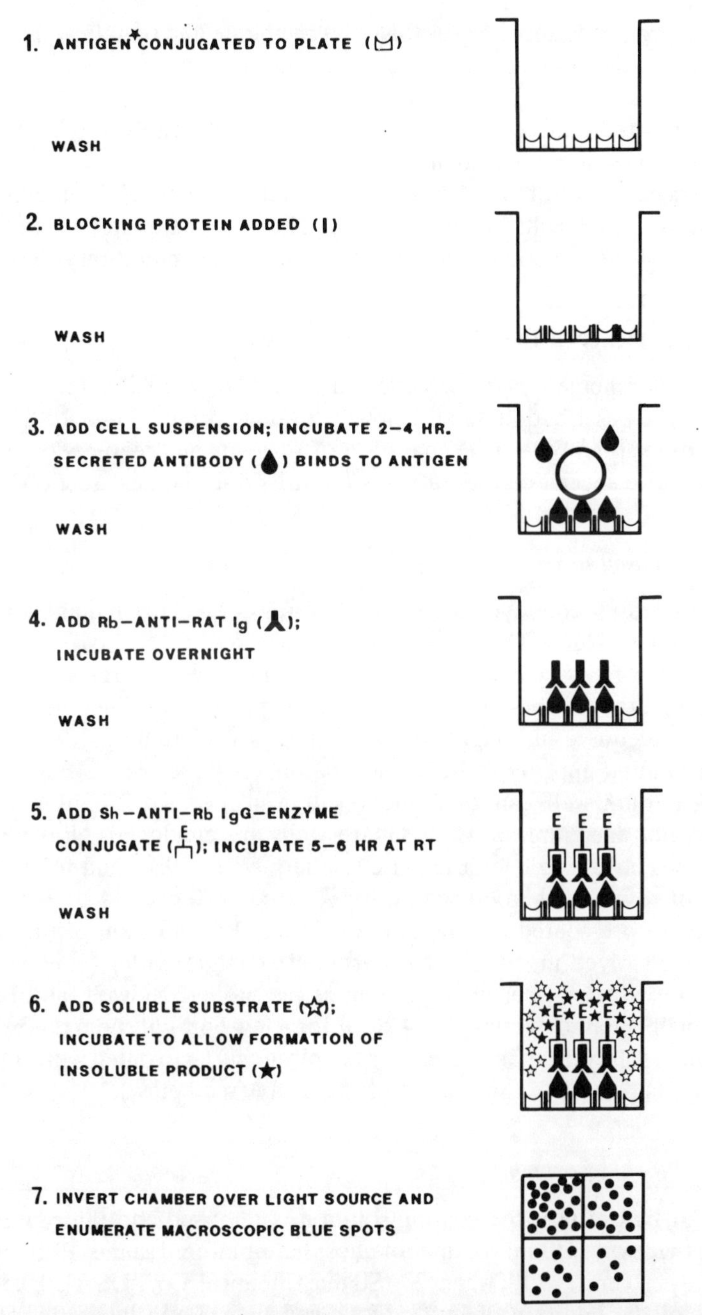

FIGURE 1 The ELISA-plaque assay for enumeration of total or antigen-specific rat ISC
* Specific antisera may be substituted for measurement of total ISC

substrate product is necessary as this will precipitate and remain at the point of formation, thus allowing the detection of discrete zones of bound antibody secreted by a single cell. The technique in principle, is relatively simple to perform but there are certain methodological peculiarities which need to be addressed to optimize sensitivity.

In the second part of this chapter we will examine in detail this methodology and outline the results of experiments which indicate the sensitivity and specificity of the technique for the detection and enumeration of ISCs.

Reaction sequence of the ELISA-plaque technique

The general scheme for the detection of rat ISC by the EP assay is shown in Figure 1 and described in detail below. The same protocol may be applied to species other than the rat. Preparation of reagents and features of the assay which require special consideration are outlined in the next section.

Step 1 Coating the plates

The assay vessels employed are either 24 round well (1.5 cm diameter) tissue culture plates (Nunc, Denmark) or 25 square well (2 × 2 cm) polystyrene replica dishes for bacteriology (e.g. from Sterilin, UK or Greiner, Germany) and it is to either of these plate types that the protocol outlined below refers. Both are adequate although the latter type is preferable as the degree of antigen binding appears to be slightly better on these plates. Furthermore, swirling in round wells can sometimes result in an uneven distribution of cells, a problem not encountered when square wells are employed. Ninety-six-well vinyl plates have also been used (Logtenberg *et al.*, 1986) and found to give consistent results with good sensitivity.

Wells are pre-coated by the addition of 0.5–1.0 ml of antigen or specific antibody dissolved in carbonate–bicarbonate coating buffer (1.59 g Na_2CO_3 plus 2.93 g $NaHCO_3$ per litre, pH 9.6; as per Voller, Bidwell, and Bartlett, 1976) and incubated overnight at 4 °C or for a few hours at room temperature. An optimal solid phase (here meaning the plastic surface coated with antigen or antibody) is central to the success of the assay (see below).

Step 2 Blocking protein

One hour before use, the coating solution is removed and the coated wells washed twice with 1.0 ml volumes of phosphate-buffered saline (PBS) containing 0.05 per cent (v/v) Tween 20 (Sigma Chemical Co., USA). Each wash consists of the addition of buffer, followed by a 3-min interval, before its removal by flicking over a sink: 0.4 ml of a 1.0 mg/ml solution of bovine serum albumin (BSA) in coating buffer is then added to each well and the plates incubated for 1 hr at 37 °C.

Step 3 Addition and incubation of cells

The antigen-coated and BSA-blocked wells are washed twice with 1.0 ml of PBS–Tween followed by a third wash with normal PBS. Slowly, from a graduated pipette or 1 ml Oxford sampler, 1.0 ml of cell suspension in a suitably iso-osmotic PBS or growth medium containing 10 per cent foetal calf serum (FCS) is then added to each well. The number of cells added depends on the size of the wells employed but for a typical 2 × 2 cm or 1.5 cm diameter well, an inoculum that yields between 20 and 150 ISC per well is usually optimal. Above this, the spots become almost confluent and it is then virtually impossible to enumerate accurately the ISC.

After the addition of cells, the plates are covered and placed in a vibration-free incubator at 37 °C for 2 hr (IgG and IgM isotypes) or 32 °C for 4 hr (IgE). The lower incubation temperature for the detection of rat IgE-ISC has been adopted as this was found to be the optimal temperature for this isotype in the standard liquid ELISA. To allow for a possible decrease in the rate of antibody secretion by IgE plasma cells at this lower temperature, the incubation time is extended to 4 hr and this has proved to be optimal for the detection of IgE-ISC, at least in the rat. We have no definite explanation for this observation but it may indicate that the IgE–antigen complex is particularly susceptible to dissociation at the higher temperature.

The incubation times described above seem to be sufficient under most circumstances, and there is little increase in the number of ISC detected if these times are extended.

After the incubation is complete, the plates are immediately emptied by flicking over a sink. Cold PBS–Tween (1.0 ml/well) is quickly added, swirled, and the wells emptied again. Three washes with 2 ml volumes of cold PBS–Tween follow.

Steps 4 and 5 Addition of antisera

All antisera are diluted in PBS–Tween containing 1.0 per cent (w/v) BSA. To each well is added 0.4 ml of an appropriate dilution of the desired anti-rat Ig (monoclonal or polyclonal; e.g. rabbit (Rb)-anti-rat IgG) and the plates incubated overnight at 4 °C. After washing, 0.4 ml of a second affinity-purified antibody (e.g. sheep (Sh)-anti-Rb IgG) conjugated to alkaline phosphatase (AP) is added and the plates again incubated at 4 °C or at room temperature. The AP conjugate is prepared exactly as detailed in Voller, Bidwell and Bartlett (1976).

The optimal working concentration of all reagents should be ascertained as excessive or limiting amounts may impair the sensitivity of the assay (see Table 1). Of paramount importance is the use of adequately absorbed antisera (or preferably monoclonal antibodies), particularly when assessing minor class or subclass ISC.

TABLE 1 Chequerboard titration of reagents for use in the ELISA-plaque assay

Dilution of Rb-anti-rat IgG	OVA concentration used to coat plates (mg/ml)					Dilution of Sh-anti-Rb IgG-AP
	4.0	2.0	1.0	0.5	0.25	
1:1000	60*	68	48	40	20	
1:500	84	84	64	48	28	1:1000
1:100		Non-specific blueing				
1:1000	82	62	86	48	30	
1:500	72	88	72	54	44	1:500
1:100		Non-specific blueing				
1:1000	58	62	70	50	50	
1:500	80	70	80	52	52	1:100
1:100		Non-specific blueing				

Wells were coated with varying concentrations of ovalbumin (OVA) in ELISA coating buffer. One million splenocytes from Brown Norway (BN) rats immunized intraperitoneally (i.p.) 28 days previously with 100 μg OVA + 10 mg aluminium hydroxide (AH) were then added to each well and incubated for 1 hr at 37 °C. Plaques were developed with three different concentrations of Rb-anti-rat IgG (Cappel Labs, USA) and Sh-anti-Rb IgG-AP conjugate, followed by the 5-BCIP substrate.

* Number of OVA-specific IgG-ISC per well.

Figures shown are means of duplicates.

Step 6 Addition of substrate

After incubation with the enzyme conjugate, the wells are again washed three time with PBS–Tween and emptied. The plates are placed on a level surface and 0.5 ml of warm (40 °C) substrate solution added to each well. The preparation and use of the AP substrate utilized here (5-bromo-4-chloro-3-indolyl phosphate; 5-BCIP), which forms a blue, insoluble product following cleavage by AP, is detailed below.

Within 15–30 minutes at room temperature, macroscopic blue spots or ‘plaques’ (Figure 2) become visible and are routinely counted after 1–2 hr. In some cases, where plaques are faint or small in size, the plates can be left overnight to allow further colour development to take place—3 M NaOH may be added to the wells to stop the AP reaction if storage is desired.

Step 7 Enumeration of immunoglobulin-secreting cells

The plates are inverted over a light source (such as an X-ray viewing box), and macroscopic blue spots (putative zones of antibody binding) scored by marking the undersurface with a fine-pointed felt pen. Similar counts are achieved with a projecting microscope. An inverted microscope can be used to distinguish occasional artefacts from zones of putative antibody.

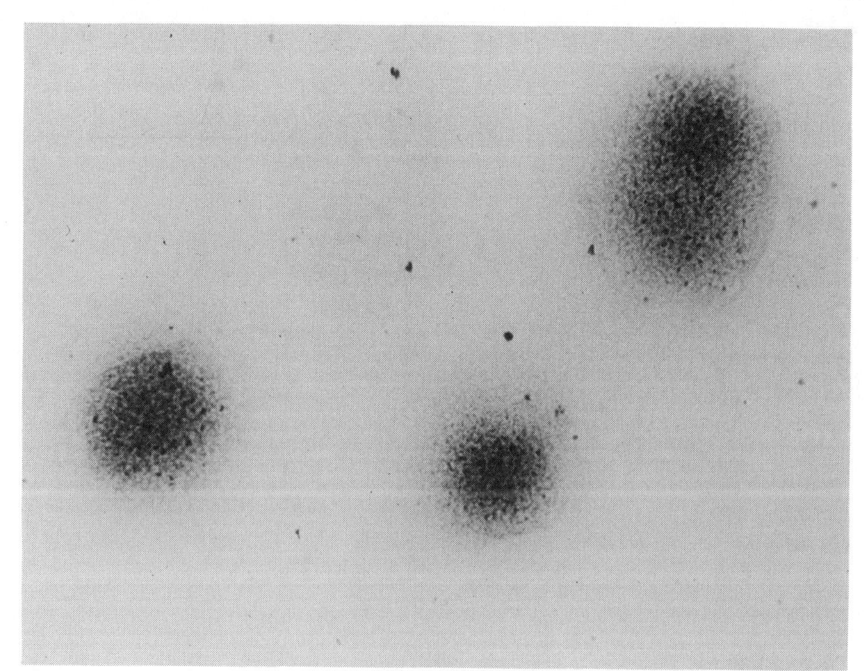

FIGURE 2 Microscopic appearance of ELISA-plaques produced by IgE-ISC. These plaques were developed after initial incubation of parathymic lymph node cells from OVA-immunized rats, on a plate coated with OVA, followed by exposure of the plate to affinity-purified Rb-anti-rat IgE. The granular appearance of the plaques is typical and the pattern is not discernibly different from that produced by cells secreting other antibody isotypes. The isolated small dots exterior to the large plaques are artefacts (see text) (Mag. × 135)

PREPARATION OF REAGENTS AND OPTIMIZATION OF THE SOLID PHASE

Concentration and nature of antigens

One of the unusual features of the EP assay is the requirement for relatively high concentrations of antigen in the coating solution. Initial studies with ovalbumin for example (OVA; Table 1) showed that up to 2 mg/ml of this protein was required for maximal sensitivity. Subsequently it was demonstrated (Table 2) that not all antigens required similarly high coating concentrations and that the amounts necessary for the preparation of an optimal solid phase dependend to a large extent on the size and nature of the antigen. Thus, larger multideterminant antigens such as viruses (Moskophidis and Lehmann-Grube, 1984), *Ascaris suum* extract, thyroglobulin (see also Logtenberg *et al.*, 1986), highly substituted haptenated proteins, or polymerized OVA and BSA, gave better results and at lower coating concentrations, than small, monomeric

TABLE 2 Efficiency of ELISA-plaque detection as a function of antigen coating of the solid phase

Antigen	Coating concentrations (mg/ml)				
	2.0	1.0	0.5	0.25	0.125
OVA*					
Native	0.5	0.42	0.11	0	—»
Monomer	0.41	0.21	0	0	—
Dimer	0.51	0.21	0	0	—
Trimer	0.68	0.13	0.26	0	—
Polymerized	1.00	0.65	0.37	0.20	—
Urea-denatured	0.32	0	0	0	—
BSA†					
Native	0.45	0.33	0	0	—
Polymerized	0.71	1.00	0.94	0.46	—
Ascaris suum‡	0.97	1.00	0.93	0.66	—
DNP-BSA§					
6.6:1	0.67	—	—	—	—
12.5:1	0.61	—	—	—	—
14.8:1	1.00	—	—	—	—
Thyroglobulin«	—	1.00	0.93	1.11	1.16

Data shown are relative numbers of antigen-specific IgG-ISC detected per well, normalized against the maximum figure obtained with each cell preparation. Wells contained 1–2 × 10^6 cells each. Cells were splenocytes from Balb/c mice, immunized 30 days previously with: * 10 μg OVA + 4 mg AH i.p.; † 10 μg BSA + 4 mg AH i.p.; ‡ 10 μg *Ascaris suum* antigen + 4 mg AH i.p.; § 10 μg DNP-OVA + 4 mg AH i.p.; or « PVG/c rats after two subcutaneous injections of thyroglobulin in Freund's complete adjuvant (for details see Holt *et al.*, 1984). Polymerized antigens were prepared as detailed in the text and monomeric, dimeric, and trimeric preparations purified from these by fractionation on a Sephacryl S-300 column (Pharmacia South Seas Pty Ltd). » Not done.

antigens. The somewhat inconsistent results that we have obtained following the coating of plates with native, unfractionated preparations of some antigens (as supplied by the manufacturer) may thus be due to variations in their content of high molecular weight aggregates (Holt *et al.*, 1981) which appear to be the important solid phase antigen in this assay (Table 2).

Polymerization of antigen

It is clear that the sensitivity of the EP assay with a number of antigens is considerably extended if they are polymerized before use (for method see below), although despite these measures the sensitivity of the assay with OVA in particular is still impaired if large amounts, even of the polymerized antigen, are not used and there is no ready explanation for this. We have noted, however, that the coating solutions can be removed and re-used a number of times without a loss in sensitivity suggesting that only a small proportion of the added antigen is actually adsorbed to the plastic surface.

A method for polymerizing antigen which has proved to be quick and simple to perform is outlined below. The product is not fractionated following polymerization so presumably a number of different molecular weight polymers are present.

(1) Dissolve protein (e.g. OVA) in phosphate-buffered saline, pH 7.4 at 20 °C.
(2) While OVA solution is stirring, add glutaraldehyde dropwise to a final molar ratio of 25 : 1 (glutaraldehyde to protein).
(3) Over ensuing 4 hr, continue stirring and repeatedly adjust the pH to 7.5.
(4) Terminate the reaction after 4 hr by adding a 50 molar excess of glycine.
(5) Remove free glycine and glutaraldehyde by extensive dialysis against 0.05 M ammonium bicarbonate buffer or, alternatively, by passage through a Sephadex G-25 column (Pharmacia, South Seas Pty Ltd) equilibrated with the same buffer.
(6) Clarify, e.g. centrifuge, at 4500 × *g* for 15 min.
(7) Lyophilize.

Anti-immunoglobulin reagents

Detection of total ISC by the EP assay is also possible simply by coating the wells with an affinity-purified anti-Ig instead of a specific antigen; however, like large, multideterminant antigens, only low coating concentrations are required. For the detection of total IgE-ISC for example (Table 3), a coating

TABLE 3 Low concentration coating solutions are adequate for the sensitive detection of total ISC

Coating solution	Concentration (μg/ml)	Cell inoculum per well			
		2×10^6	5×10^5	5×10^4	1×10^4
	1	>150*	>150	126	26
Sh-anti-rat IgE	5	>150	>150	119	34
	10	>150	>150	122	28
	20	>150	>150	131	33
OVA	2000	0	0	0	0

Mesenteric lymph nodes were removed from LOU/M rats 14 days after their infestation with 2500 third-stage *Nippostrongylus brasiliensis* larvae. Single-cell suspensions were assayed for total IgE-ISC by coating with affinity-purified Sh-anti-rat IgE at a variety of concentrations, followed by the sequential addition of cells, IgE-specific Rb-anti-rat IgE, Sh-anti-RB IgG-AP conjugate, and 5-BCIP substrate solution (see text for details). Wells coated with polymerized OVA were included as a specificity control.

* Number of IgE-ISC per well.

TABLE 4 Enhanced sensitivity by pre-coating of wells with glutaraldehyde: detection of human IgE-ISC

Total cells per well	Coating concentration of anti-IgE (μg/ml)	Number of ELISA-plaques detected per well		
		ELISA coating buffer	Glutaraldehyde coating	
1000	10	20*	>200*	(>200)†
	25	80	>200	(>200)
500	10	2	36	(84)
	25	44	156	(174)
100	10	1	11	(13)
	25	2	39	(45)
1000	blank‡	1	0	(0)
1000	Anti-IgG§	0	0	(0)

Wells were coated with affinity-purified goat-anti-human (hu) IgE (Tago Inc.) or Sh-anti-hu IgG (Cappel Labs, Inc.) in either standard carbonate–bicarbonate coating buffer or following glutaraldehyde pre-treatment (see text for details). Varying numbers of U266 human IgE myeloma cells (kindly supplied by Professor Herve Bazin, Belgium) were added to the wells followed by the sequential addition of either Rb-anti-hu IgE or Rb-anti-hu IgG, Sh-anti-Rb Ig-AP conjugate, and 5-BCIP substrate solution (see text for details).
* Means of duplicates, enumerated under × 100 magnification.
† Means of duplicates counted without magnification.
‡ Coating buffer only.
§ Coating concentration of 25 μg/ml.

solution containing 1.0 μg/ml of specific antibody is quite suffient. Similarly, total human IgE-ISC can be enumerated on plates coated with low concentrations of affinity-purified anti-human IgE although, as indicated in the data of Table 4, the detection limits can be further improved by treatment of the plastic wells with glutaraldehyde prior to addition of the anti-IgE solution (for method, see below). This procedure has been shown to increase the amounts of Ig bound to the plate, by up to 50 per cent when compared to the coating levels achieved using standard coating buffer only.

Glutaraldehyde pre-treatment of plates

A method for the coating of plates employing glutaraldehyde pre-treatment is detailed below:

(1) Add to plastic wells, 1.0 ml 0.2 per cent (v/v) glutaraldehyde (Sigma, USA) in 0.1 M phosphate buffer, pH 5.0, and leave overnight at 4 °C.
(2) Wash plates three times with 2.0 ml volumes of PBS.

(3) Add an appropriate dilution of affinity-purified anti-Ig in PBS (e.g. 25 μg/ml). Incubate for 3 hr at 37 °C.
(4) Wash plates twice with PBS.
(5) To block any remaining free glutaraldehyde binding sites add 10.0 μg/ml BSA in standard ELISA coating buffer and incubate for 1 hr at 37 °C.
(6) Wash plates twice with PBS–Tween and once with PBS.
(7) Add cells as per standard procedure.

Substrate preparation

Preparation of the 5-BCIP substrate requires the following materials, all of which are available from Sigma Chemical Co. USA:

- 2-Amino-2-methyl-1-propanol (AMP) solution
- Triton X-405
- 5-Bromo-4-chloro-3-indolyl phosphate (5-BCIP; *p*-toluidine salt)
- Agarose (36 °C gelling — Type I)

To prepare one litre of *stock AMP buffer*, proceed as follows:

(1) Warm neat AMP solution to 25–30 °C. It is solid at room temperature.
(2) Dissolve the following in 500 ml distilled water:
 (a) 150 mg $MgCl_2.6H_2O$;
 (b) 100 μl Triton X-405;
 (c) 1.0 g sodium azide.
(3) To this solution, add 95.8 ml (1.0 M) AMP solution with stirring. AMP is slow to dissolve so leave stirring for about 1 hr. Solution may still be cloudy after this time.
(4) Add distilled water to approx. 900 ml and adjust pH to 10.25 with concentrated HCl.
(5) Leave overnight at 20 °C, re-adjust pH to 10.25 with HCl, and add water to 1 litre.
(6) Filter.
(7) Stock buffer will store at 4 °C for at least 3–4 months but may require refiltering with age.

The 5-BCIP substrate solution is prepared on the day of use at a final concentration of 1.0 mg/ml. Higher concentrations (e.g. 2 mg/ml) result in greatly increased background colour and a consequent reduction in the number of detectable plaques.

(1) Take desired amount of 5-BCIP and dissolve in a small volume of stock AMP buffer then make up to correct volume.
(2) Filter the solution through a 0.22 or 0.45 μm filter. This step is to remove

any insoluble material as particles in the substrate may give false spots. The resulting solution is a clear, lightish brown liquid (colour dependent on the purity of the 5-BCIP — the Sigma product is almost colourless).

Substrate in agarose

Substrate may be added to the wells either in the form described above or mixed with agarose. The latter hardens in the wells and provides a solid matrix in which the substrate product may precipitate. Because of this, the colour intensity and size of the plaques is greater than that seen without agarose (by at least 10–20 per cent). There is the disadvantage that further preparation is necessary. To prepare the agarose/substrate mixture, proceed as follows:

(1) In advance, prepare batches of 3.0 per cent (w/v) agarose in distilled water (pre-filter water with a 0.45 μm filter to remove any particles prior to the addition of agarose).
(2) On the day of assay, warm a suitable volume of 5-BCIP substrate solution to 40 °C and melt the agarose.
(3) Allow the agarose to cool to approx. 60–70 °C, then add sufficient 3.0 per cent agarose to the 5-BCIP solution to give a final agarose concentration of 0.6 per cent. Hold the mixture at 40 °C until it is added to the assay plates.

Addition of nitro-blue tetrazolium

A further increase in the sensitivity of this assay may be achieved by the addition of nitro-blue tetrazolium (NBT; Grade III, Sigma, USA) to the 5-BCIP substrate solution (for details see Franci *et al.*, 1986). The reaction sequence involves the AP-mediated conversion of 5-BCIP to the corresponding indoxyl compound which precipitates and tautomerizes to a ketone. The ketone then undergoes oxidation and dimerizes to form an indigo which is responsible for the blue plaques described above. Hydrogen ions are released during dimerization and these can, in turn, reduce NBT salt to the corresponding diformazan which is intensely purple. Thus both a blue and purple product are precipitated at the site of AP activity. NBT should be added to the 5-BCIP substrate at a final concentration of 0.1 mg/ml. Concentrations greater than 0.3 mg/ml are inhibitory (Drs J. Vidal and C. Franci, Barcelona; personal communication).

Specificity of the ELISA-plaque assay

Experiments which attest to the specificity of the EP assay are illustrated in Table 5 where OVA-specific splenocytes were tested for their content of

TABLE 5 Specificity of the ELISA-plaque assay

Anti-IgG*	AP-conjugate†	Coating	Cells	Cell inoculum per well 2.5 × 10^6	1.25 × 10^6	0
+	+	OVA	I	60	36	0
+	−	OVA	I	0	0	0
−	+	OVA	I	0	0	0
+	+	—‡	I	0	0	0
+	+	BSA	I	0	0	0
−	−	BSA	I	0	0	0
+	+	OM	I	5	0	0
−	−	OM	I	0	0	0
+	+	OVA	NI	3	0	0
−	+	OVA	NI	0	0	0

Illustrated are the number of IgG ELISA-plaques detected in each well with the combination of reagenets shown. BN rats were immunized i.p. with 100 μg OVA + 10 mg AH and 21 days later their spleen cells removed (I) and together with splenocytes from non-immune rats (NI) assayed for IgG-ISC. Plates were coated with a 2 mg/ml solution of either OVA, BSA, or ovomucoid (OM; Type III-0, Sigma, USA) followed by a 1 h BSA blocking step.
* 1/1000 Rb-anti-rat IgG.
† 1/1000 Sh-anti-Rb IgG-AP.
‡ Not coated, received 1 h BSA blocking step only.

IgG-ISC under a variety of conditions. The results can be briefly summarized as follows:

(1) The exclusion of splenocytes, or any of the antisera, from the system abrogated development of the plaques.
(2) The use of irrelevant antigen, or no antigen, on the solid phase resulted in the development of less than 10 per cent of the plaques observed when OVA was employed.
(3) Plaque development following the parallel assay of immune versus non-immune cells occurred at a relative frequency of 20 : 1.

These studies therefore, together with the data in Table 4 with human cells, clearly demonstrate the specificity of the EP assay. Like all techniques, however, artefacts can occur and these are mostly in the form of small, intensely stained blue spots either within the agarose layer or attached to the solid phase. In general, their morphology is quite different to the typically much larger and granular ELISA plaque (Figure 2), and hence they are easily distinguished under magnification. The origin of these spots is uncertain but may result from the deposition of substrate product on to particulate impurities in the substrate solution. Removal of such contaminants is therefore desirable

TABLE 6 Frequency of OVA-specific ISC detected with the haemolytic and ELISA-plaque assays

Animal number	ELISA or haemolytic plaques per 10^6 cells			
	Direct PFC*	IgM-ISC†	Indirect PFC*	IgG-ISC†
1	1	2	0	0
2	6	18	205	212
3	0	4	194	105
4	10	5	166	140
5	7	6	12	23
6	6	50	1020	4375
7	7	13	176	474

Splenocytes from six OVA-immune BN rats (Nos. 2–7) of various immune status as determined by their levels of serum anti-OVA IgG, and one non-immune BN rat (No. 1) were examined for their content of OVA-specific IgM and IgG-ISC by both the haemolytic plaque-forming cell (PFC) assay* and EP assay†. Cells were incubated at 37 °C for 45 minutes in the PFC assay and at 37 °C for 1 h in the EP assay. PFC assays were performed exactly as described by Cunningham and Szenberg (1968) employing an optimal concentration of Rb-anti-rat IgG to detect indirect plaques and fresh OVA and sheep-erythrocyte-absorbed guinea pig serum as a source of complement. OVA was conjugated to sheep erythrocytes as described by Rector *et al.* (1980).

(see above, 'Substrate preparation'). Alternatively, adherent cells with endogenous AP activity (such as activated B cells) may be responsible.

Sensitivity of the ELISA-plaque assay

Antigen-specific IgG- and IgM-ISC: ELISA v. haemolytic plaque assay

The results of experiments comparing the sensitivity of the ELISA and haemolytic plaque techniques are shown in Tables 6 and 7. In Table 6, splenocytes from seven animals of differing immune status were analysed for their IgM- and IgG-ISC content on an individual animal basis. The data show comparable sensitivity for IgG with both methods, and indicate slightly increased sensitivity for IgM-ISC with the EP assay. In other similar experiments (not shown), where the cells were incubated in the EP assay for 2 hr rather than 1 hr, sensitivity for the IgM isotype increased still further, although the sensitivity of IgG-ISC detection was not substantially improved. The lower sensitivity of the haemolytic assay for the IgM isotype may indicate some heterogeneity in the potential of this antibody class to fix complement (Hoyer *et al.*, 1968). Additionally, the EP assay may reveal low affinity antibody which is not detected by the haemolytic technique (Peterfy, Kuusela and Makela, 1983).

TABLE 7 Detection of total human IgG-ISC by the haemolytic and ELISA-plaque assays

PWM (μl)	Subject 1	Subject 2
	Number of haemolytic plaques/10^6 cells	
0	<10	<10
5	1 050	1 350
10	5 200	2 550
	Number of ELISA-plaques/10^6 cells	
0	<10	<10
5	5 100	2 300
10	13 000	4 200

Peripheral blood leukocytes (PBL) from two subjects were prepared by Ficoll-Hypaque separation and cultured for 4 to 5 days in 24-well Nunc tissue-culture dishes. Wells contained PBL in 1.0 ml RPMI-1640 + 10 per cent FCS, with or without added pokeweed mitogen (PWM; Difco, USA). After culture, the cells were assayed for total IgG-ISC content by the ELISA- or haemolytic plaque assays. The EP assay for IgG-ISC was performed exactly as described in the legend to Table 4 where cells were incubated for 2 h at 37 °C on plates coated with affinity-purified Sh-anti-hu IgG. The protein A haemolysis-in-liquid plaque assay (Jones, 1981) employed a mixture of protein-A-coated sheep erythrocytes, cultured PBL, Rb-anti-hu IgG, and fresh guinea pig serum absorbed with sheep erythrocytes and protein A, as a source of complement. Cells were incubated for 1 h at 37 °C.

Total human IgG-ISC: ELISA v. protein A haemolytic plaque assay

To compare the ELISA and haemolytic plaque assays for measurement of total ISC, human peripheral blood leukocytes from two subjects were cultured for four to five days in the presence of pokeweed mitogen then assayed for numbers of total IgG-ISC (Table 7). These data again indicate that the EP assay was more sensitive than the haemolytic assay although in this instance it was the IgG isotype only which was assessed and hence the data contrast with that in Table 6 where the IgG isotype was detected equally well by both techniques. The use of protein A as a bridge to attach secreted IgG to the sheep erythrocytes in the experiments in Table 7 may account for this difference as it is possible that not all IgG subclasses were bound with equal efficiency to protein A. Additionally, human IgG-ISC ELISA-plaques were scored under ×100 magnification while the results shown in Table 6 were obtained without this aid.

Human ISCs are not as readily detected as those from hyperimmunized animals: in the former, plaques are often small which may possibly reflect a relatively slow rate of Ig secretion by human peripheral blood plasma cells. Others (Logtenberg *et al.*, 1986) have employed cell incubation times of 18 hr for human ISCs. An additional complication is that human ELISA-plaques may be accompanied by non-specific spots which can give falsely high counts

(Table 4). It is advisable, therefore, to check for the presence of these artefacts by examining the plaques under magnification and enumerating only those with the classical granular appearance demonstrated in Figure 2.

Antigen-specific IgE-ISC: ELISA v. the HACA assay

The sensitivity of the EP assay for the detection of antigen-specific IgE-ISC was examined by comparing it with the heterologous adoptive cutaneous anaphylaxis (HACA) assay. The latter is an *in vivo* technique which involves the injection of putative IgE-ISC intradermally into the backs of shaved albino rats. After a 24–48 hr latent period, specific antigen together with Evan's blue dye is injected i.v. and the diameter of the resultant blue spots which appear at the sites of cell injection is taken as a semi-quantitative measure of the number of IgE-ISC in the original cell inoculum. Employing splenocytes and draining lymph node cells from BN rats injected 9 days previously with OVA + AH i.p., it was evident in both assays that the draining lymph nodes, but not the spleen, contained appreciable numbers of OVA-specific IgE-ISC, a result which confirmed the specificity of the EP assay for this isotype also (data not shown; see Sedgwick and Holt, 1983). However, the sensitivity of the HACA assay was considerably less than the EP assay as in the former at least 2.5×10^6 lymph node cells were required to attain diameter values above that obtained with non-immune cells. The addition of only 5×10^5 lymph node cells to the EP assay, in contrast, revealed the presence of 50 OVA-specific IgE-ISC.

Variability in plaque size

Considerable variation in the size of plaques is a general feature of the EP assay and such variability can be seen in Figure 2 (albeit a relatively small difference in this particular case). Investigations into this effect (see Holt *et al.*, 1984 for details) have demonstrated a clear trend towards larger plaques in animals undergoing a secondary response to OVA compared with plaques from animals following a single (primary) challenge with this antigen. Furthermore, the size of OVA-specific IgG-ISC ELISA plaques has been shown to correlate closely with the affinity of anti-OVA IgG isolated from serum of the same animal where high affinity antibody is associated with larger plaques and low affinity antibody with smaller plaques.

Further competitive inhibition studies are clearly required to demonstrate conclusively the role of antibody affinity in this context. Nevertheless, the data effectively rule out the possibility that covert differences in the density of antigen coating of the solid phase are responsible for the variability in plaque size as such effects would be expected to occur randomly.

APPLICATIONS OF THE ELISA-PLAQUE ASSAY

In the preceding pages, a number of applications of the EP assay were detailed to provide examples of some important methodological aspects of the technique. In Tables 3, 4, and 7, for example, total IgG and IgE-ISC were enumerated in rat and human experimental systems while in Tables 5 and 6, the specific and sensitive detection of antigen-specific IgG and IgM-ISCs in immunized rats was illustrated. The EP assay has, however, many potential applications in addition to those described above, although it is perhaps particularly applicable to two situations: (1) where the antigen is difficult to couple to red cells. Included here are crude antigen preparations such as *Ascaris suum* extract (see Table 2), dinitrophenyl (Table 2), or some viruses (Moskophidis and Lehmann-Grube, 1984); and (2) where detection of ISC of a minor Ig class or subclass is required. With respect to this latter application, a particularly relevant example is the detection and enumeration of antigen-specific IgE-ISC. This isotype may represent less than 10 per cent of the total number of specific ISC in an immunized animal (Teale, Liu and Katz, 1981) and thus, in a haemolytic PFC assay, cannot be easily and accurately enumerated against a potentially much larger background of direct plaques.

Below are two examples of the application of the EP assay to the detection of IgE-ISC which clearly illustrate the versatility of the technique for the enumeration of cells secreting this class of antibody. Additionally, a novel application is described which will allow the detection, in the same cell inoculum, of antigen-specific ISC of more than one antibody class.

Enumeration of cells secreting the IgE isotype

Organ location of total- and antigen-specific IgE-ISC in the rat

Because of the inherent difficulties associated with the detection of IgE-ISC, issues as basic as the nature of primary sites of IgE synthesis remain controversial. Prior to the application of the EP assay to this problem, the actual locality of IgE-ISC in parenterally immunized rats for example, was not defined although early data by Mota (1966) suggested that the spleen was not involved. Employing the EP technique (Figure 3) we have confirmed Mota's prediction and shown that the bulk of IgE secretion in i.p. immunized rats occurs, not in the spleen, but in the parathymic lymph nodes which drain the peritoneum. IgG-ISC and IgM-ISC, in contrast, are found in both the draining lymph nodes and the spleen. Furthermore, after a secondary challenge with antigen, the splenic content of IgG-ISC increases still further, while the IgE-ISC numbers remain almost static and these values are roughly reflected in the serum content of each class of antibody.

In addition to mounting substantial antigen-specific IgE responses to exogenous immunogens such as OVA, high-IgE-responder rats (such as the BN) also manufacture large amounts of non-specific IgE which can be detected in the serum (see Table 8 for details). In contrast, other strains such as the Wistar Albino Glaxo (WAG) strain do not maintain these levels of IgE in the circulation. The origin of this antibody is unclear although it has been suggested that the Peyer's patches in the gut (Durkin, Bazin and Waksman, 1981) and the mesenteric lymph nodes (Waksman and Ozer, 1976) may be important in this regard. Employing the EP assay in a direct examination of the lymph nodes draining the gut and respiratory tract for their content of total (non-antigen-specific) IgE-ISC (Table 8), it was clearly demonstrated that in BN rats, the lymph nodes draining the respiratory tract comprised the major source of mature IgE-ISC and not the mesenteric lymph nodes or Peyer's patches in the gut, although the latter were also a rich source of these cells compared with other organs. Taken together, the lymph nodes which drain these two mucosal sites, and not the spleen, bone marrow or blood, were the predominant sites for IgE-secreting plasma cells in the rat. The WAG rat, in contrast, contained only minimal numbers of IgE-ISC in the lymph nodes of the respiratory tract.

The data from these two experiments are important as for the first time it can be clearly stated that IgE secretion in the rat, whether antigen specific (Figure 3) or background (Table 8), principally occurs in the draining lymph nodes. Moreover, based on this premise, it is apparent that environmental antigen at the mucosal surface may provide the stimulus for the generation of background IgE-ISC. The IgG isotype, in contrast, is secreted from a number of sites including the spleen, and other results (not shown) have demonstrated high numbers of IgG but not IgE-ISC, in the bone marrow also.

Immunoprinting

Sequential readout of cells secreting different antibody isotypes on individual cell monolayers

A novel application of the EP assay is illustrated in Figure 4 where, instead of adding cells to the solid phase, the solid phase is added to the cell monolayer. This variation allows the application of a number of different solid phases and, thus, the detection in a given cell inoculum of:

(1) More than one class of antigen-X-specific ISC by the sequential addition of different solid phases, each coated with X; or
(2) more than one class of total ISC by the sequential addition of different solid phases, each coated with a different anti-Ig; or
(3) ISC of more than one antigenic specificity by the sequential addition of different solid phases, each coated with a different antigen.

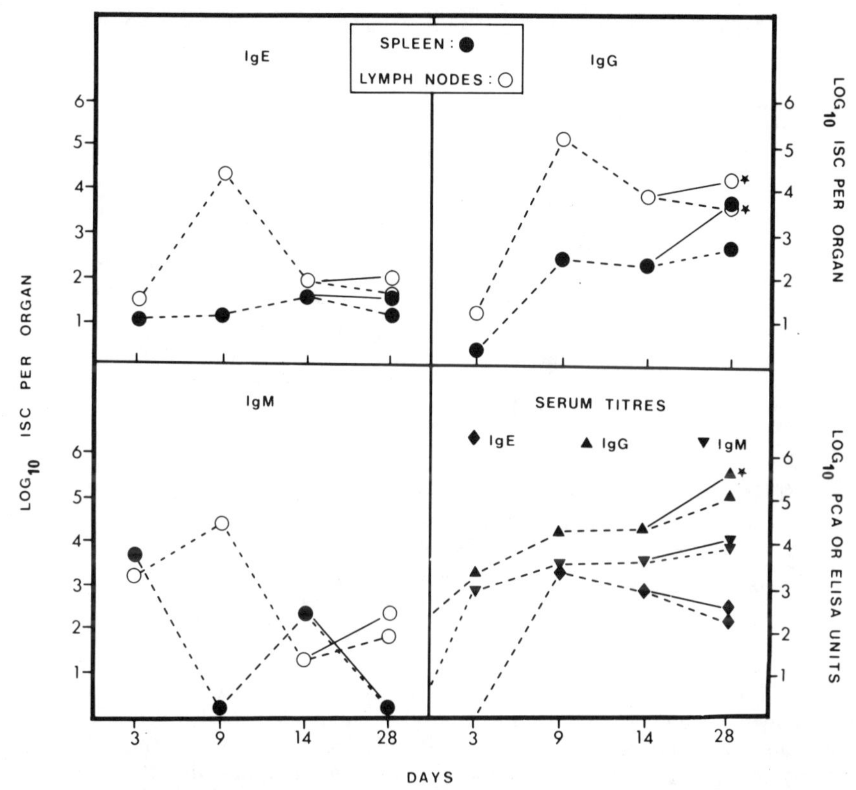

FIGURE 3 The kinetics and distribution of IgE-ISC in parenterally immunized rats. Two groups of 4 BN rats were immunized i.p. on day 0 with 100 μg OVA + 10 mg AH, and the primary anti-OVA response of individual animals from one group followed up to day 28 (- - -). The second group was given a secondary challenge with OVA + AH on day 14 (—). The data represent the mean number of anti-OVA IgE, IgG, and IgM-ISC found in the spleen and parathymic lymph nodes of these animals, together with mean serum anti-OVA IgE (determined by passive cutaneous anaphylaxis, PCA; Ovary, (1964) and IgG and IgM titres (determined by ELISA; see Sedgwick and Holt, 1985) detected at the times shown

Represents significant secondary responses (t-test, $p < 0.01$) comparing secondarily challenged with non-challenged animals

In this technique (for details, see legend to Figure 4) cells are first adhered to the bottom of the plastic wells with poly-L-lysine, an agent which does not appear to adversely affect the viability of the cells, even over an extended time period. Coated plastic cover-slips are then added and the cells incubated for 1–2 hr to allow time for antibody secretion to occur. The cover-slip is then removed, replaced with a second, and the process repeated up to four time.

In a number of studies employing both an anti-DNP IgE secreting hybri-

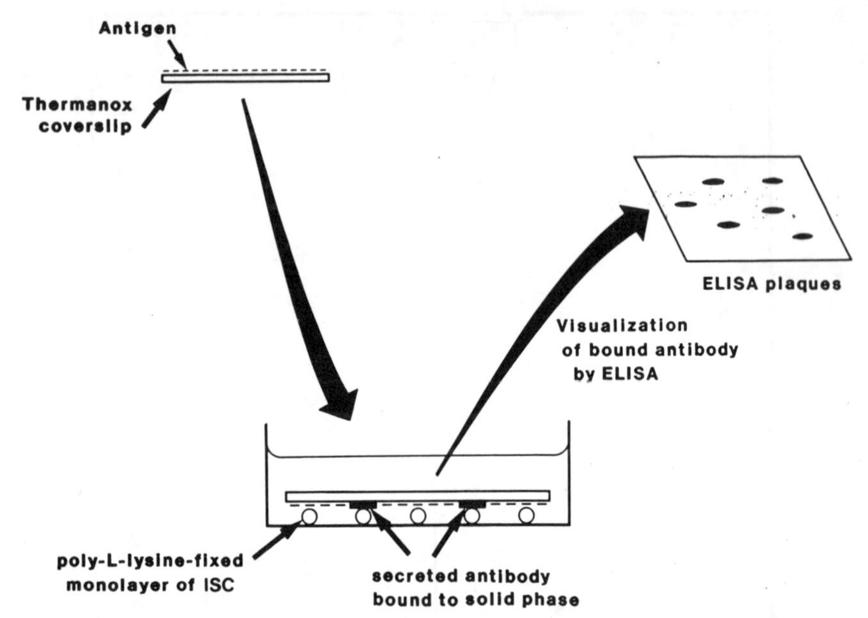

FIGURE 4 Immunoprint modification of the ELISA-plaque assay. Eight-well multi-plates (Lux, USA; well size 26 × 33 mm) are pre-coated with 1 mg/ml poly-*l*-lysine (Sigma, USA) in PBS for 60 min at 37 °C. After washing with PBS–Tween and RPMI-1640 medium + 10 per cent FCS, 2 ml of putative ISC in RPMI/FCS are gently added and incubated for 90 min at 37 °C. After incubation, the fixed cell monolayers are washed carefully with RPMI-FCS and 2 ml fresh medium added back to the wells. Thermanox cover-slips (Lux, USA; 24 × 30 mm), pre-coated with antigen or affinity-purified antibody as previously described, are then lowered into the wells and the plates incubated for 2 hr at 37 °C. This process is then repeated as required with new coated cover-slips. Immediately after removal from the wells, each cover-slip is washed three times in PBS–Tween and incubated with appropriate developing reagents as described in the text

doma and hyperimmune splenocytes from OVA-immunized animals, it has been shown, employing DNP- or OVA-coated cover-slips, respectively, that at least two consecutive cycles of immunoprinting may be taken from an individual ISC monolayer, before the detection efficiency (relative to a standard EP assay) falls below 90 per cent, and that a third cycle still detects in excess of 75 per cent of ISC in the cell inoculum.

In principle, therefore, at least two and possibly three distinct ISC variants may be determined in the one monolayer. Furthermore, by superimposing the cover-slips over the stained or otherwise specifically labelled cell monolayer in the orientation originally employed while immunoprinting, additional information may possibly be obtained, particularly with respect to the ISC itself.

TABLE 8 Total IgE-ISC in rats of high (BN)- and low (WAG)- IgE-responder phenotype

Cell source	Total IgE-ISC per 10^7 mononuclear cells (per organ)	
	BN*	WAG*
Respiratory tract lymph nodes	1252 (4726)	11 (10)
Spleen	53 (1120)	—†
Mesenteric lymph nodes	146 (1466)	—
Peyer's patches	10 (9)	—
Bone marrow	0 (–)	—
Peripheral blood leukocytes	4 (–)	—

Groups of three BN and WAG rats were sacrificed, and lymphoid cells obtained from a variety of sources. Total IgE-secreting cells were then determined on pooled samples, employing the EP assay. The respiratory tract lymph nodes consisted of the superficial cervical, internal jugular, posterior mediastinal, and parathymic groups. Blood leukocytes were obtained as a buffy coat following centrifugation of heparinized blood.
* Mean total IgE levels detected by a solid phase radioimmune assay in serum from 33 adult BN and WAG rats was 21 492 and 157 ng IgE/ml serum respectively (data kindly provided by Assoc. Prof. K.J. Turner, Perth, Western Australia).
† Not determined.

CONCLUSION

Following recent advances in the methodology employed for the measurement of antibody levels in solution, it was perhaps inevitable that the technologies encompassed by ELISA would find application elsewhere. The assay described in this chapter is such an example.

In providing as it does a sensitive and specific method for the detection of ISC without the technical constraints of haemolytic PFC assays, the EP assay should in time become the technique of choice for the determination of numbers of ISC in a given cell population. Moreover, it is not unrealistic to envisage the use of the EP assay in the detection of cells which secrete biologically active molecules other than immunoglobulin. For example, it is now feasible in principle, to detect and enumerate cells secreting factors such as λ interferon and interleukin 1, simply by applying the appropriate cells to plates coated with antibody specific for these molecules. The recent development and increasing availability of monoclonal antibodies for these and other immunologically important factors, now means that such investigations may be pursued.

ACKNOWLEDGEMENTS

The authors would like to thank Dr G.A. Stewart (Perth, WA) for assistance with, and advice on, immunochemical procedures, Mr Stan Buckingham

(Oxford, UK) and Catherine Lee (Oxford, UK) for preparing photographs, and Mrs Caroline Griffin (Oxford, UK) for typing the manuscript. JS is currently supported by a post-doctoral Fellowship from the Multiple Sclerosis Society of Great Britain and Northern Ireland.

REFERENCES

Cunningham, A.J., and Szenberg, A. (1968). Further improvements in the plaque technique for detecting single antibody-forming cells. *Immunology*, **14,** 599.

Durkin, H.G., Bazin, H., and Waksman, B.H. (1981). Origin and fate of IgE-bearing lymphocytes. I. Peyer's patches as differentiation site of cells simultaneously bearing IgA and IgE. *J. Exp. Med.*, **154,** 640.

Franci, C., Ingles, J., Castro, R., and Vidal, J. (1986). Further studies on the ELISA-spot technique: Its application to particulate antigens and a potential improvement in sensitivity. *J. Imm. Methods*, **88,** 225.

Holt, P.G., Rose, A.H., Batty, J.E., and Turner, K.J. (1981). Induction of adjuvant-independent IgE responses in inbred mice: primary, secondary and persistent IgE responses to ovalbumin and ovomucoid. *Int. Arch. All. Appl. Immun.*, **65,** 42.

Holt, P.G., Sedgwick, J.D., Stewart, G.A., O'Leary, C., and Krska, K. (1984). ELISA–plaque assay for the detection of antibody-secreting cells: observations on the nature of the solid phase and on variations in plaque diameter. *J. Imm. Methods*, **74,** 1.

Hoyer, L.W., Borsos, T., Rapp, H.J., and Vannier, W.E. (1968). Heterogeneity of rabbit IgM antibody as detected by C′1a fixation. *J. Exp. Med.*, **127,** 589.

Jerne, N.K., Henry, C., Nordin, A.A., Fuji, H., Koros, A.M.C., and Lefkovits, I. (1974). Plaque forming cells: Methodology and theory. *Transplant. Rev.*, **18,** 130.

Jones, B.M. (1981). Clinical evaluation of B cell and T-regulator cell function using a protein A haemolytic plaque assay. *Clin. Exp. Immunol.*, **46,** 196.

Logtenberg, R., Kroon, A., Gmelig-Meyling, F.H.J., and Ballieux, R.E. (1986). Production of anti-thyroglobulin antibody by blood lymphocytes from patients with autoimmune thyroiditis, induced by the insolubilized antoantigen. *J. Immunol.*, **136,** 1236.

Moskophidis, D., and Lehmann-Grube, F. (1984). The immune response of the mouse to lymphocytic choriomeningitis virus. IV. Enumeration of antibody-producing cells in spleens during acute and persistent infection. *J. Immunol.*, **133,** 3366.

Mota, I. (1966). On the site of rat reagin formation. *Immunology*, **11,** 137.

Ovary, Z. (1964). *Immunological Methods*, CIOMS Symposium (ed J.F. Ackroyd). Blackwell Scientific Publications, Oxford, p. 259.

Peterfy, F., Kuusela, P., and Makela, O. (1983). Affinity requirements for antibody assays mapped by monoclonal antibodies. *J. Immunol.*, **130,** 1809.

Rector, E.S., Lang, G.M., Carter, B.G., Kelly, K.A., Bundesen, P.G., Bottcher, I., and Sehon, A.H. (1980). The enumeration of mouse IgE-secreting cells using plaque-forming cell assays. *Eur. J. Immunol.*, **10,** 944.

Sedgwick, J.D., and Holt, P.G. (1983). Kinetics and distribution of antigen-specific IgE-secreting cells during the primary antibody response in the rat. *J. Exp. Med.*, **157,** 2178.

Sedgwick, J.D., and Holt, P.G. (1985) Down-regulation of immune responses to inhaled antigen: studies on the mechanism of induced suppression. *Immunology*, **56,** 635.

Sedgwick, J.D., and Holt, P.G. (1986). The ELISA–plaque assay for the detection and enumeration of antibody-secreting cells: An overview. *J. Imm. Methods*, **87,** 37.

Teale, J.M., Liu, F.-T., and Katz, D.H. (1981). A clonal analysis of the IgE response and its implications with regard to isotype commitment. *J. Exp. Med.*, **153,** 783.

Voller, A., Bidwell, D., and Bartlett, A. (1976). *Manual of Clinical Microbiology* (eds N.R. Rose and W. Friedman). Amer. Soc. Microbiology. Washington, DC, P. 506.

Waksman, B.H., and Ozer, H. (1976). Specialized amplification elements in the immune system. *Prog. Allergy*, **21,** 1.

Wortis, H.H., Dresser, D.W., and Anderson, H.R. (1969). Antibody production studied by means of the localized haemolysis in gel (LHG) assay. III. Mouse cells producing five different classes of antibody. *Immunology*, **17,** 93.

ELISA and Other Solid Phase Immunoassays
Edited by D.M. Kemeny and S.J. Challacombe

CHAPTER 12

Chemiluminescence Immunoassay

Ian Weeks and **J.S. Woodhead**
University of Wales College of Medicine, Cardiff

CONTENTS

INTRODUCTION

Currently there exists a plethora of immunochemical techniques for the specific quantitation of a wide variety of analytes, particularly those of biological interest. The range of analytes measurable using such techniques is incredibly diverse both in terms of molecular complexity and molar concentration. The former consideration does not impose restrictions on the type of immunochemical system used for quantitation. For example, both ferritin (MW $>$ 650 000) and dinitrobenzene (MW = 168) can be quantified, using competitive binding radioimmunoassay (RIA). Further, albumin can be quantified using immunoelectrophoresis or enzyme-linked immunosorbent assays (ELISA). In such situations the choice of method will depend on the whim of the investigator or, more importantly, on the superior precision or technical convenience that one method in particular may exhibit.

Of more fundmental importance in the choice of immunoassay system is the sensitivity of detection of the method and the ability to quantify precisely and accurately the analyte within the relevant concentration range.

In the clinical laboratory there is a requirement for high sensitivity assays for many analytes, e.g. polypeptide hormones. In this situation there are other considerations to be borne in mind such as requirements for assay robustness, technical simplicity, ease of quality assurance and preference for stable, non-radioactive reagents. Here then the number of options is considerably diminished and has stimulated research into the development of suitable non-isotopic immunoassay systems. Few of the systems described to date are suitable for routine application in the clinical laboratory and only enzyme amplification, time-resolved fluorescence and chemiluminescence assays, exhibit the necessary level of performance to yield high sensitivity immunoassays. Of these methods only certain types of chemiluminescent immunoassay can offer improved performance over radioassays without the need for complex instrumentation or reagent chemistries.

CHEMILUMINESCENCE

The phenomenon of chemiluminescence is widespread though this is rarely appreciated since many forms of the emission are weak and not observed by the naked eye. Familiarity is thus more with bioluminescence such as is observed in the firefly. Further, confusion frequently arises between the term chemi-

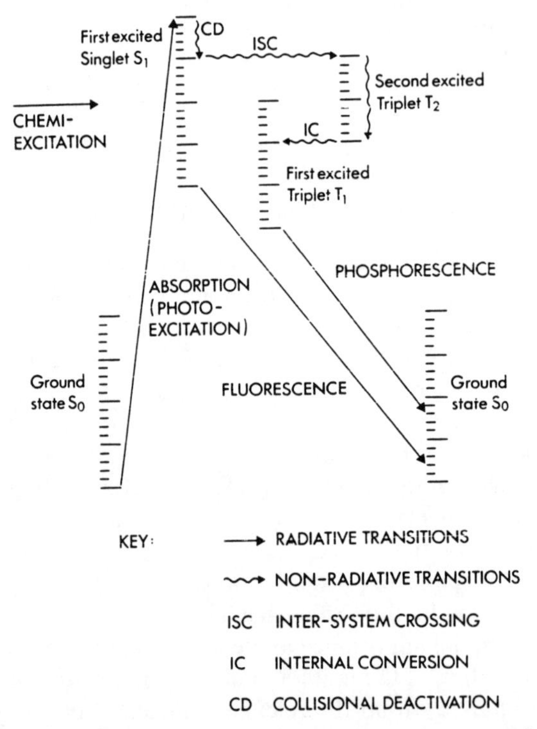

FIGURE 1 Jablonski diagram. Chemiluminescence involves formation of excited states as the products of a highly energetic chemical reaction

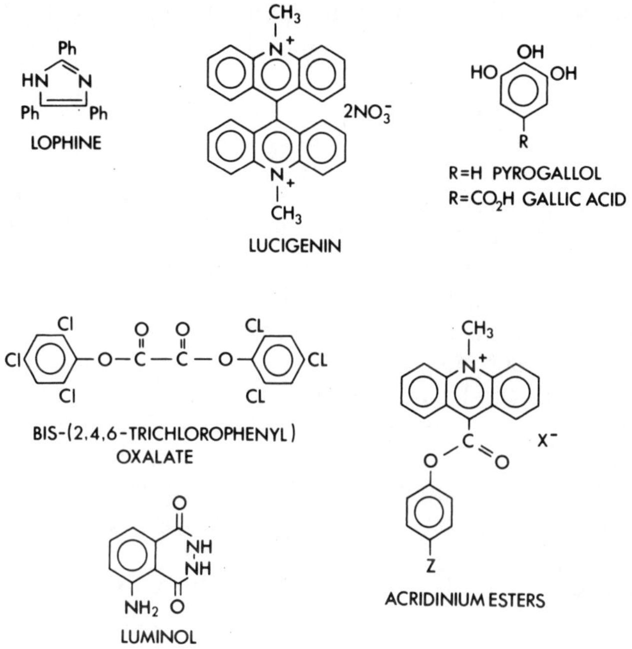

FIGURE 2 Chemiluminescent molecules

luminescence and other commonly encountered forms of luminescence, namely fluorescence and phosphorescence. All types of luminescence involve the transition of excited atomic or molecular states to a ground state. In relaxing from a highly energetic state to a stable state, energy must be dissipated and in luminescence it is emitted as photons (Figure 1). The wavelength of emission need not correspond to the visible region of the spectrum but may also be generated in the infra-red or ultra-violent regions.

The terminology of luminescence phenomena derives from the way in which the excited states are produced. Hence fluorescence and phosphorescence, both of which involve excited state production by absorption of light by a particular chemical species, are examples of photoluminescence. Therefore, it

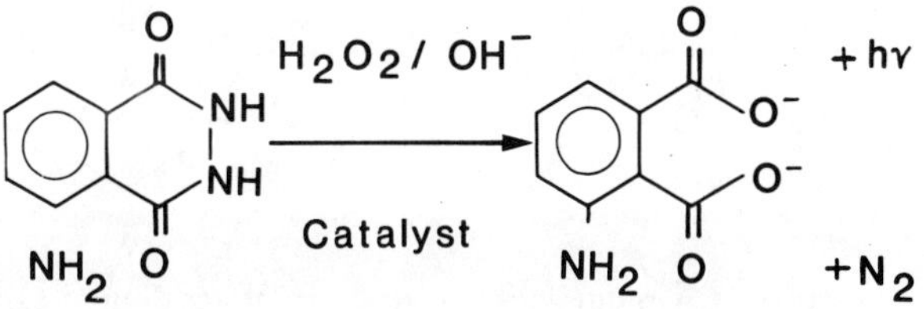

FIGURE 3 Chemiluminescent reaction of luminol

follows that chemiluminescence involves the production of excited states as the result of a chemical reaction. The emission of heat during chemical reactions (approximately 1–10 kJ mol^{-1}) is well known in the context of exothermic reactions where vibrationally excited states are involved. The emission of visible light (> 100 kJ mol^{-1}) requires the initial generation of electronically excited states which, if done chemically, requires a highly energetic reaction usually involving energy 'channelling' in a highly coordinated fashion. In short, chemiluminescence is the emission of light from a chemical reaction which produces product molecules in an electronically excited state.

Several classes of molecule are capable of exhibiting 'bright' chemiluminescence (Figure 2) and many have been extensively studied. The first synthetic chemiluminescent compound to be studied was luminol (Albrecht, 1928) after which the bis-acridinium molecule, lucigenin, became the subject of much study (Gleu and Petsch, 1935).

The chemical conditions which cause these various molecules to undergo chemiluminescent reactions vary considerably and may be formulated in either organic solvent or aqueous systems. Under aqueous conditions the phthalhydrazide compounds (luminol, isoluminol, etc.) require a catalyst, hydrogen peroxide, and alkaline pH for high intensity light emission (Figure 3). A wide variety of chemical species are capable of catalysing the reaction, from simple transition metal cations to macromolecules such as horseradish peroxidase. The emitting species is the dicarboxylate anion though there is still some doubt as to the nature of the reaction mechanism (McCapra, 1974).

The acridinium species, e.g. aryl acridinium esters and lucigenin, have no catalytic requirement and undergo a chemiluminescent reaction in the presence of dilute alkaline hydrogen peroxide. In the case of lucigenin, emission can be seen at two wavelengths depending upon physical conditions. The

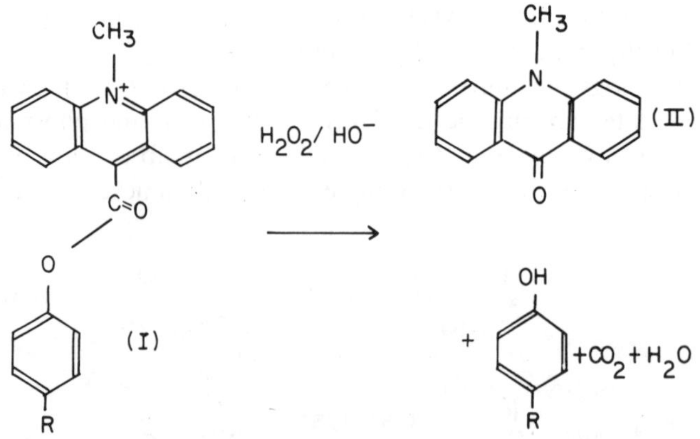

FIGURE 4 Chemiluminescent reaction of acridinium esters

primary emitter is *N*-methylacridone which emits photons in the blue region of the spectrum but in certain circumstances a green emission is observed due to energy transfer to the native lucigenin molecule. The mechanism of the chemiluminescent reaction of aryl acridinium esters involves a concerted multiple bond cleavage via a dioxetanone intermediate to yield an electronically excited molecule of *N*-methylacridone (Figure 4) which subsequently relaxes to its ground state with the emission of photons at 430 nm.

Low levels of chemiluminescence can be precisely measured using photon counters (luminometers), several of which are now commercially available. It is well established that digital photon counting offers advantages over photocurrent measurement at low luminescent intensity. A wide range of digital instruments is available from Laboratorium Prof. Dr. Berthold, Wildbad, FRG.

Chemiluminescent reaction mechanisms in immunoassay

In common with other non-isotopic immunoassay probes the behaviour of chemiluminescent molecules is dependent on their chemical environment. This fact means that it is important to appreciate the possible effects that species present in the biological matrix may have on quantitation. For immunoassays employing solid phase reagent techniques for the separation of immune complexes these effects are minimized by extensive washing procedures prior to quantitation. Non-separation (homogeneous) immunoassays thus have the greatest potential for interference.

Luminol and its various derivatives utilize a vigorous oxidation system together with a catalyst to achieve high intensity emission. Recent studies have also shown that certain molecules are capable of 'enhancing' the light emission when horseradish peroxidase is used as a catalyst (Whitehead *et al.*, 1983). The mechanism by which this occurs is unknown and is as ubiquitous as the luminol reaction itself. In absolute terms the light emission from the horseradish-peroxidase-catalysed oxidation of luminol is not as intense as the microperoxidase-catalysed reaction and the use of molecules such as benzothiazoles, iodophenols, and hydroxycinnamic acid increases the relatively low light intensity to approximately the same level as the microperoxidase system. What is evident is that the latter system is not improved by the presence of these 'enhancer' molecules.

The fact that a large number of diverse chemical species can catalyse or otherwise affect the chemisty of the phthalhydrazide reactions, coupled with the observation that several 'active oxygen' species are capable of initiating chemiluminescence (Campbell, Hallett and Weeks, 1986), gives some indication of the potential for interference and high background that exists in analytical techniques based on this class of molecule.

Other chemiluminescent systems

A number of other chemiluminescent molecules, which do not require the presence of a catalyst, have also been used for immunoassay purposes. It has been observed that fluorescein isothiocyanate is involved in a chemiluminescent process in the presence of hypochlorite anion and that the intensity of emission is increased when the label is covalently coupled to protein. Use has been made of this property in a coated-tube assay for rubella antibody (Shapiro *et al.*, 1984). However, this system is unlikely to be widely used since high concentrations of sodium or calcium hypochlorite are required which are rather harsh reagents, cause non-specific chemiluminescence, and can damage luminometer pumping systems.

Another chemiluminescent system which has been used involves bis (2,4,6-trichlorophenyl oxalate) (Arakawa, Maeda and Tsuji, 1985). The endpoint quantitation is performed in a non-aqueous solvent containing a 'fluor'. This is necessary since the oxalate ester itself does not exhibit visible chemiluminescence emission and requires that energy transfer occurs to a suitable acceptor molecule which then emits light in the visible region of the spectrum.

The most widely used non-catalytic chemiluminescence reaction is that of the acridinium esters, which proceeds in the presence of dilute alkaline hydrogen peroxide. Though this does not suffer from the disadvantages of the other systems, problems arise due to lack of suitable labelling materials which necessitates the undertaking of a multi-step synthesis (Weeks *et al.*, 1983).

Interface between immunochemical and chemiluminescent reactions

There are a number of ways in which the immunochemical and chemiluminescent systems can be coupled:

(1) Covalent coupling of the chemiluminescent molecule to antigen or antibody.
(2) Covalent coupling of catalyst or cofactor of the chemiluminescent reaction to antigen or antibody.
(3) Antigen or antibody labelled with a system capable of generating a component of the chemiluminescent reaction.

Several variations of the above schemes can be envisaged which utilize indirect coupling. For example, labelled second antibody or antibody labelled with biotin or avidin which is then linked into the chemiluminescent system by the appropriate biotin or avidin coupled component.

Only the first of the schemes listed above is directly analogous to radioisotopic assays, in that visible photons rather than high energy photons are generated directly from a labelled component of the immunochemical reaction. The other options merely represent enzyme-driven systems with their associated disadvantages of complex reaction chemistry.

Several systems representative of schemes 2 and 3 have been reported. An early example is an immunoassay for cortisol (Arakawa, Maeda and Tsuji, 1979) in which a peroxidase–cortisol conjugate bound to antibody was then quantified by its ability to catalyse luminol chemiluminescence. More recently this basic system has been used in the presence of benzothiazoles (luciferins) or other molecules found to increase the light output from the basic system (see earlier).

An example of scheme 3 is an immunoassay for thyroxine based on oxalate ester chemiluminescence (Arakawa, Maeda and Tsuji, 1985). Here thyroxine is labelled with glucose oxidase and used as the labelled antigen. Immune complexes formed by incubation with anti-thyroxine antibodies are then separated using a second antibody coated onto a 'macrobead'. Following washing, the bead is incubated with glucose solution such that hydrogen peroxide is formed if thyroxine–glucose-oxidase conjugate is present in the immune complex. An aliquot of the incubation mixture is then added to solutions of bis (2,4,6-trichlorophenyl oxalate) and 8-anilino-1-naphthalenesulphonic acid (ANS) in ethyl acetate. Any hydrogen peroxide present causes the oxalate ester to undergo a chemiluminescent reaction with energy transfer to the ANS such that ultimately the intensity of light emission is proportional to the amount of antibody-bound thyroxine–glucose-oxidase conjugate and hence inversely proportional to the concentration of thyroxine present in a sample introduced into the immunochemical system.

These latter assays are enzyme immunoassays utilizing chemiluminescent endpoints rather than more conventional colorimetric or fluorimetric endpoints.

USE OF CHEMILUMINESCENT MOLECULES AS LABELS IN IMMUNOASSAY

Phthalhydrazide derivatives

The first reported use of a chemiluminescent molecule in labelling a component of a competitive binding reaction was that published by Schroeder *et al.* (1979). This was an assay for biotin based on avidin binding and used a biotin–isoluminol conjugate as the tracer. The assay was also the first example of a homogeneous chemiluminescent assay (see later). Shortly after this a number of heterogeneous chemiluminescent immunoassays appeared, based upon the use of both lumionl and isoluminol derivatives as labels. At the same time Simpson *et al.* (1979) reported the preparation of chemiluminescent labelled antibodies which were capable of exploiting the advantages of immunometric assay procedures.

These early studies demonstrated the feasibility of immunoassays based on chemiluminescent labels but together with later work highlighted problems in the use of luminol and isoluminol derivatives. The high sensitivity predicted was not achieved possibly for two major reasons. Firstly, the complex oxidation system was susceptible to background and interference effects. Secondly, it was observed that the chemiluminescence quantum yield was critically dependent on the chemical structure of the molecule and that the modifications required to covalently couple the molecule to components of an immunochemical reaction reduced the efficiency of light emission. The extent to which this occurred was dependent on the nature of the luminol derivative used and also the coupling chemistry. Schroeder *et al.* (1981) reported a fourteenfold decrease in light output upon coupling aminohexylethylisoluminol to immunoglobulin G using an *N*-succinimidyl ester derivative. Using luminol Simpson *et al.* (1979) and Pratt, Woldring, and Vilterius (1978) observed loss of light output to levels of 0.7 per cent and 20 per cent following diazotization and periodate oxidation respectively.

Problems in early assays

The problems of high background and non-specific interference were to a certain extent obviated by the use of an organic solvent extraction step and several steroid immunoassays based upon steriod–aminobutylethylisoluminol (ABEI) conjugates were reported. A more convenient approach was subsequently developed for several steroid immunoassays which utilized antibody-coated tubes to permit extensive washing of the immune complexes prior to quantitation. Using this format, assays were developed for certain steroids which compared favourably with their tritium-labelled counterparts and did not require scintillation counting (Collins *et al.*, 1983). The major problem with these assays was that they required exposure to high pH at about 60 °C prior to quantitation of chemiluminescence. In the absence of this treatment light emission was low, presumably due to some form of quenching due to the nature of the immune complex. It was suggested that the treatment released not only the labelled antigen from the immune complex but also the aminobutylethylisoluminol label from the steriod. Subsequent modification of this technique involved the use of a low temperature releasing agent based on organic solvents or detergent. There is confusion as to whether such ‘releasing’ solutions are necessary since many ABEI–hapten immunoassays have been reported which do not require such procedures in order to produce adequate performance (Wood, 1984).

Recent assay developments

More recently a large number of immunoassays have been reported using both labelled antigen and labelled antibody techniques in conjunction with coated macrospheres (6.4 mm diameter) to facilitate assay separation. These assays have used derivatives of aminobutylethylisoluminol or aminobutylethyl-naphthalhydrazide as labels and microperoxidase or haemin as catalyst (Wood, 1984).

Two types of reaction have been widely used for the ABEI labelling of steroid and protein molecules. The first of these involves the modification of the steriod nucleus with an appropriate bridging group, e.g. hemisuccinate or carboxymethyloxime followed by derivatization to yield the active N-succinimidyl ester. The product is then allowed to react with the primary amine moiety of ABEI to produce the steroid–label conjugate via an amide bond (Collins *et al.*, 1983).

The second method involves the modification of ABEI to yield a carboxyl derivative in the form of a hemisuccinamide (commerically available from LKB as ABEI-H) which can be further modified to yield an *N*-succinimidyl ester. This reacts with the ϵ-amino group of lysine residues in the protein under mild alkaline conditions. Labelling methods are not restricted to these options though they are by far the most common. Other methods which have been reported have mainly involved the use of mixed anhydride or carbodiimide coupling procedures.

Acridinium esters

More recently acridinium esters have been used as immunoassay labels both in labelled antigen and labelled antibody systems. Competitive binding immunoassays have been reported for 17 beta-oestradiol (Patel *et al.*, 1982) and progesterone. (Richardson *et al.*, 1985) based on steroid–acridinium ester conjugates. Though the detection limits of the conjugates themselves were extremely low (5×10^{-20} mol for the latter) sub-optimal immunochemistry resulted in assay sensitivities merely equivalent to their radioisotopic assay counterparts. This observation highlights the fact that careful assay design is of fundamental importance in governing sensitivity irrespective of the choice of label. It is only when the immunochemistry is optimal that full advantage can be taken of a very high specific activity label to yield assays of higher sensitivity than are achieved with conventional labels.

A further disadvantage in the use of label–hapten conjugates lies with the fact that the production of such conjugates will require the use of different chemistries depending on the nature of the hapten. The use of labelled antibody techniques for the measurement of both large and small molecules therefore represents a more universal approach to immunoassay, particularly in view of the rapid expansion in monoclonal antibody technology.

The first description of a clinically relevant immunoassay based on acridinium ester-labelled monoclonal antibodies was that of a two-site immunoche-

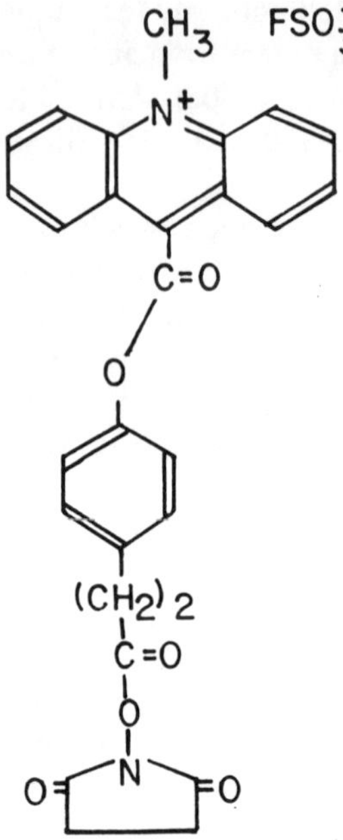

FIGURE 5 Acridinium ester label for the production of chemiluminescent species from precursor molecules containing primary or secondary aliphatic amines

miluminometric assay for alpha-fetoprotein (Weeks *et al.*, 1983). Here antibodies were labelled with a *N*-succinimidyl derivative of a chemiluminescent acridinium ester (Figure 5). Using this derivative it has proved possible to produce stable labelled antibodies with high specific activity, in a reproducible manner.

The chemistry of acridinium ester chemiluminescence has important implications for its exploitation in immunoassays. Firstly, the reaction proceeds in the presence of dilute alkaline hydrogen peroxide without any catalytic requirement, which means that background and interference effects are minimal. Secondly, the reaction consists of a concerted multiple bond cleavage mechanism via a dioxetanone intermediate to yield an electronically excited molecule of *N*-methylacridone which is dissociated from the rest of the original molecule (Figure 6). Thus, linkage to an antibody via the 'tail' of the molecule has little effect on the light emission, resulting in a consistently high quantum yield. Both of these features are in marked contrast to the situations encountered with the phthalhydrazides. Following this, there were reports of a variety of two-site ICMAs using both polyclonal and monoclonal antibodies for a variety of analytes including ferritin, thyrotrophin, complement component C9 and intact parathyroid hormone (Weeks, Sturgess and Woodhead, 1986; Brown *et al.*, 1986).

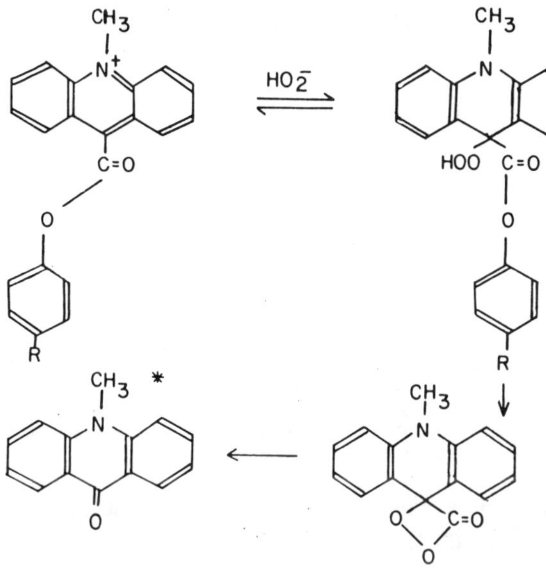

FIGURE 6 Mechanism of acridinium ester chemiluminescence

Use of magnetic particles

The use of magnetizable particles as solid phase reagents permits all the advantages of particulate materials in terms of rapid assay kinetics and ease of assay optimization, together with a rapid separation system which obviates the need for centrifugation or filtration. A two-site ICMA utilizing a magnetizable particle-linked antibody was subsequently engineered to provide a rapid assay for human growth hormone (Weeks and Woodhead, 1986).

Chemiluminescent acridinium ester-labelled antibodies have also been used for the measurement of small molecules using competitive binding immunochemiluminometric assays. The first such assay reported was an ICMA for serum thyroxine (Figure 7) using acridinium ester-labelled monoclonal antibodies and a magnetic separation procedure (Sturgess *et al.*, 1986). Here the serum sample (100 μl) was incubated for 30 min at room temperature with the labelled antibody 8-anilinonaphthalenesulphonic acid (ANS) and a T4–RIgG conjugate. Magnetizable particle sheep (anti-RIgG) antibody was then added and the reaction allowed to proceed for 10 min in order to permit separation of labelled antibody/conjugate immune complexes. The intensity of chemiluminescence emission associated with the solid phase was thus inversely proportional to the concentration of T4 in the serum samples. Recently this technique has also been applied to the measurement of free thyroxine in serum (Sturgess *et al.*, 1988). This latter system has important implications in thyroid function testing since there is no interference due to abnormal protein binding

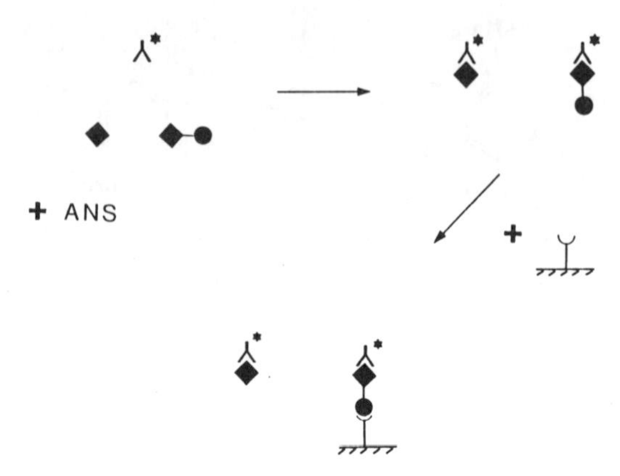

FIGURE 7 Immunochemiluminometric assay for thyroxine (see text)

status, a problem which compromises the diagnostic accuracy of immunoassays using radiolabelled analogue tracers.

HOMOGENEOUS ASSAYS

As with other non-radioactive immunoassay systems chemiluminescence has the potential to yield assays without the requirement for immune complex isolation prior to quantitation, so-called homogeneous assays. Two types of homogeneous immunoassay have been reported, based on two phenomena. The earliest systems described essentially made use of the enhancement phenomenon observed in some cases when a labelled hapten was bound by an antibody (Kohen *et al.*, 1979). It is apparent that this is a somewhat variable phenomenon in that it is not universally observed; further, it has only been demonstrated for some small molecular weight analytes. Thus, although it is undoubtedly an interesting observation, it is perhaps not possible to apply it to a wide range of analytes.

The other system described makes use of radiationless energy transfer purported to occur by the Forster mechanism. Here the antibody is labelled with a fluorphore and is partitioned between the analyte of interest and a tracer antigen with a chemiluminescent label (Patel *et al.*, 1983). When the latter is complexed with the antibody, initiation of the chemiluminescent reaction results in light emission at the wavelength of the fluorophore emission, this is in contrast to the native chemiluminescent emission wavelength seen when the tracer antigen is unbound. Thus, if the two wavelengths are monitored simultaneously, the ratio of intensities is a measure of the concentration of analyte in the system. In order for the technique to be successful, it is necessary that the overlap between the chemiluminescent emission spectrum and the absorption spectrum of the fluorophore (the Joule integral) is substantial. Also the molar extinction coefficient of the fluorophore should be large and, finally, the distance between the donor and acceptor molecules upon immune complex formation must be close enough to permit Forster energy transfer.

Both these homogeneous systems utilize luminol derivatives and both are insensitive in common with other homogeneous assay systems. At present, high sensitivity assays are only obtained by the use of heterogeneous assay systems.

CONCLUSION

The development of chemiluminescence over the last decade has resulted in the production of commercially available reagents and instrumentation. Undoubtedly this will mean the gradual replacement of radioimmunoassay with this major non-isotopic assay technology.

REFERENCES

Albrecht, H.O. (1928). Über die Chemiluminescenz des Aminophthalsåurehydrazids. *Zeitschrift der physikålischen Chemie*, **136,** 321.

Arakawa, H., Maeda, M., and Tsuji, A. (1977). (Enzyme immunoassay of cortisol by chemiluminescence reaction of luminol–peroxidise). *Bunseki Kagaku*, **26,** 322.

Arakawa, H., Maeda, M., and Tsuji, A., (1985). Chemiluminescence enzyme immunoassay for thyroxin with use of glucose oxidase and a bis (2, 4, 6 -trichlorophenyl) oxalate — fluorescent dye system. *Clin. Chem.*, **31,** 430.

Brown, R.C., Aston, J.P., Weeks, I., and Woodhead, J.S. (1986). Measurement of circulating intact parathyroid hormone using a two-site immunochemiluminometric assay. *J. Clin. Endocrinol. Metabl.*(in press).

Campbell, A.K., Hallett, M.B., and Weeks, I. (1986). Chemiluminescence as an analytical tool in cell biology and medicine. In: *Methods of Biochemical Analysis*, Vol. 31 (ed. D. Glick,) Wiley, Chichester, p. 317.

Collins, W.P., Barnard, G.J., Kim, J.B. *et al.* (1983) Chemiluminescence immunoassay of plasma steroids and urinary steroid glucuronides. In: *Immunoassays for Clinial Chemistry* (eds. W.M. Hunter and J.E.T. Corrie), 2nd edn., Churchill Livingstone, Edinburgh, p. 373.

Gleu, K., and Petsch, W. (1935). The chemiluminscence of the dimethylbiacridilium salts. *Angew. Chem.*, **48,** 57.

Kohen, F., Pazzagli, M., Kim, J.B., Lindner, H.R., and Boguslaski, R.C. (1979). An assay procedure for plasma progesterone based on antibody enhanced chemiluminescence. *FEBS Lett.*, **104,** 201.

McCapra, F. (1974). Chemiluminescence of organic compounds. *Prog. Org. Chem.*, **8,** 231.

Patel, A., Woodhead, J.S., Campbell, A.K., Hart, R.C., and McCapra, F. (1982). The preparation and properties of a chemiluminescent derivative of 17β-estradiol. In: *Luminescent Assays: Perspectives in Endorcinology and Clinical Chemistry* (eds M. Serio and M. Pazzagli), Raven Press, New York, p. 181.

Patel, A., Davies, C.J., Campbell, A.K., and McCapra, F. (1983). Chemiluminescence energy transfer: A new technique applicable to the study of ligand–ligand interactions in living systems. *Anal. Biochem.*, **129,** 162.

Pratt, J.J., Woldring, M.G., and Vilterius, L. (1978). Chemiluminescence-linked immunoassay. *J. Imm. Methods*, **21,** 179.

Richardson, A.P., Kim, J.B., Barnard, G.J., Collins, W.P., and McCapra, F. (1985). Chemiluminescence immunoassay of plasma progesterone with progesterone – acridinium ester used as the labeled antigen. *Clin. Chem.*, **31,** 1664.

Schroder, H.R., Vogelhut, P.O., Carrico, R.J., *et al* (1976). Competitive protein binding assay for biotin monitored by chemiluminescence. *Anal. Chem.*, **48,** 1933.

Schroder, H.R., Hines, C.M., Osborn, D.D., *et al.* (1981). Immunochemiluminometric assay for hepatitis B surface antigen. *Clin. Chem.*, **27,** 1378.

Shapiro, R., Chan, J., Pierson, A., Vaccaro, K., and Quick, J. (1984). Protein-enhanced fluorescein chemiluminescence used in an immunoassay for rubella antibody in serum. *Clin. Chem.*, **30,** 889.

Simpson, J.S.A., Campbell, A.K., Ryall, M.E.T., and Woodhead, J.S. (1979). A stable chemiluminescent-labelled antibody for immunological assays. *Nature* (*Lond.*), **279,** 646.

Sturgess, M.L., Weeks, I., Mpoko, C.N., Laing, I., and Woodhead, J.S. (1986). Chemiluminescent labeled-antibody assay for thyroxin in serum, with magnetic separation of the solid-phase. *Clin. Chem.*, **32,** 532.

Sturgess, M.L., Weeks, I., Mpoko, C.N., Laing, I., and Woodhead, J.S. (1988). An immunochemiluminometric assay for serum free thyroxine. *Clin. Endocrinol.*, (in press).

Weeks, I., Campbell, A.K., and Woodhead, J.S. (1983). Two site immunochemiluminometric assay of α_1-fetoprotein. *Clin. Chem.*, **29,** 1480.

Weeks, I., Sturgess, M.L., and Woodhead, J.S. (1986). Chemiluminescence immunoassay: an overview. *Clin. Sci.*, **70,** 403.

Weeks, I., and Woodhead, J.S. (1986). Measurement of human growth hormone (hGH) using a rapid immunochemiluminometric assay. *Clin. Chim. Acta*, **159,** 139.

Weeks, I., Beheshti, I., McCapra, F., Campbell, A.K., and Woodhead, J.S. (1983). Acridinium esters as high specific activity labels in immunoassay. *Clin. Chem.*, **29,** 1474.

Whitehead, T.P., Thorpe, G.H.G., Carter, T.J.N., Groucutt, C., and Kricka, L.J. (1983) Enhanced luminescence procedure for sensitive determination of peroxidase-labelled conjugates in immunoassy. *Nature* (*Lond.*), **305,** 158.

Wood, W.G. (1984). Routine luminescence immunoassay for haptens and proteins. In: *Analytical Applications of Bioluminescence and Chemiluminescence* (eds. L.J. Kricka *et al.*). Academic Press, London, p. 189.

ELISA and Other Solid Phase Immunoassays
Edited by D.M. Kemeny and S.J. Challacombe

CHAPTER 13

The Use of ELISA for Rapid Viral Diagnosis: Viral Antigen Detection in Clinical Specimens

Shireen M. Chantler and **Anne-Louise Clayton**
Wellcome Research Laboratories, Beckenham, Kent, UK

CONTENTS

INTRODUCTION

Reliable diagnosis of viral infections is essential for differential clinical diagnosis, the implementation of selective therapy, epidemiological studies, and the control of disease spread. The laboratory diagnosis is usually made by demonstrating the presence of a virus or viral component and/or virus-specific antibody in the appropriate clinical specimen. In some instances, such as hepatitis B infection where the clinical and virological course of disease is well documented, an analysis of several markers is required to establish clinical status with accuracy while in others the detection of either antigen or antibody alone may be sufficient for diagnosis. The failure to demonstrate a virus does not exclude a viral aetiology but its presence in samples taken from symptomatic or asymptomatic patients provides indisputable evidence of current infection or carrier state and allows appropriate clinical decisions to be made. Hence, there is increasing interest in rapid and sensitive tests for the detection and identification of viruses in clinical specimens.

The detection of a virus is achieved traditionally by isolating the pathogen in cell culture. This method is reliant upon achieving virus replication in monolayer cell cultures and observing secondary cytopathic or transformation effects on the cells by microscopy. Virus isolation in cell culture, with the inherent amplification achieved by cell to cell spread, provides an exquisitely sensitive procedure for detection of viable virus but several factors limit its usefulness for rapid diagnosis. It is slow and success is dependent upon retention of virus infectivity during transport of the specimen to the laboratory, the continuous availability of susceptible cell lines, and the existence of cell culture facilities and expertise. Several important human virus pathogens either cannot be cultivated or are cultivated with difficulty in existing cell culture systems. Finally, definitive identification rather than detection necessitates additional procedures which identify the agent on the basis of characteristic morphology, antigenicity, genomic organization, or biological properties.

Electron microscopy procedures which define viruses on the basis of characteristic morphology (Almeida, 1980) have contributed greatly to the diagnosis of viral infections, particularly for those refractory to cell culture isolation and in mixed infections. While the procedure is excellent for rapid demonstration of virus particles in single 'clean' specimens such as vesicular fluid, it is time consuming and impractical for analysis of large numbers of specimens containing extraneous materials such as faeces. Furthermore, the equipment and specialist expertise are not available in most diagnostic laboratories.

A plethora of alternative procedures such as agglutination, immunofluorescence, immunoperoxidase, radio- and enzyme-immunoassay and DNA hybridization have been described for the direct detection and identification of viruses in clinical specimens. With the exception of DNA hybridization (Highfield and Dougan, 1985; Kingsbury and Falkow, 1985), all rely upon immunological interaction between labelled antibody and virus-specific antigen determinants and differ mainly in the choice of antibody label used to visualize or quantitate this. Many of these techniques have been exploited with considerable success for the routine immunodiagnosis of viruses not amenable to isolation, notably, hepatitis B and enteric viruses such as rotavirus and adenovirus, but their wider application for viruses which can be readily isolated has been limited by the failure to achieve equivalent sensitivity to cell culture. If these methods are to provide a practical adjunct or alternative to cell culture they must offer either the benefits of simplicity, economy, speed, convenience, and specificity such that subsequent culture of assay-negative specimens provides a real saving of effort, or preferably, exhibit a level of specificity and sensitivity equivalent to virus isolation. It follows that the acceptable sensitivity of a new test will depend upon the prevalence of disease in the population under study and the ease and efficacy of the existing method.

Enzymeimmunoassays offer the potential for assays of considerable sensitivity by combining the specificity of antigen–antibody interaction with the amplification of catalytic enzyme reactions. One molecule of enzyme catalyses the conversion of many molecules of substrate resulting in amplification of the signal which, with selected chromogens, produces a coloured reaction product detectable by the naked eye or in a colorimeter. Solid phase assays using these principles are extensively used for viral pathogens refractory to culture or present in high concentration in body fluids. Their adoption for the routine diagnosis of other viruses has not occurred because available assays have been insufficiently sensitive to detect viruses at the concentration present in clinical specimens during the course of infection. However, several recent technical advances have had a considerable impact on the potential role of ELISA in immunodiagnosis. These relate to the improvements in defined antigen preparation by recombinant DNA technology (Eisenstein and Engleberg, 1986) and peptide synthesis (Walter, 1986), the availability of selected monoclonal antibodies of high specificity and affinity, and the introduction of novel enzyme detection systems of increased sensitivity. Judicious exploitation of these technical aspects now offers a realistic possibility of establishing ELISA tests for viral antigen which have a comparable sensitivity to traditional methods and the advantage of speed and flexibility in use.

This chapter outlines the principles and practical problems encountered in the establishment and application of ELISA tests for viral antigen detection and draws heavily upon our own experience and that of others closely involved in this field.

PRINCIPLE OF ASSAYS

A variety of approaches, outlined below, are available for the detection and identification of viruses in clinical specimens by enzyme immunoassays based upon specific interaction of viral antigen with the corresponding antibody. Some of these are applicable to assays incorporating nucleic acid hybridization for specific virus detection.

ELISA systems may be conveniently divided into competitive and non-competitive assays.

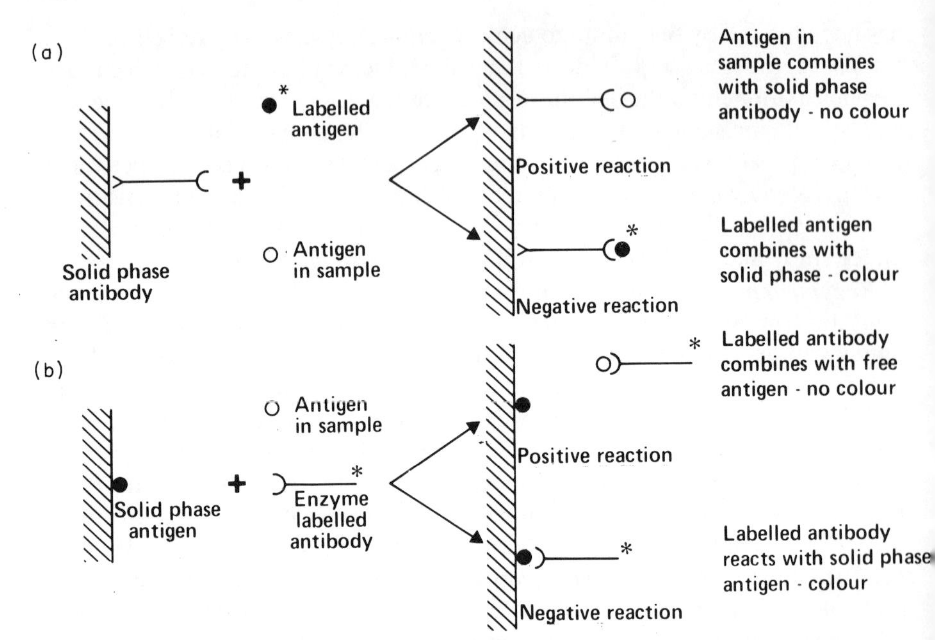

FIGURE 1 Schematic representation of competitive assays

Competitive assays

In these, one of the components of the immune interaction is immobilized on the solid phase and the other is labelled with enzyme (Figure 1a,b). If antigen is present in the specimen it will inhibit or decrease binding of the labelled component to the solid phase reactant by combining with the immobilized antibody (Figure 1a) or with labelled antibody (Figure 1b). This results in reduced colour development on subsequent addition of enzyme substrate and chromogen. While this type of assay has the advantage of a single immune incubation step it has several practical and theoretical limitations. The assay range is small, a positive reaction gives a decrease in colour, and antibody in the specimen will give the same result as antigen. Furthermore, the assay assumes a similar pattern and affinity of interaction between the test and reagent antigen with antibody. This may not be achieved consistently by the use of tissue-derived antigens, containing a heterogeneous mixture of antigen components, or monoclonal antibodies of defined but restricted specificity. This approach has not been exploited for viral diagnosis for the reasons above and because of the practical difficulties of obtaining purified antigens of consistent quality for labelling. The provision of defined antigens by DNA recombinant techniques or peptide synthesis is likely to stimulate renewed interest in this approach.

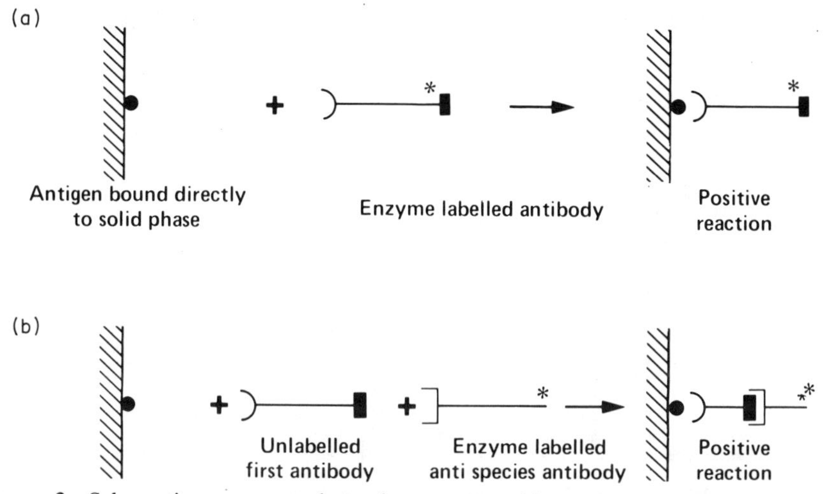

FIGURE 2 Schematic representation of non-competitive assays — direct antigen immunobilization

Non-competitive assays

These are almost exclusively used for viral antigen detection and may involve direct or indirect immobilization of test antigen. In the former, antigen in the sample is directly absorbed on the solid phase and later identified by a single incubation with enzyme-labelled antibody (Figure 2a) or by sequential incubations with unlabelled specific antibody followed by enzyme-labelled anti-species immunoglobulin (Figure 2b), procedures analogous to direct and indirect immunofluorescence tests performed on microscope slide preparations of clinical specimens. The use of anti-species conjugate will amplify the immune interaction but the likelihood of irrelevant specific interaction, due to natural antibodies in the conjugate reacting with other antigens and debris immobilized with the test antigen, is increased. The method does have the advantage, however, of enabling a single labelled reagent to be used for the identification of several antigens provided that the first antibodies are derived from a single species. It also permits the use of primary antibodies which are denatured during conjugation. Direct immobilization of the relevant antigen in a clinical specimen containing other extraneous materials is dependent on the efficacy of adsorption. The presence of other highly adsorptive substances or added protein in diluents will decrease efficacy. Hence, whenever possible, specimens should be suspended in protein-free buffer prior to direct immobilization to avoid poor test antigen adsorption due to competitive adsorption effects.

In the indirect antigen immobilization procedure, alternatively described as

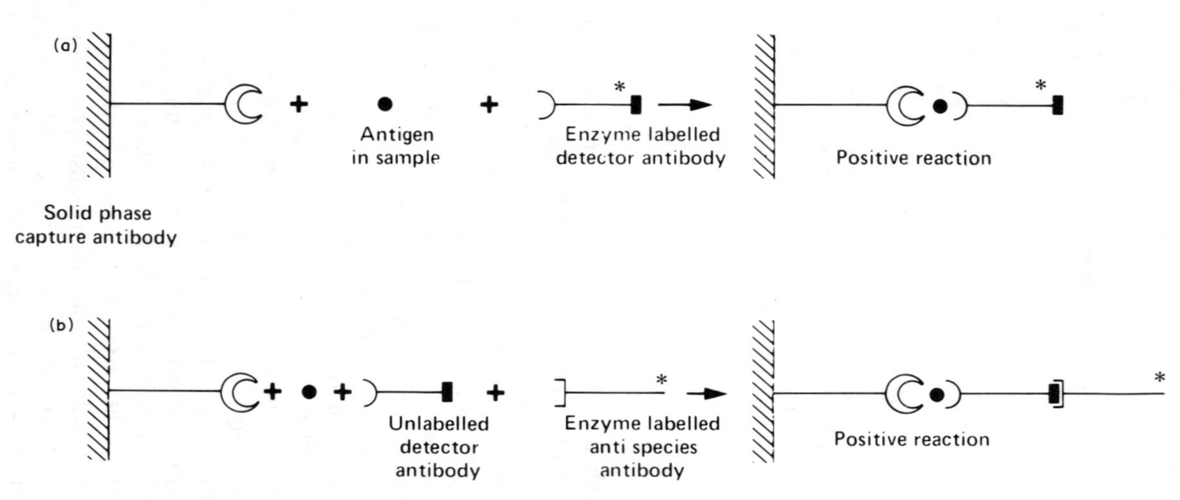

FIGURE 3 Schematic representation of non-competitive assays — antigen cpature

antigen sandwich or antigen capture, specific antibody immobilized on the solid phase is used to bind selectively the pertinent antigen from the specimen and the contaminants are removed by washing. The bound antigen is detected by incubating with enzyme-labelled specific antibody (Figure 3a) or, as described above, by sequential addition of unlabelled antibody and enzyme-labelled anti-species immunoglobulin (Figure 3b). If the former procedure is employed, antibody from a single species may be used as capture and detector. Utilization of unlabelled specific antibody and an anti-species conjugate, on the other hand, necessitates the use of capture and detector antibodies derived from different species to prevent specific interaction with capture antibody in the absence of viral antigen. Several groups have overcome this potential problem by using $(Fab)_2$ fragments of specific antibody as capture in combination with anti-Fc specific anti-species conjugates. In our experience, any non-essential manipulation of virus-specific antibodies is best avoided if maximal activity is to be retained.

Antigen capture procedures, unlike direct antigen immobilization, have the advantage of selectively binding the relevant antigen from a heterogeneous mixture of components commonly present in clinical specimens which are later removed by washing. The efficacy with which antigen is bound and retained during subsequent manipulations will be dependent upon the properties of the antibodies used. Antibodies of the same or different specificity may be used as capture and detector. The choice will depend on the distribution density of the antigens present on the virus or viral component, the affinity and the multiple or restricted epitope specificity of the antibodies used. The use of monoclonal antibodies directed against different eptiopes (i.e. two-site assay) may confer a practical advantage by allowing the simultaneous rather than sequential addition of test sample and conjugate, thereby eliminating one incubation step. This procedure has been used successfully in our laboratory for the detection of

TABLE 1 Relative sensitivity of HSV antigen detection in antigen capture ELISA using sequential or simultaneous addition of antigen and labelled monoclonal detector antibody

HSV antigen dilution	Reactivity in:			
	Sequential assay		Simultaneous assay	
	OD_{450}	P/N	OD_{450}	P/N
1:40	0.953	19.4	1.876	42.6
1:160	0.309	6.3	1.555	26.2
1:640	0.114	2.3	0.403	9.2
1:2560	0.069	1.4	0.146	3.3
1:10240	0.053	1.08	0.069	1.57
Negative	0.049		0.044	

P/N = Ratio of the absorbance of test sample/negative cell control.

TABLE 2 Detection of HSV antigen in ELISA using direct antigen immobilization and antigen capture

HSV antigen dilution	Reactivity in:			
	Direct assay*		Capture assay†	
	OD_{450}	P/N	OD_{450}	P/N
1:40	<2.0	22.5	1.876	42.6
1:160	1.134	12.7	1.155	26.3
1:640	0.361	4.1	0.402	9.2
1:2560	0.154	1.7	0.146	3.3
1:10240	0.083	0.9	0.069	1.6
Negative	0.089		0.044	

* Antigen immobilized on an uncoated solid phase was incubated with HRP-labelled monoclonal antibody.
† Solid phase was coated with anti-HSV antibody and then incubated with antigen and HRP-labelled monoclonal antibody.
P/N = Ratio absorbance of test sample/negative cell control.

hepatitis B(HBV), herpes simplex (HSV), and respiratory syncytical (RS) viruses. Contrary to expectation, the sensitivity with which antigen is detected is at least as good as that of a sequential assay using the same reagents (Table 1) but caution should be exercised in assuming this is so with all systems.

The choice of non-competitive assay employed will be determined by a variety of factors including the spectrum of antibodies available, their physiochemical properties, and the nature of the clinical specimen to be investigated. We have found the sensitivity of HSV detection by direct antigen immobilization and antigen capture ELISA systems using the same labelled monoclonal anti-HSV detector antibody to be comparable (Table 2). However, this may not be universally true as both the ability of the antibody to discriminate between test and contaminating antigens and variations in the efficacy of test antigen adsorption on the solid phase in the presence of other highly adsorptive materials will affect assay sensitivity.

Alternative detector systems

Several systems other than labelled anti-species conjugate have been described for the detection of bound specific antibody. These rely upon specific non-immunological interaction between antibody immunoglobulin and protein A or, alternatively, avidin/streptavidin and biotin. Protein A is isolated from *Staphylococcus aureus* (usually Cowan I strain) and reacts with the Fc portion of immunoglobulins from most mammalian species. Enzyme-labelled protein A is frequently used as a universal tracer in place of an anti-species conjugate and has been found to give equivalent sensitivity with lower background

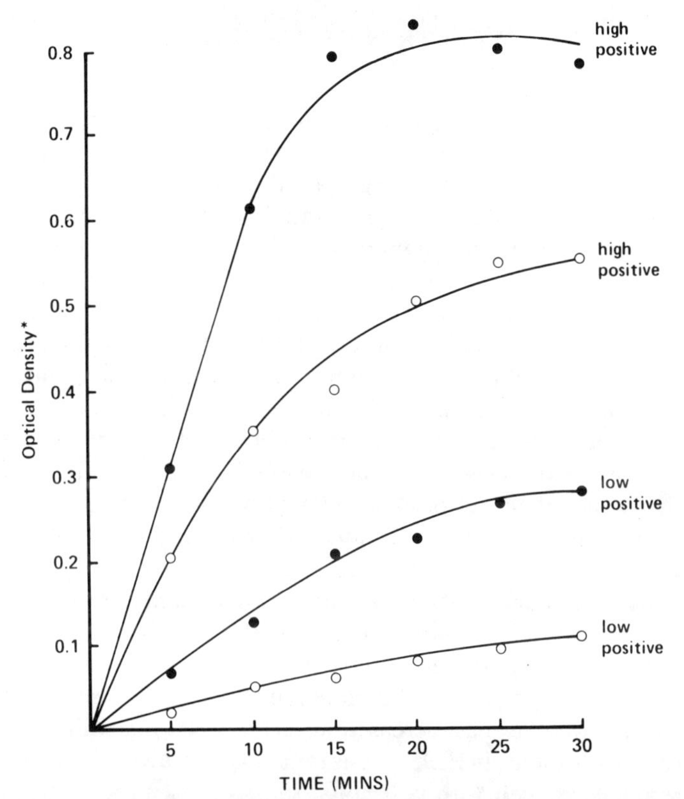

FIGURE 4 Time course of HSV antigen capture assays: comparison of enzyme-labelled detector antibody with biotinylated detector used with streptavidin-biotin-enzyme complex. Both assays employed the same monoclonal capture and detector antibodies, enzyme, and chromogen for enzyme detection

* Corrected optical density, i.e. OD test — OD negative cell control. (*closed circles*) Simultaneous addition of antigen and enzyme-labelled detector antibody. (*open circles*) Simultaneous addition of antigen and biotin-labelled detector antibody followed by sequential additional of streptavidin–biotin–enzyme complex

colour. Other detection systems have been used for amplification of initial binding. The small vitamin biotin has a very high affinity for proteins avidin and streptavidin. Biotinylated detector antibody used in conjuction with avidin/ streptavidin enzyme conjugates is reported to be more sensitive than detector antibody directly labelled with enzyme. Unfortunately, few comparative studies have convincingly excluded the contribution of the differing conjugation procedures and/or variation in batches of antibody used on the results. In a comparative study using monoclonal antibody to HSV coupled to horseradish peroxidase or biotin by analogous conjugation procedures, we have failed to confirm an increase in sensitivity (Figure 4).

SELECTION OF ANTIBODIES

Highly potent antibodies with the appropriate specificity, spectrum of reactivity, and potency are required for the detection of viral antigen in ELISA. Readers familiar with the art of raising polyclonal antisera will know that different preparations, although prepared in an identical manner, rarely exhibit the same qualitative and quantitative properties on subsequent investigation. The reproducible preparation of high affinity specific polyclonal antisera is particularly difficult due to the antigenic strain variability of micro-organisms, the presence of existing antimicrobial antibodies in animals used for immunization, and the problems associated with obtaining adequate supplies of highly purified virus for immunization. Undesirable immunological cross-reactions can be minimized by the use of gnotobiotic or pathogen-free animals but the antisera, once prepared, will contain a mixture of antibodies heterogeneous with regard to immunoglobulin class, avidity, and epitope specificity, hence their performance may vary.

The advent of hybridization techniques for monoclonal antibody preparation (Kohler and Milstein, 1975) has overcome many of the problems associated with the consistent production of high quality anti-viral polyclonal reagents. Relatively impure viral antigens may be used as immunogen and, once selected, a continuous supply of antibody of defined specificity is available without the need for additional immunopurification. The range of anti-viral monoclonal antibodies has increased in recent years but while many may be used to great advantage in ELISA tests for antigen, readers must recognize that the heterogeneity seen with polyclonal sera is present between the specific products of different hybridoma clones. For these reasons, preconceived notions on the superiority of polyclonal or monoclonal antibodies are untenable in our view. Their suitability for use in ELISA can only be assessed by a systematic analysis of their immunological and functional properties. Particular emphasis should be placed on potency, specificity, and those physiochemical characteristics which may result in denaturation during immunoglobulin purification and enzyme labelling procedures or inefficiency following immobilization on a solid phase.

Potency

The relative potency of candidate antibodies may be assessed in ELISA utilizing cell-culture-derived antigens as target, directly immobilized on the solid phase or, indirectly, by the use of a broad spectrum polyclonal capture antibody. Dilutions of hyperimmune serum, ascitic fluid or hybridoma culture supernatant containing test antibody should be added followed by the appropriate enzyme-labelled anti-species conjugate. The test should include parallel titrations of known antibody-positive and -negative preparations and test and control target antigen substrates which contain or lack the relevant viral antigens. Endpoint titres obtained on different occasions may vary in this type of assay but the inclusion of reference sera allows the selection of preparations with the highest relative potency. Indirect immunofluorescence using suitable virus-infected slide preparations and a range of dilutions of test and control antibodies as the intermediary layer, followed by the appropriate labelled anti-species conjugate, is a suitable alternative or additional system. This procedure has the advantage of confirming the presence of specific anti-viral activity and absence of anti-host cell reactivity by the pattern of intracellular virus staining observed and is satisfactory for analysis of antibody activity when antigens suitable for ELISA are unavailable.

Specificity and spectrum of reactivity

Antibodies used for viral antigen detection should be directed against group-common determinants to ensure the detection of all strain variants likely to be present in clinical specimens. It is imperative that this epitope is unique to the organism under test and not shared by other viruses or infectious agents to exclude identification of other unrelated pathogens. For example, the lipopolysaccharide of *Chlamydia trachomatis* has been shown to contain immunodeterminants shared by other Gram-negative bacteria (Brade *et al.*, 1985). Preliminary specificity analysis of anti-chlamydia antibodies should be designed to exclude selection of antibodies exhibiting this type of cross-reaction.

It is clear that a knowledge of the antigenic differences between strains of a particular virus and the likelihood of immunological interactions between different organisms is required to establish the most appropriate tests for specificity analysis of antibodies. These may be performed in test systems, similar to those described above but utilizing a spectrum of appropriate target antigens. Specificity assessments, when done, are usually designed to exclude the possibility of expected immunological cross-reactions and relatively little attention is paid to establishing whether an adequate reactivity spectrum for strain variants is present. This is particularily important when monoclonal antibodies of restricted epitope specificity are to be used. The spectrum of reactivity obtained in our laboratory with monoclonal and polyclonal anti-respiratory syncytial virus (RS) antibodies against a range of known RS-positive and -negative culture supernatants (Table 3) illustrates the occurrence of restricted activity with some monoclonal antibodies and emphasizes the importance of selecting antibodies with the appropriate spectrum of reactivity. If the number of available antibodies is large, it may be useful for practical purposes to identify those with similar epitope reactivity by competitive inhibition experiments at this stage. Representative antibodies within the various groups can then be investigated initially.

Time spent on initial antibody selection minimizes wasted effort and the

TABLE 3 Spectrum of reactivity of monoclonal and polyclonal anti-RS antibodies with RS-positive and -negative culture supernatants

Isolation	Number positive with enzyme-labelled detector antibody							
	n	MCA 1	MCA 2	MCA 3	MCA 4	MCA 5	MCA 6	PCA
+	27	27	27	21	20	19	25	27
–	10	0	0	0	0	0	0	0

MCA — monoclonal antibody.
PCA — polyclonal antibody.
All assays employed a broad spectrum polyclonal capture antibody and one of the above enzyme-labelled detector antibodies. All detector antibodies were of similar potency, when tested by indirect immunofluorescence, but some reacted with only a restricted number of RS strain variants.

likelihood of erroneous results at a later stage. However, in defining the specificity criteria for antibody selection, readers should be aware that the absence of cross-reactivity with unrelated organisms in preliminary evaluations may not exclude its occurrence in the final assay adopted for antigen detection, particularly if the sensitivity of the two systems differs considerably.

Functional utility

The results of preliminary potency and specificity tests will indicate whether candidate antibodies are unsuitable or are likely to be useful in antigen detection assays. Their practical utility, however, can only be determined by experiment. It is an ironic twist of fate that some monoclonal antibodies fulfilling the best selection criteria, are unstable after conventional immunoglobulin fractionation or enzyme labelling or are unsuitable when immobilized

TABLE 4 Reactivity of combinations of selected anti-HSV antibodies in an antigen capture ELISA

Capture antibody:	Detector antibody:				
	MCA 1	MCA 2	MCA 3	MCA 4	PCA
MCA 1	–	+	–	–	++
MCA 2	+++	–	NT	–	–
MCA 3	–	+	–	–	NT
MCA 4	–	+	–	–	NT
PCA	+++	+	–	–	++

NT — not tested.
Globulin preparations of all antibodies were derived from ascitic fluid or hyperimmune serum and were labelled with enzyme. Each unlabelled antibody was tested as a capture in combination with each of the enzyme-labelled reagents. The visual discrimination obtained between positive and negative control antigen for each combination is shown in the table.

on a solid phase. While this may preclude their use in the two-site antigen capture ELISA, they may be suitable in alternative assay procedures outlined previously. In view of this potential problem, it is prudent to monitor retention of antibody activity by appropriate tests after each major preparative step.

Once immunoglobulin preparations and enzyme conjugates of selected antibodies are available, the best combination of single antibodies or mixtures of antibodies for use as capture or detector should be determined. In our experience, non-competing antibodies reactive against repetitive eptiopes provide the most effective combinations. However, each theoretical combination should be examined with control-positive and -negative antigens. The results obtained with different combinations of four anti-HSV antibodies, fulfilling the initial selection criteria, indicate that very few were suitable in an antigen capture ELISA (Table 4). Antibody combinations giving the best discrimination should be selected for use.

ASSAY DEVELOPMENT AND OPTIMIZATION

Selection of solid phase

The choice of solid phase used will depend upon the requirements of individual laboratories. A wide range of high quality polystyrene microtitre plates or strips are available which are suitable for most applications. However, a preliminary investigation to confirm suitability in use should be undertaken with the reagents selected.

Choice of enzyme label and conjugation procedure

The choice of enzyme label used and the conjugation procedure adopted will depend upon individual preferences and the availability of suitable chromogenic substances. Good conventional non-carcinogenic chromogens are available for a number of enzymes and are suitable for most applications. However, the use of amplified enzyme detection with alkaline phosphatase as label offers the potential for increased assay sensitivity—an important requirement for viral antigen detection.

A variety of methods are available for coupling enzymes to antibody. The simplest method utilizes glutaraldehyde to cross-link amino groups on the enzyme with those on the protein to be labelled. An alternative procedure involves the generation of aldehyde groups of horseradish peroxidase, by periodate oxidation of glycol groups, which then interact with amino groups of the protein. Other methods involve the introduction of sulphydryl–maleimide groups which result in the formation of stable amide and thioether bonds between enzyme and antibody. Conjugates derived by all these procedures have been used in ELISA tests. A variation of the sulphydryl–maleimide method (Duncan, Weston and Wrigglesworth, 1983) is routinely used in our laboratory. It is suitable for labelling with both horseradish peroxidase and alkaline phosphatase and allows greater control of the level of enzyme incorporation resulting in conjugates with consistent enzyme–antibody molar ratios.

Selection of optimal assay conditions

Once reagent antibodies and solid phase have been selected, the optimal dilutions of reagents required to give maximal discrimination between positive and negative antigen samples for the preferred assay conditions of incubation time and temperature are determined by chequerboard titration of the test components. The work pattern, and hence the requirements of individual laboratories, will differ and should be considered prior to undertaking extensive assay optimization. The practical approach outlined below relates principally to the development and optimization of antigen capture ELISA but the technical aspects and the principles involved are analogous for ELISA antigen assays adopting other assay principles.

When establishing an antigen capture ELISA we have found it convenient to use polystyrene wells coated with a single concentration of capture antibody (10 μg-ml is suitable for most purposes) in carbonate buffer (0.1 M, pH 9.0). These are tested with a range of antigen concentrations, selected to represent samples of high, moderate, and low reactivity and a broad but limited number of conjugate dilutions. While the use of different coating buffers and capture antibody concentrations is mandatory for some antibodies, in our experience a capture antibody found suitable by the above procedure is rarely improved by these manipulations. The derivation of antigen-positive and -negative samples should be matched whenever possible; if crude cell-culture-derived viral lysates are used as reference antigen, a similar preparation of uninfected cells is the most appropriate control. The addition of detergents in the antigen diluent may have a beneficial effect but the choice of additive, its concentration, and its effect in a particular assay must be assessed.

Close observation of the rate of colour development immediately after the addition of enzyme substrate and chromogen often provides information on the potential suitability or otherwise of reagent dilutions which is lost, due to increased background colour, if the initial observation is delayed. The optimal conditions are then established by testing a more extensive range of reagent dilutions under assay conditions of maximal sensitivity and convenience, achieved by manipulation of incubation time and temperature, relative volume of reactants, and judicious use of additives.

Once a functional assay is established, it is prudent to confirm the specificity of the system by demonstrating the absence of reactivity, firstly, with known unrelated antigens and, secondly, with the test antigen when the capture or detector antibodies are replaced with antibodies of unrelated specificity derived from the same species as the test reagents.

ENZYME DETECTION SYSTEMS

Numerous studies investigating ELISA with conventional chromogenic substrates for the direct detection of viruses in clinical specimens have shown that the sensitivity is low relative to cell culture isolation. We have obtained analogous results with an antigen capture ELISA utilizing monoclonal antibodies, horseradish peroxidase (HRP) as antibody label, and tetramethylbenzidine dihydrochloride (TMB)/H_2O_2 as substrate for HSV detection in genital specimens (Clayton *et al.*, 1985). However, the use of alkaline phosphatase as label and an amplified enzyme detection system (Self, 1985) significantly increased the sensitivity of HSV antigen detection with no decrease in specificity (Clayton *et al.*, 1986). This procedure utilizes a 'cyclical' system for enzyme detection. During the first phase, alkaline phosphatase converts the substrate nicotinamide adenine dinucleotide phosphate (NADP) to nicotinamide adenine dinucleotide (NAD). On addition of an 'amplifier' containing a second enzyme alcohol dehydrogenase and a chromogen, NAD is reduced to NADH which in turn reduces the chromogen to give a coloured tetrazolium salt. During this latter reduction, NADH is converted back to NAD which is available for re-utilization in the system.

We have compared the relative sensitivity of an antigen capture simultaneous ELISA for HSV using the enzyme amplification described above and conventional detection systems for horseradish peroxidase (Figure 5a) and for alkaline phosphatase (Figure 5b). The dose–response curves clearly show that in both cases lower levels of HSV were detected by the use of enzyme amplification when measurements were performed at an equivalent period after addition of substrates. Reference to Figure 5b shows a wide difference in the antigen detection levels with the amplified and conventional enzyme detection systems for alkaline phosphatase when readings were performed at 40 min. However, a comparable result was achieved with *p*-nitrophenyl phosphate (*p*-NPP) after considerably prolonged incubation. The implementation of these sensitive detection systems not only allows lower levels of antigen to be detected within a given time period but, in situations where virus levels are sufficienctly high to eliminate this need, its use offers the possibility of decreasing assay time.

ASSESSMENT OF ASSAY

An investigation of the assay performance must be undertaken to define the criteria for a positive reaction, and to establish its specificity and sensitivity in use. For viruses which can be isolated by cell culture, testing a range of known isolation-positive and -negative supernatants can provide useful information on the quantitative discrimination of positive and negative samples and on the ability of the test to detect a range of strain variants. As the concentration of virus in cell culture supernatants is high relative to that in body fluids or

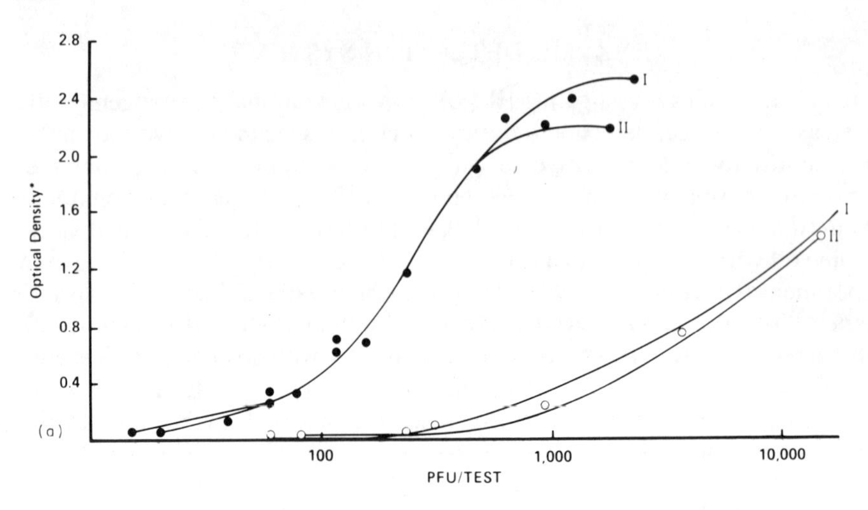

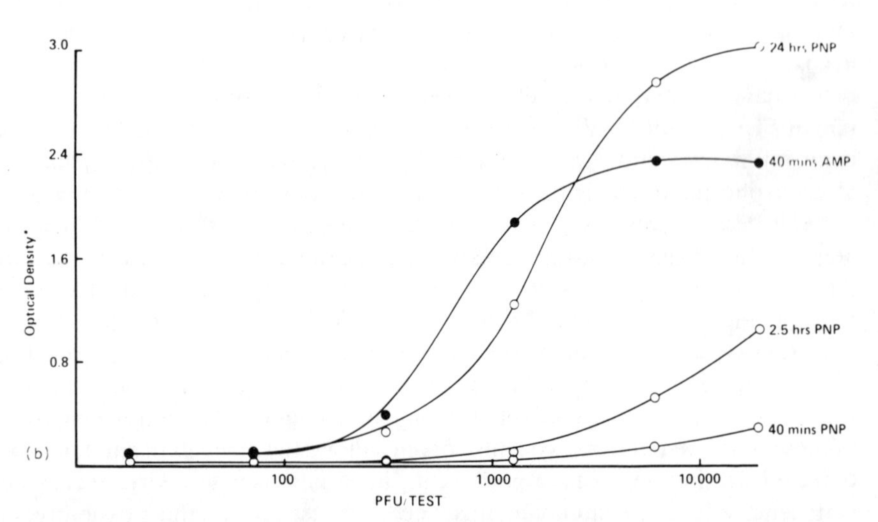

FIGURE 5 Titration of two HSV isolates of defined infectivity in HSV antigen-capture assays using the same monoclonal capture and detector antibodies and either enzyme-amplified or conventional chromogenic detection of enzyme

* Corrected optical density, i.e. OD test — OD negative cell control. I — HSV type I isolate. II — HSV type II isolate.

(a) (*closed circles*) Enzyme-amplified detection of alkaline phosphatase. (*open circles*) Conventional chromogenic (TMB/H_2O_2) detection of horseradish peroxidase.

(b) (*closed circles*) Enzyme-amplified detection of alkaline phosphatase. Total reaction time with substrate and amplifier was 40 min. (*open circles*) Conventional chromogenic (*p*-nitrophenolphosphate) detection of alkaline phosphatase. The test was read at 40 min, 2.5 hr, and 24 hr after addition of substrate specimens taken from lesions, the failure to detect a known positive isolate probably indicates a deficiency in assay specificity rather than sensitivity, a potential problem in assays utilizing antibodies of extremely restricted specificity. Missed, putatively positive isolates should be checked whenever possible by an independent test.

If the results of these tests show good agreement then the absence of reactivity with other unrelated organisms, particularly those most likely to be present in the clinical material ultimately to be examined, should be established. The results of these tests will determine whether the assay possesses the correct specificity and spectrum of immunological reactivity. Subsequent evaluation of assay sensitivity is best performed on clinical samples in direct comparison with the most efficient existing diagnostic procedure. If the reference method is isolation in cell culture, investigators should be aware that isolation-negative samples may contain non-viable virus. Samples which are isolation-negative but positive in ELISA should be confirmed by the use of appropriate controls and confirmatory tests.

Differential neutralization with known seropositive and seronegative sera is an effective confirmatory procedure for analysis of ELISA-positive, isolation-negative specimens. Test specimens are pre-incubated with the relevant sera before testing in ELISA. Abolition of ELISA reactivity with seropositive serum alone indicates the presence of non-viable virus in the specimen. A demonstrable reduction in activity may not be observed with highly positive samples and in these conditions a diluted sample should be re-investigated. Weakly positive discrepant specimens may present interpretative difficulties due to interassay variations observed in absolute absorbance values. The inclusion of a known low positive control similarly treated with seropositive and negative sera as a 'cut-off' control is helpful.

Certain viruses, known to generate Fc receptors on infected cells, protein-A-containing micro-organisms, and rheumatoid-factor-like components in body fluids may give rise to false-positive results due to non-immunological interaction with immunoglobulin reagents in the assay. Parallel testing on test wells coated with specific capture antibody, and on control wells coated with unrelated antibody, is a useful way of confirming the immunological specificity of assay results, despite the increased workload. Non-specific Fc interference is independent of the specificity of the reagents used and, if present, will occur with samples tested on both test and control wells.

Specimens which are isolation-positive and ELISA-negative may result from a deficiency in assay sensitivity or specificity. Testing matched cell culture supernatant, when available, will assist in identifying the problem. The absence of reactivity, with isolates containing elevated levels of virus, usually indicates that the assay specificity is too narrow and that additional antibodies are required. However, readers should be aware that cell culture is not infallible as the occurrence of cytopathic effect (CPE) does not necessarily identify a particular virus.

FACTORS INFLUENCING VIRAL ANTIGEN DETECTION IN CLINICAL SPECIMENS

A variety of factors can influence the ease with which immunoassays detect viral antigen in clinical specimens even with assays satisfying the most stringent specificity and sensitivity criteria. These relate principally to the nature of the antibodies used, the accessibility and concentration of viral antigen present in specimens taken during the course of infection, and the presence of interfering substances in body fluids. An awareness of these potential limitations will decrease the possibility of erroneous conclusions regarding the efficacy of particular ELISA tests for viral antigen detection.

Monoclonal antibodies of defined but restricted specificity are increasingly used in ELISA for antigen detection. The exquisite specificity of selected high affinity products undoubtedly exceeds that attainable by many, but not all, polyclonal antibodies and the practical benefits are undeniable. However, immunological limitations imposed by their use often only became apparent after prolonged usage. Aberrant results both false-positive, due to previously unknown common epitopes on unrelated pathogens, and false-negative, resulting from minor strain variations, may be due to this cause.

Non-specific reactions due to non-immunological interaction between constituents in clinical specimens and test reagents may also occur. For instance anti-immunoglobulins, commonly generated during acute viral infection (Salonen *et al.*, 1980), can react with the Fc region of reagent antibodies (Yolken and Stopa, 1979).

The inaccessibility of viral antigen in a specimen due to its intracellular localization or to masking by complexing with host antibody or other inhibitory substances such as β2 microglobulin (McKeating *et al.*, 1986) may reduce the efficacy of antigen detection. A variety of detergents may be used to release intracellular viral components but in identifying a suitable treatment procedure, care should be taken to ensure that there is no detrimental effect on virus antigenicity or on the efficacy of immune interaction in the assay.

In our experience the requirement to dissociate irreversibly the host antibody complexed with test antigen presents a greater practical problem. Classical procedures for dissociating antigen–antibody complexes require physical separation of the components and reversal of the conditions prior to testing if re-association of the complexes or interference with the immune reaction implicit in the ELISA procedure is to be avoided. Hence, the potential interference by local antibody in the detection of viral antigen by ELISA and the problems associated with its simple, effective removal present an important theoretical limitation for immunoassays. While the problem has been shown to influence the detection of rotavirus in faeces late in the course of infection (Watanabe, Gust, and Holmes, 1978), and hepatitis B antigen when antibody to hepatitis surface antigen is produced (Deinhardt, 1980), the extent of this problem in other body fluids needs further investigation. We have investigated the incidence of anti-HSV antibody in a small number of genital specimens which were isolation-positive but ELISA-negative for HSV. The specimens were tested for secretory antibody in ELISA or for cell-bound antibody in washed cytocentrifuge preparations by immunofluorescence staining with $F(ab')_2$ anti-human immunoglobulin conjugate. Four out of eighteen of these specimens contained specific antibody which may have interfered with antigen detection in ELISA. It is likely that the affinity of host antibody as well as its concentration will influence the extent of interference.

The rate of viral antigen detection in ELISA relative to cell culture will depend upon the efficacy of virus isolation and the absolute level of virus present. The excretion of virus, with notable exceptions, following initial infection is of limited duration. The concentration may be elevated early after onset of infection and thereafter decreases progressively. Hence, the presence of virus and the level of excretion will vary with clinical stage, duration, type, and site of the lesion sampled. The ease of viral antigen detection by ELISA will similarly vary. We have found the rate of HSV antigen detection to be higher with samples taken from vesicular than encrusted lesions and, within each type, to vary with the site of lesion sampled (Clayton *et al.*, 1986). The apparent sensitivity of an ELISA test will, therefore, depend upon the spectrum of samples used for evaluation. It is imperative, therefore, that an assay sensitivity evaluation includes samples taken from a range of clinical categories and that tests to determine the relative sensitivity of different assays are performed on the same panel of specimens. Interpanel differences in the composition of specimens may give widely differing results even within a single study in which factors such as the efficacy of isolation, sampling procedure, retention of viral infectivity, and test procedure are standardized (Table 5).

TABLE 5 Variations in the observed sensitivity of an HSV antigen capture ELISA with different panels of genital specimens

Panel	n	Number ELISA positive	% ELISA positive
1	15	14	93
2	15	11	73
3	15	9	60

All swabs were collected by a standard procedure and placed in 3 ml of transport medium. A single isolation procedure was employed throughout the study. A simultaneous assay employing capture and alkaline-phosphatase-labelled monoclonal antibodies with enzyme-amplified detection was used.

As the concentration of viral antigen in a specimen influences the ease with which it is detected in ELISA, the specimen treatment/collection procedure may be of critical importance in determining whether a sample, particularly those with low levels of antigen, is detected. For this reason it is prudent to collect specimens in the smallest volume of collection buffer/medium containing additives, where appropriate, to ensure maximum virus release. We have investigated the influence of specimen processing on the sensitivity of HSV antigen detection in an ELISA using enzyme amplification. Sequential swab specimens taken from genital lesions were collected in transport medium or directly processed in a small volume of diluent containing detergents. The latter procedure significantly increased the rate of antigen detection giving a result equivalent to isolation (Clayton *et al.*, 1986).

A variety of technical advances can be exploited to achieve ELISA tests giving a sensitivity comparable to cell culture. However, several biological factors associated with the type of specimen to be investigated may limit the sensitivity of antigen detection attained in practice, irrespective of the type of diagnostic procedure used. Recognition of these factors and the implementation of technical manipulations which minimize these effects are essential to achieve assays of acceptable sensitivity and specificity for viral antigen detection. The level of acceptable sensitivity and specificity of non-culture assays will depend upon the availability, ease and expense of the corresponding culture method, and the intended use of the test. Readers should be aware that an assay performance profile which is acceptable for use in a population with a high prevalence of disease may not be so in a low prevalance setting as the predictive values may differ markedly.

FUTURE PROSPECTS

Recent improvements in enzymeimmunoassays for viral antigen detection have made a considerable impact on the sensitivity, specificity, and speed of assay systems. These diagnostic tests are at an advanced stage of development but a chapter on detection of virus infections would be incomplete without reference to nucleic acid hybridization techniques which are inceasingly being applied in this field (Kinsbury and Falkow, 1985). This procedure relies upon hybridization of single-stranded viral nucleic acid with a complementary sequence present in the appropriate labelled viral DNA probe and subsequent detection of bound probe. The technique is likely to be particularly useful where the viral antigenic composition is not defined, e.g. human papilloma virus (Gissmann *et al.*, 1983) and parvovirus (Anderson, Jones and Minson, 1985); when antigen determinants are poorly expressed or masked (Mckeating *et al.*, 1986) or when information on the virulence of an infecting virus is sought. Recent evidence suggests that circulating DNA in hepatitis is a more reliable indicator of infectivity than antigenic markers (Harrison *et al.*, 1985).

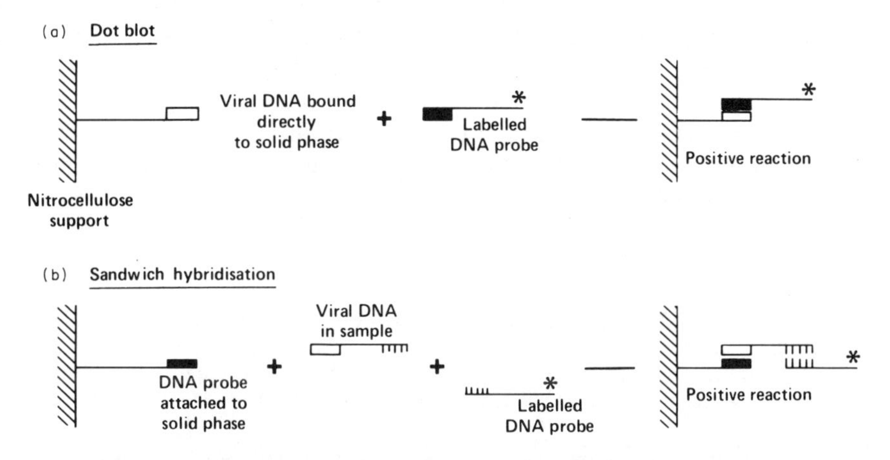

FIGURE 6 Schematic representation of DNA hybridization assays

The ability to identify and select unique nucleotide sequences which only hybridize to the gene of interest confers a high level of specificity unlikely to be matched by monoclonal antibodies which rely on configurational effects. The sensitivity of the technique is limited, however, by the low copy number of the genes present in an infected cell. It remains to be seen whether assays based on nucleic acid hybridization can approach the sensitivity of immunoassays which take advantage of high replication of antigenic epitopes.

The choice of DNA probe used is subjected to selection criteria analogous to those used for antibody selection; regions unique for the test agent and conserved in all strain variants are required for diagnostic tests. The principles of assays are similarly analogous to ELISA systems (Figure 6). Direct immobilization on a solid support (usually nitrocellulose), alternatively described as dot blot, or capture hybridization procedures have been used for virus detection (Ranki *et al.*, 1983; Virtanen *et al.*, 1984; McKeating *et al.*, 1985). The latter procedure involves the use of two different probes as capture and detector respectively which are complementary to different non-overlapping regions of the viral nucleic acid. The viral nucleic acid in the sample forms a bridge between the probes and the bound labelled DNA is then measured. This approach is more specific than dot blot and allows the removal of extraneous material which might interfere in the assay.

The major methodological limitations of the DNA hybridization system relate to the need for the rather tedious pre-treatment of the sample to extract the nucleic acid and the requirement to use a radioactive label, ^{32}P, to achieve sensitive detection. A number of alternative labels including enzymes are being actively investigated and it is likely that suitable methodological refinements will become available in the near future and extend the role of DNA probe assays in the diagnosis of virus infections.

ACKNOWLEDGEMENTS

We would like to thank Drs Godley, Thin, Sullivan, and Watkins and Mr Bertrand for supplying specimens used in these studies; Jill Dale for performing the assays of secretory HSV antibody; Jackie Mitchell and Pamela Thurbin for typing the manuscript.

REFERENCES

Almeida, J.D. (1980). Practical aspects of diagnostic electron microscopy. *Yale J. Biol. Med.*, **53**, 5.

Anderson, M.J., Jones, S.E., and Minson, A.C. (1985). Diagnosis of human parvovirus infection by dot-blot hybridisation using cloned viral DNA. *J. Med. Virol.*, **15** (2), 163.

Brade, L., Nurminen, M., Makela, H., and Brade, M. (1985). Antigenic properties of *Chlamydia trachomatis* lipopolysaccharide. *Infect. Immun.*, **48** (2), 569.

Clayton, A.L., Beckford, U., Roberts, C., Sutherland, S., Druce, A., Best, J., and Chantler, S. (1985). Factors influencing the sensitivity of herpes simplex virus detection in clinical specimens in a simultaneous enzyme-lined immunosorbent assay using monoclonal antibodies. *J. Med. Virol.*, **17**, 275.

Clayton, A.L., Roberts, C., Godley, M., Best, J.M., and Chantler, S. (1986). Herpes simplex virus detection by ELISA: Effect of enzyme amplification, nature of lesion sampled and specimen treatment. *J. Med. Virol.* **20**, 89.

Deinhardt, F. (1980). Predictive value of markers of hepatitis virus infections. *J. Infect. Dis.*, **141**, 299.

Duncan, R.J.S., Weston, P.D., and Wrigglesworth, R. (1983). A new reagent which may be used to introduce sulphydryl groups into proteins, and its use in the preparation of conjugates for immunoassay. *Anal. Biochem.*, **132**, 68.

Eisenstein, B. I., and Engleberg, N.C. (1986). Applied molecular genetics. New tools for microbiolgists and clinicians. *J. Infect. Dis.*, **153** (3), 416.

Gissmann, L., Wolnik, L., Ikenberg, H., Koldovsky, U., Schnurch, H.G., and Zur Hauséen, H. (1983). Human papillomavirus types 6 and 11 DNA sequences in genital and laryngeal papillomas and in some cervical cancers. *Proc. Natl. Acad. Sci.*, **80**, 560.

Harrison, T.J., Bal, V., Wheeler, E.G., Meacock, T.J., Harrison. J.F., and Zuckerman, A.J. (1985). Hepatitis B virus DNA and e antigen in serum from blood donors in the United Kingdom positive for hepatitis B surface antigen. *Brit. Med. J.*, **290**, 663.

Highfield, P.E., and Dougan, G. (1985). DNA probes for microbial diagnosis. *Med. Lab. Sci.*, **42**, 352.

Kingsbury, D.T., and Falkow, S. (1985). *Rapid Detection and Identification of Infectious Agents*. Academic Press, New York.

Köhler, G., and Milstein, C. (1975). Continuous culture of fused cells secreting antibody of predefined specificity. *Nature (Lond.)*, **256**, 495.

McKeating, J.A., Al-Nakib, W., Greenaway, P.J., and Griffiths, P.D. (1985). Detection of cytomegalovirus by DNA-DNA hybridisation employing probes labelled with 32 phosphorus or biotin. *J. Virol. Methods*, **11**, 207.

McKeating, J.A., Grundy, J.E., Varghese, Z., and Griffiths, P.D. (1986). Detection of cytomegalovirus by ELISA in urine samples is inhibited by β2 microglobulin. *J. Med. Virol.*, **18**, 341.

Ranki, M., Virtanen, M., Palva, M., Loaksonen, M., Pattersson, R.F., Kaariainen, L. Halonen, P.E., and Soderlund H. (1983). Nucleic acid sandwich hybridisation in adenovirus diagnostics. *Curr. Top. Microbiol. Immunol.*, **104**, 309.

Salonen, E.M., Vaheri, A., Suni, J., and Wager, O. (1980). Rheumatoid factor in acute viral infections; Interference with determination of IgM, IgG and IgA antibodies in an enzymeimmunoassay. *J. Infect. Dis.*, **142**, 250.

Self, C.H. (1985). Enzyme amplification — a general method applied to provide an immunoassisted assay for placental alkaline phosphatase. *J. Imm. Methods*, **76**, 389.

Virtanen, M., Syvanen, C.A., Oram, J., Soderlund, H., and Ranki, M. (1984). Cytomegalovirus in urine: detection of viral DNA by sandwich hybridisation. *J. Clin. Microbiol.*, **20** (6), 1083.

Walter, G. (1986). Production and use of antibodies against synthetic peptides. *J. Imm. Methods*, **88**, 149.

Watanabe, H., Gust, I.D., and Holmes, I.H. (1978). Human rotavirus and its antibody: their coexistence in faeces of infants. *J. Clin. Microbiol.*, **7** (5), 405.

Yolken, R.H., and Stopa, P.J. (1979). Analysis of nonspecific reactions in enzyme linked immunoabsorbent assay testing for human rotavirus. *J. Clin. Microbiol.*, **10**, 703.

ADDITIONAL READING

Hames, B.D., and Higgins, S.J. (1985). *Nucleic Acid Hybridisation, A Practical Approach*. Irl Press. Oxford.

Kurstak, E. (1985). Progress in enzyme immunoassays: production of reagents, experimental design, and interpretation. *Bull. Wld. Hlth. Org.*, **63** (4), 793.

Meinkoth, J., and Wahl, G. (1984). Hybridisation of nucleic acids immobilized on solid supports. *Anal. Biochem.*, **138**, 267.

Richman, D.D., Cleveland, P.H., Redfield, D.C., Oxman, M.B., and Wahl, G.M. (1984). Rapid viral diagnosis. *J. Infect. Dis.*, **149** (3), 298.

Richman, D.D., Schmidt, N., Plattein, S., Yolken, R., Cherensky, M., McIntosk, K., and Matteis, M. (1984). Summary of a workshop on new and useful methods in rapid viral diagnosis. *J. Infect. Dis.*, **160** (6), 941.

Schuurs, A.H.W.M., and Van Weeman, B.K. (1977). Enzymeimmunoassay. *Clin. Chim. Acta*, **81**, 1.

Yolken, R.H. (1982). Enzyme immunoassays for the detection of infectious agents in body fluids: current limitations and future prospects. *Rev. Infect. Dis.*, **4**, 35.

ELISA and Other Solid Phase Immunoassays
Edited by D.M. Kemeny and S.J. Challacombe

Published by John Wiley & Sons Ltd

CHAPTER 14

The Use of ELISA for Rapid Viral Diagnosis: Antibody Detection

R.J.S. Duncan

Wellcome Research Laboratories, Beckenham

CONTENTS

INTRODUCTION

Developing an ELISA to detect antibodies directed against a virus is more onerous than developing an ELISA for the corresponding virus because viral particles present many different epitopes. Each or all of these could be used for detecting the virus, but for diagnostic purposes antibodies to all of the epitopes might have to be detected. In addition, specific antibodies may reside in a particular class of antibody. The sensitivity of an antibody test is, however, usually less than that required of an antigen test.

An ELISA for antibody detection involves capture (with varying degrees of specificity) of the antibody on to a solid phase, and subsequent or simultaneous identification of the bound antibody, again with varying degrees of specificity. Methods of labelling with an enzyme, and detection of the enzyme, are common to all ELISAs. However, for some systems of antibody detection it is necessary to conjugate viral proteins with an enzyme, which may require more advanced chemistry than labelling an antibody, and will require purified or cloned viral antigen.

The three most commonly used ELISAs for antibody detection are:

(1) A competitive test in which test antibody and enzyme-labelled antibody compete for immobilized antigen.
(2) Capture of the antibody on to the solid phase by immobilized antigen, with subsequent detection of the captured antibody with either a labelled anti-species antiglobulin (which may be class specific) or with labelled antigen.
(3) Capture of the antibody by an immobilized anti-species antibody (possibly class specific) with detection by labelled antigen.

Other methods have been used more rarely, for example antibody capture or detection with labelled protein A, labelled immune complex capture by rheumatoid factor (Sachers *et al.*, 1985), and variations of the competitive system. Western blotting is also an ELISA in the broadest sense, but will not be discussed here. Several recent reviews of antibody testing are available (Bergmeyer, 1986) and so this chapter will be concerned with practical aspects of the various tests. Most of the examples and references will be based on detection of antibodies against the AIDS virus (anti-HIV) because these ELISAs have possibly been examined more thoroughly than any other ELISA for antibody (Mortimer, Parry, and Mortimer, 1985; Carlson *et al.*, 1985) and illustrate the difficulties and advantages of each of the possible techniques in both commercial and laboratory settings.

SUMMARY OF GENERAL METHODS

Immobilization of antigen

The choice of the solid phase (microtitre wells, tubes, dipsticks, etc.) depends on the way the assay will be used (mass screening, single patient assay, experimental work) and on the equipment available, but makes little difference to the development of the assay. The material of the solid phase is more important but there appear to be no general rules for its selection. Similarly, while chemical activation of the solid phase has been recommended (Von Klitzing *et al.*, 1982) and activated materials are commercially available, the advantages, if any, can only be determined experimentally.

Experience shows that passive adsorption to polystyrene surfaces is difficult to surpass when the physical conditions for adsorption (pH, ionic strength, dielectric constant) have been optimized. To ascertain these conditions it is necessary to measure the amount of antigen bound in a functional state and this is best done with a labelled antibody of the type to be detected. Binding of protein and peptide antigens occurs with a broad pH optimum usually in the pH range 6–10 and it is normally sufficient to examine coating at pH 9.6 (carbonate or glycine buffer), pH 8 (triethanolamine buffer), and pH 6 (citrate buffer). The ionic strength can be critical for antigen binding, and it may be possible to 'salt-out' the antigen with ammonium sulphate or phosphate in concentrations just below that at which it would precipitate from solution. Other antigens may be forced on to a solid phase by the slight denaturation and change in hydrophilicity induced by incorporating into the coating solution organic solvents such as ethanediol or methanol in concentrations up to 20 per cent.

An antigen which is loosely bound to a surface may be fixed by treatment with glutaraldehyde (up to 0.5 per cent) either during or after the actual coating reaction, but this treatment usually leads to unaccaptable losses of activity of the antigen. On very rare occasions it has been observed that treatment of an antigen with detergent (e.g. 0.5 per cent sodium dodecylsulphate) will greatly increase its hydrophobicity (and hence binding to polystyrene) without loss of antigenicity. The temperature and time of coating can also influence the extent of antigen binding, but overnight at 37 °C is a normal condition with dilute and valuable antigens. If the assay design to be used requires immobilization of an antibody either to capture an antigen or to bind selectively a class of immunoglobulins the techniques described above will allow this to be done readily, but efficiency may be improved on occasion by momentarily acidifying the antibody (Conradie, Govender, and Visser, 1983).

Antigen purity

High antigen purity is not usually a particular requirement for its immobilization on a solid phase, as long as precautions are taken so that any impurities do not interfere with the ultimate assay, see below. However, should the antigen prove either too impure or too intractable to immobilize directly it may be possible to coat the solid phase with a specific antibody (often a monoclonal) which will capture the required antigen from solution. Advantages of using a 'capture antibody' are simplicity (it is easy to immobilize an antibody), the possibility of using very impure and dilute antigen preparations because of the affinity purification offered, and the possibility that the antigen may be bound in a more active conformation than that forced upon it by direct contact with a hydrophobic plastic (Ouldridge *et al.*, 1984). Disadvantages of using a capture antibody are that detection by labelled antigen methods is ruled out, the antibody selected (polyclonal or monoclonal) may not capture all the epitopes of a virus or may mask rare sites, anti-immunoglobulins in sera to be tested may interfere with specific detection of antibodies against the antigen, and at least one extra coating stage is required.

Blocking plates

Having immobilized the antigen it is usual to treat the coated solid phase for a few hours with a strong protein solution (0.2 per cent bovine albumin in the coating buffer, or 0.75 per cent casein in Tris, pH 7.8) so blocking any remaining binding sites on the solid phase to prevent non-specific conjugate attachment during the subsequent assay. The value of this blocking is open to question, however, because variable dissolution of the blocking coat may lead to increased sample-to-sample variability, and the conjugate is normally presented in a proteinaceous solution which suffices to prevent non-specific attachment. Coated surfaces may be air-dried and stored for long periods with no loss of efficacy.

Capture of target antibody

The prepared solid phase is used to capture specifically the antibody of interest from a sample. This reaction is straightforward, the only difficulties being associated with non-specific binding in assays with labelled anti-species detection or with lack of sensitivity for antibodies to rare or non-immobilized epitopes. Non-specific binding can be overcome by dilution of the sample with a relatively concentrated protein solution (e.g. 1 per cent bovine albumin or up to 20 per cent serum from the species used to prepare the conjugate or any capture antibody), and by including in the diluent material from disrupted cells or medium of the type in which the viral antigens were cultured. Detergents and mildly chaotropic salts in concentrations sufficient to disrupt non-specific binding but not immune interactions may also be useful. Lack of sensitivity to particular epitopes is more difficult to deal with, but can be of importance (Carlson *et al.*, 1985), and usually requires a different approach to antigen preparation or immobilization.

Conjugate capture

In the competitive assay design antibody and conjugate capture are normally added simultaneously; all the other assay designs usually involve a sequential capture of antibody followed in a separate stage by conjugate. Non-specific binding of the conjugate must be prevented in either case, and for this reason the conjugates are usually used in a solution containing specific proteins, salts and detergents as described above. The nature of the conjugate is important in minimizing non-specific binding. Ishikawa (1980) and his colleagues have shown the relevance of preparing defined, polymer-free conjugates. The enzyme used too can have a great effect, the larger enzymes — galactosidase, urease, and alkaline phosphatase — giving, all other things being equal, larger background variability than the smaller enzymes such as peroxidase (Imagawa *et al.*, 1984).

It is easy to show that in a typical ELISA only about 1/5000 of the conjugate added to the reaction vessel is actually specifically bound in the presence of a weakly positive sample. For this reason the final wash is crucial to the performance of an assay, and although in some cases a wash using merely water proves satisfactory (and is very desirable for a commercial assay), it is more usual to use detergents and salts to remove completely the excess conjugate.

SPECIFIC EXAMPLES OF ELISA FOR ANTIBODY DETECTION

Competitive ELISA

The basis of this type of ELISA is shown in Figure 1. The sample to be tested and a labelled antibody are added together to a prepared solid phase bearing immobilized antigen. During a period of incubation competition occurs between any specific antibodies in the sample and the conjugate. If there are large amounts of such antibodies in the sample, the conjugate will be prevented from binding and so strongly positive samples will have little or no enzyme bound to the solid phase. Negative samples will not prevent the enzyme from binding, and intermediate samples will prevent the binding of a fraction of the conju-

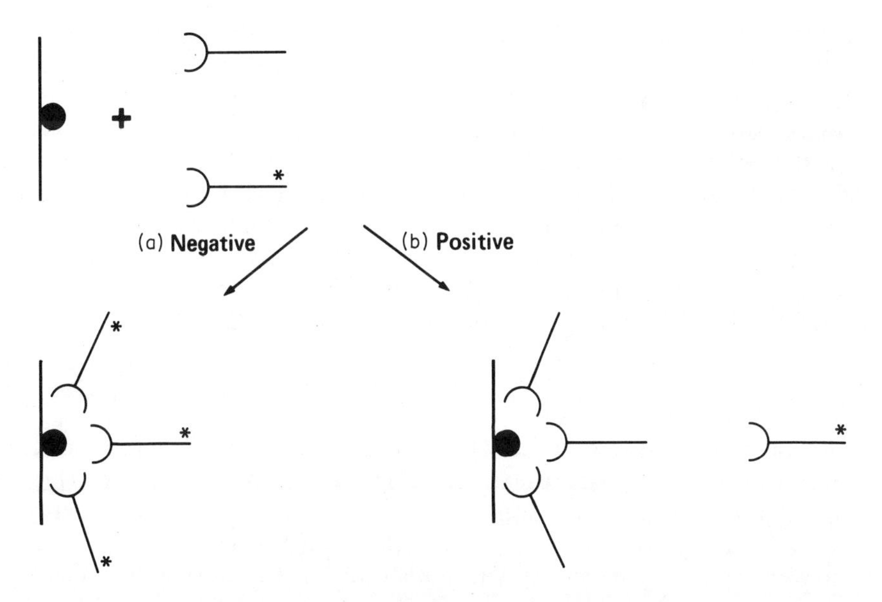

FIGURE 1 *Competitive ELISA*. Competition between antibodies in the sample and labelled antibodies for immobilized antigen

gate. This assay format is attractive in that it requires conjugated antibodies, not labelled antigen.

Disadvantages of competitive assays

Impure antigen on the solid phase is not a problem so long as the conjugate does not react with the impurities — and this factor is under the control of the assayist by selection of a suitable antibody source. Another useful feature of this system is that only a single incubation and washing stage are required, a valuable feature in a screening laboratory. There are essentially four disadvantages of the system: 1) Since positive samples prevent binding of the conjugate, the washing stage must be thorough in order that no traces of conjugate are left to convert a positive to a false negative. In practice this is not a real problem because failed washing leads to gross excess of conjugate remaining, and the assay is obviously void. 2) Of more practical importance is the presumed necessity to detect antibodies to all the viral epitopes. In the case of the AIDS virus there is some evidence that individuals may have antibodies only to certain epitopes, and not to all (Carlson *et al.*, 1985). This makes it necessary to ensure that all the epitopes are present on the solid phase, that the conjugated antibody used recognizes all those epitopes, and that the presence in the sample to be tested of antibody to a single epitope will prevent binding of sufficient of the conjugate to be detected. Although it is easy to design an assay in the competitive format, it is much more difficult to validate than many of the other assays because of the problem of weak competition when only antibodies to a rare epitope occur in the sample. 3) The third major problem of the competitive format lies in the reproducible preparation of the components of the system; it is important that the relationship between the antigen on the solid phase and the conjugate remains constant or both specificity and sensitivity will change. When different batches of either of the components are used this may be difficult to ensure: titrations of the various components must be undertaken whenever anything is changed. 4) Finally, it is not possible to design a competitive assay to detect antibodies of a specified class.

Sensitivity of competitive assays

The competitive assay is unique in that negative samples result in enzyme being attached to the solid phase, usually meaning that negative samples give a higher signal in the detection system than positive samples. Distinguishing between positive and negative then depends on deciding what signal is lower than that of a typical negative. This can be done statistically, assuming a Gaussian distribution of values of a group of negative samples (say 20) run at the same time as the test samples, or by comparison with a cut-off standard which is made from a dilution of a positive sample and set to give a signal which is, on average, different from negative by some statistical test (see below). As the cut-off is related to both the amount of antigen on the solid phase and the concentration of the conjugate, a change in either will require a new cut-off to be defined to maintain the same statistical sensitivity.

Practical details of competitive assays

Developing a competitive assay involves binding antigen to the solid phase, and then titrating the conjugate in the presence and absence of a weak positive (at the limit of what it is hoped to be able to detect) to obtain roughly the correct conjugate dilution. The coating antigen is then titrated at the chosen conjugate concentration, again in the presence and absence of the weak positive, and with successsive rounds of titration the best concentrations can be found. A cut-off standard is then selected, either by analysis of negatives along with the sample or by titration of a positive along with may negatives (using several batches of reagents) to find a dilution which is at the chosen level of difference from the negatives. If a variety of conjugated antibodies are available, selection can be made by choosing the one which gives the least colour, on average, with several strong positive samples as well as a satisfactory titre and sensitivity.

Capture ELISA

Capture of immobilized antigen

The basis of this type of ELISA is shown in Figure 2. The sample to be assayed is incubated with a solid phase to which an antigen has been bound, either by direct adsorption or through a capture antibody. Deciding upon the method of immobilization depends on the antigen; direct coating is preferable, but is not always chemically possible and may cause conformational changes in the antigen. During the assay any specific antibody will itself become bound to the antigen on the solid phase, and then in a second incubation is detected with either a labelled anti-species antibody or with labelled antigen. Detection by labelled antibody is straightforward, but the use of labelled antigen is less obvious. The basis of the latter is that antibodies are at least bivalent, one valence is used in attaching the antibody to the immobilized antigen leaving the other(s) free to bind to the labelled antigen (see Chapter 6). Using labelled antigen, both capture and detection of the target antibody depend on its specificity towards the antigen, so if the antigen is correctly chosen and purified the assay can be made very specific overall.

On the other hand, detection with labelled anti-species antibody makes use of the specificity of the target antibody only once, so the overall assay is inherently less specific; this is the case in practice, labelled antibody detection

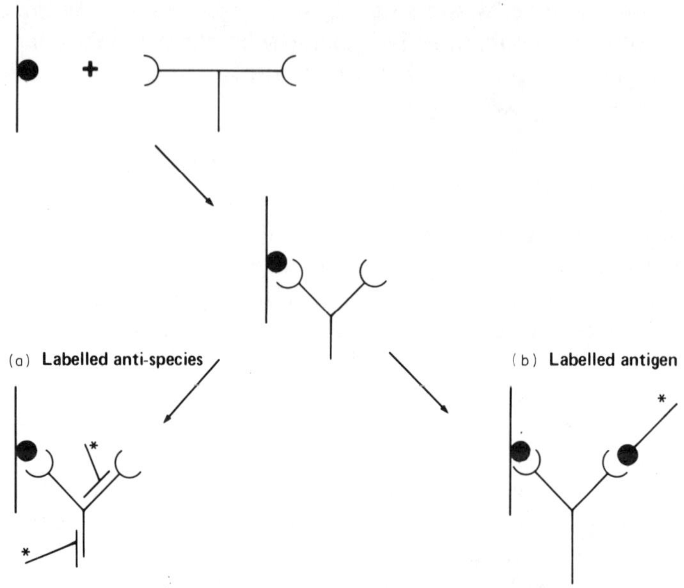

(a) Labelled anti-species

(b) Labelled antigen

FIGURE 2 *Capture ELISA I.* Antibody is captured from the sample and detected with either labelled anti-species (which could be class specific) or with labelled antigen

gives falsely positive results whenever non-target antibody becomes attached to the solid phase by any mechanism. The choice of using labelled anti-species or labelled antigen usually depends on the availability of a labelled antigen which will detect all the epitopes of the virus which are of diagnostic importance. If the antigen was immobilized through an antibody, labelled antigen detection will be impractical because the conjugate will be bound to the capture antibody and cause high backgrounds.

Factors causing non-specificity in capture assays

One of the most important problems met during the development of this type of ELISA is caused by impurities in the antigen which react immunologically with antibodies in some of the samples to be tested. The most common source of impurity is the cell line in which the antigen is cultured, but affinity purification columns and proteins used in diluents can each introduce reactive material. Any antibody which is captured from the sample will be detected by labelled anti-species antibody; whether it is detected by labelled antigen will depend on the extent of labelling of the homologous impurity in the antigen. Similarly, any antibody which is non-specifically adsorbed on to the solid phase or binds to any capture antibody will also be detected by labelled anti-species antibody although usually not by labelled antigen.

Non-specific adsorption could arise because of imperfectly blocked binding sites on the solid phase, because of maltreatment of the sample (heating, freeze-thawing), or from the nature of the sample. Apart from the technical problem of inadequately blocked binding sites it is difficult to solve these problems, and a change in the assay design or the source and purification methods of the antigen could be the easiest solution if many samples give incorrectly positive results of this kind. As in the competitive assay format, it is important to ensure that all of the epitopes of the viral antigens are immobilized, but in this assay design it is not as important to ensure a critical balance between the epitopes — as long as the epitope is on the solid phase a capture assay should detect antibodies to it. Again, unlike the competitive format it is unnecessary to maintain a critical relationship between the conjugate and the immobilized antigen, in general the more antigen bound to the solid phase the better, and changing batches of reagent is not as difficult as in the competitive system. Capture by immobilized antigen followed by detection with a labelled class-specific anti-immunoglobulin (e.g. anti-IgM) is possible in this assay format, but in general better results are obtained using immobilized antibody capture as described below.

Sensitivity of capture assays

Distinguishing between positive and negative samples is a matter of deciding what optical density is different from the low value given by negative samples. This can be done statistically from a number of negative samples, but as usual a cut-off standard (a dilution of a positive sample) or an absolute optical density value can be used instead, with greater economy but less sensitivity. The absolute sensitivity of assays involving capture on to an immobolized antigen is theoretically the greatest available by ELISA for antibody detection, but the useful sensitivity can occasionally be lost because of the equally great sensitivity for non-specifically-bound antibody.

Development of capture assays

Developing an ELISA using an immobilized antigen requires above all a satisfactory method for the immobilization (see Chapter 2). The methods outlined above should allow this to be done relatively easily, but verifying that the procedure has been performed satisfactorily is necessary. It may be possible to use monoclonal antibodies to test that all important epitopes have been immobilized but as many 'positive' sera as possible, from various stages of the viral infection, should be assayed to make sure that all stages and expected antibody classes are detected by the ELISA. Preparation of a labelled antigen is more difficult than preparation of a labelled antibody, but the specificity improvement obtained by its use is normally sufficient to merit the effort.

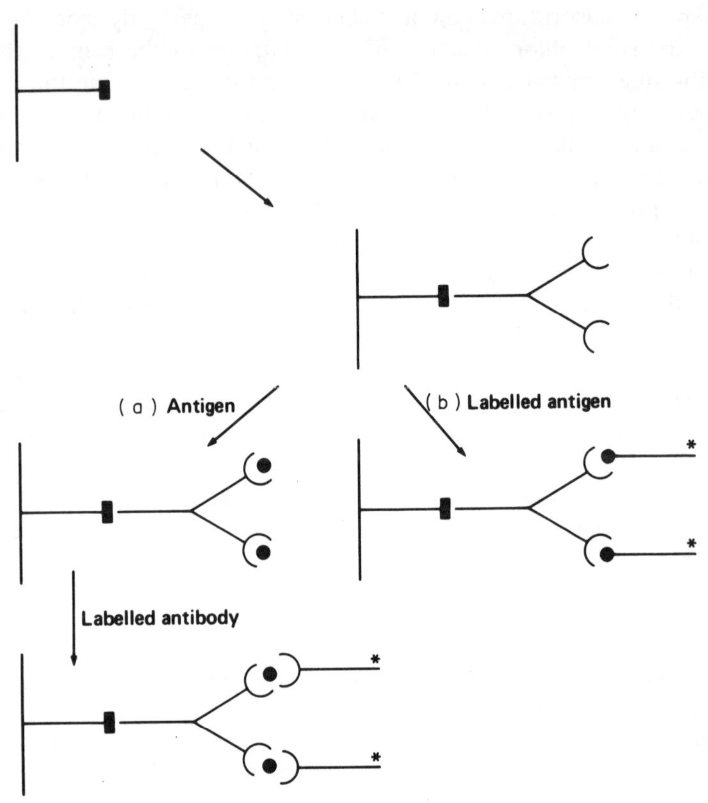

FIGURE 3 *Capture ELISA II.* Antibody is captured by anti-species antibody (which might be class specific) and detected with labelled antigen or with antigen and a labelled antibody

Capture by immobilized anti-species antibodies

The basis of this type of ELISA is shown in Figure 3. The sample to be assayed is incubated with a solid phase to which has been adsorbed an anti-species antibody, usually directed against a particular class of immunoglobulin, e.g. anti-human IgM. Immunoglobulins from the sample are bound to the solid phase and, usually in a subsequent incubation, are detected with a labelled antigen. For this particular assay format the antigen is most commonly labelled in a yet further incubation, with a conjugated antibody raised in the same species as the initial capture antibody and directed against the antigen. There is no reason why antigen which has been directly labelled cannot be used, and with the advent of cloned antigens this will become more common.

Sensitivity of assay

The most important problem met in this type of assay is that of sensitivity. Antibodies directed against the particular viral antigen of interest will only form a small proportion of the total antibody present in the sample, and will be bound in competition with the total present: resulting in the low sensitivity. This is in contrast with the situation described earlier, where only specific antibody is captured and then detected. However, experience has shown that assays of a particular class of immunoglobulin are more likely to be successful if the capture is with immobilized class-specific anti-species rather than with immobilized antigen. Although the reasons for this are not completely clear, it is probable that since IgG will normally be present in overwhelming concentrations and also with a higher avidity, it may exclude the class of interest if initial capture is on the basis of antibody specificity rather than class. In practice anti-species capture, because of the lack of sensitivity, is used almost solely for assays of immunoglobulins of a specific class, where the advantages outweigh the disadvantage.

Specificity of assay

Problems of specificity are not commonly met because labelled antigen is used in the detection system, but unless precautions are taken to prevent non-specific adsorption of an excess of cross-reacting IgG, it may on occasion be difficult to ensure class specificity even though the capture antibody be of high selectivity. Of course, the difficulty of ensuring detection of antibodies to all the epitopes of the virus has to be overcome, but this is usually a problem only if directly labelled antigen is used. The practice of designing a capture ELISA of this type is similar to that described above, the principle difference being that the specificity of the antibody used is critical.

ANALYSIS

The information which can be obtained from ELISA for antibodies is fourfold: whether a sample contains any antibodies to a particular virus; what particular classes of antibody are present; what epitopes they recognize; and, finally, how much of the antibody is present.

Defining positive results

Information about the presence or absence of antibodies to a virus is related to the sensitivity of the assay and, in turn, to the method of defining 'positivity'. The only true method of achieving this is statistical: there is some level of statistical certainty at which the signal generated by a sample in an ELISA can be said to be different from the signal arising from the general population of negative samples, but what level of difference is allowed before declaring a sample as positive depends on the purpose of the assay and the use to which the results will be put (Galen and Gambino, 1975). Detailed analysis is beyond the scope of this chapter but as a guide for the initial validation of an assay, it is common to select a difference of two to three standard deviations from the mean of a number of different negative samples as implying that a particular sample is different from those negatives. Whether the difference is due to the presence of the antibody which it is hoped to detect or to some other interfering factor will depend on how well the ELISA has been developed.

Statistical problems

There are at least two problems associated with adopting this statistical approach. One is that unless the signals produced by the general population of negative samples are normally distributed the level of confidence associated with a particular cut-off will be incorrect, but the loss of accuracy is usually negligible in practice. A second problem is that it is usually inconvenient to measure several different negative samples along with the sample under test, and a single negative measured repeatedly is obviously not a valid indicator of the general negative result. In these cases it is usual to include a 'cut-off' standard which has been defined beforehand as being at the appropriate statistical difference from the population of negatives.

The chosen level of statistical confidence (making the assumption of normality) is then applied to the mean of those negatives and the dilution of the positive consistently giving the result closest to that required is chosen as the appropriate standard. Several different batches of prepared solid phase and conjugates should be used in defining the cut-off so that a value representative of likely future preparations is selected. Comparison with a cut-off standard determined from earlier assays is less efficient than using the information available within a particular assay, but saves work in that fewer negative samples need to run along with a presumed positive.

Non-statistical methods

Methods other than statistical analysis or comparison with a standard are used to allow distinction between positive and negative, for example absolute optical density differences from a negative value, or taking as positive a sample reading which is some multiple of the negative reading (the positive-to-negative ratio method). All such methods are based on the prior experience and judgement of the assayist, and would be more valid if supported by analysis of the data which must have been obtained to give the subjective criteria. The positive-to-negative ratio method is the least satisfactory of these methods when applied to ELISA because the negative value is strongly influenced by any variation in the background caused by the substrate or the operating parameters of the reading device.

Sometimes a sample may be tested in two ways, for example with a test well to which antigen has been adsorbed and with another well which acts as a control, identically treated but containing no antigen. A two-well format such as this may be adopted to overcome the problem of samples which are actually negative giving rise to a signal because of some form of non-specific binding. The treatment of this situation is very complex because absolute rather than statistical differences are involved, and as the difference between the two values will most probably change from assay to assay a great deal of sensitivity will inevitably be lost, unless both test and control are performed repeatedly and then statistically analysed sample by sample. This is a cumbersome procedure and it is probably best to avoid a two-well format (unless the loss of sensitivity can be tolerated), to improve the assay to reduce the non-specific effects, and to devise a confirmatory test which will screen out the non-specific results at a later stage.

Obtaining information about the class and epitopic specificity of the antibodies which may be present in a sample is identical, statistically, to that of deciding whether any antibody is present, although the assay may have a different form.

Quantitation

Quantitating the antibody in a sample rather than merely detecting its presence introduces additional difficulties (see Chapter 3). It is not correct to conclude that a sample contains only a small amount of antibody because it gives a signal little different from the negatives; a large amount of low affinity antibody and a small amount of high affinity antibody may give rise to the same signal in most assay formats (see Chapter 6). Moreover, different samples may give high signals in one assay format and low signals in another because of differences in the affinity of the detecting system. For this reason it is not at present possible to quantitate antibodies (other than monoclonals) in absolute units, the best that can be done is some sort of comparative evalution. (For *antigen* detection the signal generated bears a direct relationship to the amount present.)

The easiest comparison is between the signals generated by a known set of antibody concentrations and the sample. Unfortunately, the linear range of the usual ELISA detectors is restricted to about two orders of magnitude, i.e. from an absorbance of about 0.02 to 2.0, while the anti-viral antibody concentration in a sample could range over five or six orders of magnitude (de Savigny and Voller, 1980), so the sample may have to be diluted to bring a strong signal within the range of the detector, or the assay stopped at different times after adding the substrate for the enzyme label.

Dilution series

A single assay of antibody concentration, even if carried out in replicate, is not reliable because of the lack of linearity of ELISA results. It is more informative to prepare a series of dilutions of the unknown sample and to compare the titration curve so obtained with that of the standard, either the last dilution different from the negative standard or a direct comparison with the whole of the standard curve being used to give the antibody concentration by application of the methods set out in Colquhoun (1971). The estimate of an antibody concentration from endpoint dilution is less satisfactory than that obtained using the information available from the whole of the titration curve because dilutions are notoriously difficult to reproduce, an error of one serial dilution being typical, and samples containing a preponderance of high affinity antibodies will give a markedly skewed titration curve which is difficult to analyse. Even comparisons of antibody concentration are assay-format dependent, and it is very unlikely that two different assays will have the same range of epitopic specificity or the same specificity in the detection system. This has been proved to be the case in the evalution of anti-AIDS test kits where kits perform differently with different sets of samples and it is not possible to give a ranking of antibody content which is consistent between the various kits.

ACKNOWLEDGEMENTS

I would like to thank my colleagues at Wellcome Diagnostics for their patience in discussing many of the propositions presented here.

REFERENCES

Bergmeyer. H.-U. (1986). *Methods of Enxymatic Analysis*, Vol. 10, *Antigens and Antibodies*. VCH Press, Berlin.

Carlson, J.R., Hinrichs, S.H., Levy, N.B., Gardner, M.B., Holland, P., and Pedersen, N.C. (1985). Evaluation of commercial AIDS screening kits. *Lancet*, **1**, 1388.

Colquhoun, D. (1971). *Lectures of Biostatistics*, Clarendon Press, Oxford.

Conradie, J.D., Govender, M., and Visser, L. (1983). ELISA solid phase: partial denaturation of coating antibody yields a more efficient solid phase. *J. Imm. Methods*, **59**, 289.

de Savigny D., and Voller, A. (1980). The communication of ELISA data from laboratory to clinician. *J. Immunoassay*, **1**, 105.

Galen, R.S., and Gambino, S.R. (1975). *Beyond Normality: the predictive value and efficiency of medical diagnoses*. Wiley, Chichester.

Imagawa, M., Hashida, S., Ishikawa, E., Niitsu, Y., Urushizake, I., Kanazawa, R., Tachibana, S., Nakazawa, N., and Ogawa, H. (1984). Comparison of β-Å- galactosidase from *E. coli* and horse radish peroxidase as labels of anti-human ferritin FAB by sandwich enzyme immunoassay technique. *Jap. J. Biochem.*, **96**, 659.

Ishikawa, E. (1980). Enzyme labelling of antigens and antibodies for quantitative enzyme immunoassay. *Immunoassay* (Suppl.), **1**, 1.

Mortimer, P.P., Parry, J.V., and Mortimer, J.Y. (1985). Which anti HTLV-III/LAV assays for screening and confirmatory testing? *Lancet*, **ii**, 873.

Ouldridge, E.J., Barnett, P.V., Parry, N.R., Syred, A., Head, M., and Rweyemamu, M.M. (1984). Demonstration of neutralizing and non-neutralizing epitopes on the trypsin-sensitive site of foot-and-mouth disease virus. *J. Gen. Virol.*, **65**, 203.

Sachers, M., Emmerich, P., Mohr, H., and Schmitz, H. (1985). Simple detection of antibodies to different viruses using rheumatoid factor and enzyme labelled antigen. *J. Virol. Methods*, **10**, 99.

Von Klitzing, L., Schultek, T., Strasburger, C.J., Fricke, H., and Wood, W.G. (1982). Comparison between adsorption and covalent coupling of proteins to solid phases using different polymers as support. *Radioimmunoassay Relat. Proced. Med., Proc. Int. Symp.* p. 57.

ELISA and Other Solid Phase Immunoassays
Edited by D.M. Kemeny and S.J. Challacombe

CHAPTER 15

Application of ELISA to Microbiology

S.J. Challacombe
UMDS, Guy's Hospital, London

CONTENTS

INTRODUCTION

Cultivation of the infecting agent is the mainstay in the diagnosis of infectious diseases. However, there are many diseases where cultivation of the infecting agent is not generally available, and in particular several viral agents that cause medically important diseases cannot be easily grown in a tissue culture or animal systems. The latter include rotavirus, hepatitis A and hepatitis B virus, and Epstein–Barr viruses. In other diseases, the infective agent is not known, and contributions to the diagnosis must be made from serological investigations of various body fluids. In diseases of uncertain aetiology, the study of naturally induced antibodies to micro-organisms is necessary to elucidate their pathological significance. This may be particularly important for those diseases in which a number of different bacteria may be implicated.

Thus solid phase assays have two quite distinct applications with regard to microbiology. The first is the detection of a bacterial antigen in various body fluids, which may help in the direct diagnosis of various diseases, and the second is in the detection of antibodies to bacteria, which may help in the indirect diagnosis of disease, and in the elucidation of pathogenic mechanisms.

THE DETECTION OF MICROBIAL ANTIGENS

Detection of bacterial antigens in various body fluids as an aid to diagnosis of an infectious disease is useful where traditional methods of cultivation are for some reason inadequate. As discussed above, this is particularly applicable to various viral infections and these are discussed in more detail in Chapter 13. However, there are a number of other situations in which techniques other than cultivation are necessary for the rapid diagnosis of an infectious disease. These include situations such as systemic fungal diseases where the replication time of the organism might be long (McIntosh *et al.*, 1978) and cases where specimens are obtained after partial antibiotic treatment (Winkelstein, 1970).

A prerequisitive of the usefulness of this type of test is that the microbial particles are antigenic and that the antigens are available for binding to the detection antibody. The test should be able to be completed in a short period of time and should be applicable to a wide range of infectious agents in a range of body fluids.

In recent years, many assays have been developed to attempt to detect infectious agents. These include immunodiffusion (Prince, 1968), agglutination (Ward *et al.*, 1978), counterimmunoelectrophoresis (Thirumoorthi and Dajani, 1979), immunofluorescence (Fulton and Middleton, 1974) and, more recently, radioimmunoassay (Felber, 1978). Many of these assays have been developed primarily for the detection of viral antigens, but are also applicable to bacterial and fungal antigens.

Radioimmunoassay

This form of immunoassay has achieved widespread usage because of its sensitivity and because the antigen–antibody reaction can be measured objectively by the use of gamma counters, rather than subjectively. It has proved very useful in situations where screening of blood samples for antigens is required. However, radioimmunoassays have a number of disadvantages, which include the short half-life of common gamma emitting isotopes, the requirement for expensive gamma counting equipment, and the potential radiation hazard of the radioactive reagents.

The limitations have led to the development of other solid phase assay systems, which have the same benefits of being sensitive and objective, but do not require the use of radioactive reagents.

ELISA

The benefits of the ELISA system, radioimmunoassay (RIA), and other immunoassays have been discussed in earlier chapters. An assay system based on colour has the advantage over other tests, such as agglutination, fluorescent or radioactivity in that the antigen–antibody reaction can be measured objectively in simple colorimeters (Engvall and Perlmann, 1972) and can also be seen with the naked eye (Bullock and Walls, 1977). In addition, the use of ELISA microtitre plate readers enables a large number of reactions to be read in a short period of time. Thus ELISAs can be adapted to large-scale testing for infectious agents (Voller, Bartlett, and Bidwell, 1976). In principle, the ELISA system can be used either for large-scale determination of the presence or absence of an agent or, in relation to standards, the amount of antigen can be quantitated (Engvall and Perlmann, 1972).

Applications of ELISAs

In recent years, ELISAs have been developed to detect antigens in a variety of fields. As well as widespread use in the field of virology (reviewed in Chapter 13) ELISAs have proved useful for the quantitation of the type-specific antigens of *Haemophilus influenzae* (Pepple, Moxon, and Yolken, 1980), parasitic antigens (Voller, Bartlett, and Bidwell, 1976), and for the diagnosis of invasive candidiasis by the measurement of serum antigen (Warren *et al.*, 1977).

Many bacteria are pathogenic because of the production of bacterial toxins, and enzyme immunoassays have proved useful for the measurement of such toxins as the heat labile toxin of *Escherichia coli* (Yolken *et al.*, 1977; Merson *et al.*, 1980). The assay has also proved useful for the rapid and simple identification of such heat labile enterotoxin-producing *E. coli* (Czerkinsky and Svennerholm, 1983). In this example, the assay is used to detect and identify whole bacteria from fluids which may contain a multitude of non-pathogenic bacteria.

Sputum samples. ELISA systems are capable of detecting antigens from *Haemophilus* and other lower respiratory tract infections in clinical specimens where culture might prove difficult, especially those which have become negative because of previous treatment with antibiotics. Thus Drow, Maki, and Manning (1979) have used the sandwich enzyme-linked immunoassay for rapid detection of *Haemophilus influenza* type B infections, and Harding *et al.* (1979) have used the assay for the detection of *Streptococcus pneumoniae* antigen.

Comparison of ELISA with other methods of antigen detection

ELISA sytems are undoubtedly much more sensitive than other immunoassay systems such as immunofluorescence, counterimmunoelectrophoresis, and latex agglutination. However, this advantage in terms of sensitivity is counterbalanced by the disadvantage of a longer assay time (usually 4 hr or greater) which might often be too long for urgent therapeutic decisions.

It is probable that, though the ELISA systems are sensitive, they are less reliable than culture for detection of certain organisms. Yolken (1982) reported that ELISA systems detected less than half of the clinical specimens proved positive for *Mycoplasma pneumoniae* and *Chlamydia trachomatis*. This meant that, while the ELISA was much more rapid and less expensive than tissue culture isolation, the system required the culture assays were always performed on negative specimens to ensure accurate diagnosis. The detection may be particularly capricious where specimens are obtained after partial antibiotic treatment (Winkelstein, 1970).

ELISA systems for the detection of antigen

A number of different ELISA systems have been developed for the direct measurement of infectious agents and their antigens in body fluids. Theoretically it is possible to use either enzyme-labelled antigen in a competitive assay or enzyme-labelled antibody in a direct measurement assay. Since, with most infectious diseases, sufficient quantitites of purified antigen are not readily available, it has been much more practical to devise ELISA systems that make use of enzyme-labelled antibody rather than antigen. In essence these assays may be direct or indirect.

Direct ELISA for antigen measurement

The principle of this test is shown in Figure 1. It is the simplest form of the non-competitive assay, and in this assay an unlabelled antibody is bound to the solid phase, which could be either a microtitre plate or a plastic bead. The binding of immunoglobulin or purified antibody to the solid phase by simple absorption or covalent linkage is discussed in Chapter 2 (Herrmann and Collins, 1976). After the removal of unbound antibody by washing, the plates can be stored until needed. The clinical sample is added then at a series of appropriate dilutions to the solid phase in a buffer containing 0.1 per cent of non-ionic detergent such as Tween 20, and 0.5 per cent protein such as foetal calf serum or gelatine. This buffer can be added at an earlier stage to saturate unbound sites on the solid phase and to decrease the rate of non-specific absorption (Kato *et al.*, 1979). After incubation of the clinical specimen and washing, enzyme-labelled antibody is added. This enzyme-labelled antibody will of course react with antigen bound to the solid phase. The amount of this second antibody bound is directly proportional to the amount of antigen, and is determined by the addition of substrate after the removal of any free enzyme-labelled antibody. Quantitation of antigen can be determined by comparison

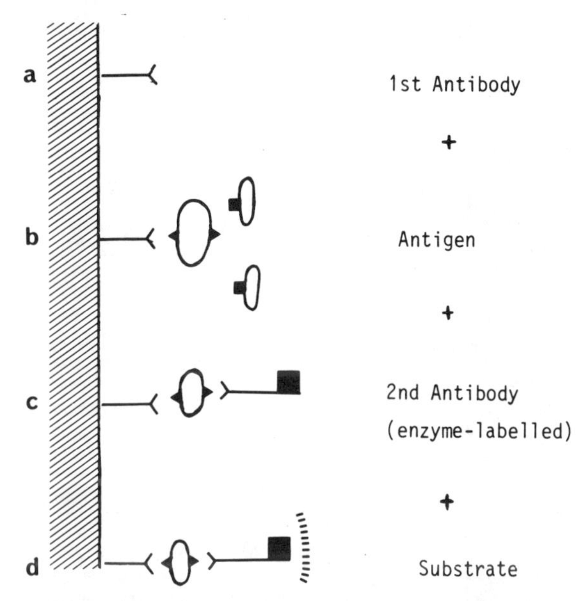

FIGURE 1 Direct ELISA for antigen measurement. The plate is coated with antibody to the antigen (a). The test fluid is added and any specific antigen will bind to the antibody (b). After washing, enzyme-labelled antibody is added which will bind only if specific antigen has been 'captured' (c). The conversion of substrate to colour is proportional to the amount of antigen captured (d)

with known positive test specimens, and with the inclusion of appropriate positive and negative controls.

A limitation of this technique is that for every antigen to be measured a specific antibody–enzyme conjugate is required. Sometimes the conjugation of antigen–enzyme is difficult, and this places limitations on the technique. This is a particular problem where a number of different antigens need to be assessed.

Indirect ELISA systems for antigen measurement

An advantage of indirect ELISA systems is that one of the limitations for the direct system discussed above, which requires a distinct antibody–enzyme conjugate for every antigen, is overcome. In indirect systems, any antigen in the clinical samples is bound to the solid phase just as in the direct system (Figure 2). At the second stage, instead of adding enzyme-labelled antibody, unlabelled antibody is added and reacted with bound antigen. After washing, any antibody bound to the antigen on the solid phase is reacted with an enzyme-labelled antiglobulin directed at the species of the second antibody. After a further incubation step, substrate is added and the reaction is quantified by relation to known positive controls.

The advantages of the indirect system is that a single labelled antiglobulin

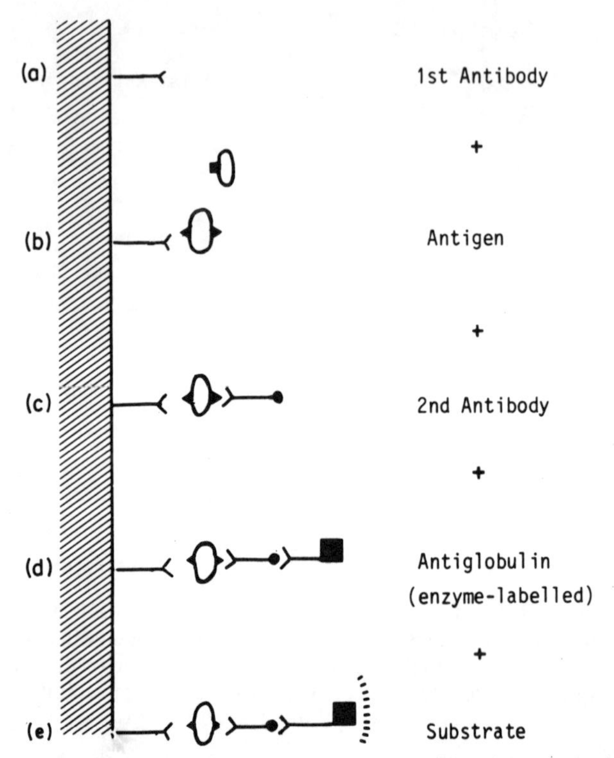

FIGURE 2 Indirect ELISA for antigen measurement. The plate is coated with specific antibody and antigen 'captured' as in the direct assay (Figure 1 a, b). Unlabelled antibody of a different species from that used for coating is added and binds only to specific 'captured' antigen (c). Enzyme-labelled antibody directed against the immunoglobulin of the species-specific antibody is added (d). After addition of a substrate the degree of colour conversion is proportional to the amount of antigen present (e)

reagent can be used for detection of a large number of different antigens, providing that the second antibody is paired in the appropriate species. The system is thus more flexible than the direct system.

Most authors have also found that the system is more sensitive than the equivalent direct system (Yolken, 1978; Yolken *et al.*, 1977). It is assumed that the increased sensitivity it due to the fact that several molecules of antiglobulin can react with a single molecule of second antibody.

The advantages of increased sensitivity can be offset by the disadvantage that the system requires an extra incubation stage with washing, and therefore takes longer (Yolken and Leister, 1981). In addition, the system requires antibody prepared in two different animal species with antigenically dissimilar immunoglobulins, to prevent non-specific interactions between the enzyme-labelled antiglobulin and the original antibody utilized for coating the solid phase.

THE USE OF ELISA FOR DETECTION OF ANTIBODIES TO BACTERIA

In many instances, the study of naturally induced antibodies to whole bacteria as well as to individual antigens is necessary to elucidate their pathological significance. This may be particularly important for those diseases in which a number of different bacteria may be implicated as well as for certain allergies. Antibodies to whole bacteria have often been detected using the fluorescent antibody test, or by agglutination. However, these assays are not very sensitive and are not strictly quantitative. The use of solid phase immunoassays offers an attractive approach since they combine sensitivity and simplicity and can be rendered isotype specific. However, whereas these assays have proved very sensitive with soluble antigens, which combine directly to the solid phase (Catt and Tregear, 1967), assays using whole bacteria have not proved so reliable, mainly because of the problem of instability of the bacterial attachment to the solid phase.

A whole bacterium contains many antigens on the cell surface, and can be regarded as a complex antigen. In many diseases, the relevant 'pathogenic' antigen has not been identified or purified, and whole bacteria can be used to give an indication of whether that particular bacterium is associated. However, an obvious disadvantage is that the use of whole cells increases the chance of cross-reactivity with other organisms, and thus specificity for the bacterial complex must be shown in any such assays. This is usually achieved by absoption of the sera with the homologous organism and results compared with the absorption with heterologous antigens. This is discussed in more detail below.

Attachment of whole bacteria to the solid phase

One of the major problems with whole bacteria, or with other particulate antigens, is to achieve consistent and persistent binding of the antigen to the solid phase. This can be a problem in assays where the total incubation times exceed 2 hr, as in most indirect assays. We have shown in ELISA experiments, where *Streptococcus mutans* was labelled with carbon-14 and bound to the plates using alkaline buffer, that desorption of bacteria from the solid phase occurred after a total incubation time of 2 hr.

Many different ways to enhance the attachment of bacteria or other particulate antigens to the solid phase have been reported. These include: (a) a direct method in an appropriate alkaline buffer, (Carlsson, Hurvell, and Lindberg, 1976); (b) the use of glutaraldehyde or methylglyoxal as an agent to fix the bacteria to the solid pahse (Czerkinsky *et al.*, 1983; Challacombe *et al.*, 1986); (c) an indirect or sandwich method where antibody to the bacteria is first coated on to the solid phase, which then results in attachment of capture bacteria (Challacombe, Czerkinsky, and Rees, 1983); (d) centrifugation and ethanol–methanol fixation (Barka, Tomasi, and Stadtsbaeder, 1986); and (e) Tris–HCl buffer with EDTA (Hancock and Tsang, 1986); and (f) poly-L-lysine; and (g) glutaraldehyde. Some of these attachment methods are covered elsewhere in this issue.

Barka, Tomasi, and Stadsbaeder (1985) have used whole *Streptococcus pneumoniae* cells as a solid phase absorbent to assay C-reactive protein in ELISA. In this instance bacteria were attached to the solid phase using a centrifugation at 350×*g* for 10 min and then fixation with ethanol followed by methanol. The same group of workers have extended the method of attachment to *Legionella pneumophilia* cells and report that the ELISA is more sensitive for the serodiagnosis of legionellosis than indirect fluorescence (Barka, Tomasi, and Stadtsbaeder, 1986).

Kinetic ELISA in microbiology

Kinetic-based enzyme-linked assays are thought to be very quantitative since the initial velocity of substrate conversion is directly proportional to the antibody concentration. In contrast, conventional ELISAs, which depend on doubling dilutions of serum and an endpoint determination, have substrate times which far exceed the initial linear reaction. These assays are thus thought by some to be only semi-quantitative (Hancock and Tsang, 1986). The latter authors have developed a K-ELISA for antibodies to *Schistosoma mansoni* where polystyrene beads are coated with antigen and the beads are attached to sticks moulded to the lid of microtitration plates. Reagents and sera are placed in the plates and the beads exposed by immersion. The exposure time for a single dilution of serum, conjugate, and substrate is 5 min each and timed interval repeat readings can be made. This method seems to overcome the main problem of K-ELISAs of elaborate reaction rate analysers.

Comparison of methylglyoxal and antibody attachment techniques

Bacteria

Both methods can be applied to the ELISA for a variety of Gram-positive and Gram-negative bacteria (Czerkinsky *et al.*, 1983). In our experience little difference between species in the attachment has been found except that optimal concentrations for each species may differ. Strains of *Streptococcus mutans*, *Actinomyces viscosus* (WVU 371) *E. coli*, and *Veillonella alcalescens* were selected as representatitve Gram-positive and -negative strains. *S. mutans* and *E. coli* were cultured in a semi-defined medium (Bowden, Hardie, and Fillery, 1976) at 37 °C for 18 hr under anaerobic conditions. *A. viscosus* and *V. alcalescens* were cultured in BBL Actino broth (BBL Ltd) for 3 days and 5 days respectively at 37 °C under anaerobic conditions. Bacteria were harvested, washed four times in PBS, pH 7.4, 0.1 M containing 0.1 per cent NaN_3, and resuspended in this buffer and stored at −20 °C (stock).

It should be noted that bacteria may absorb antigens from the culture medium. This may lead to erroneous conclusions regarding specificity. For example *S. mutans* has been claimed to cross-react with cardiac muscle based on immunization of rabbits, but it seems that some of this cross-reactivity was due to antigens absorbed from the broth (brain heart infusion) (Russell, 1985). Thus depending on the circumstance, defined or semi-defined media should be considered.

Prior to use in the radioimmunoassay, suspensions should be ultra-sonicated for 1 min at 20 KHz (Ultrasonics Ltd, Rapidis A180G) in order to dissociate bacterial chains and/or clumps. Shorter periods may be satisfactory with most Gram-negative bacteria but longer periods should be avoided as damage to the bacteria may occur. Dissociation should be confirmed by staining.

Antisera

Preparation of antisera. Antisera to each bacterium can easily be prepared in rabbits or mice. For mice we found that excellent IgG titres (approximately 1 in 10^7) can be obtained by immunization with 2×10^8 bacteria in 100 μl of H37Ra adjuvant (Difco Ltd) intraperitoneally at 0 and 14 days with antisera taken at 21 days.

Commercial antisera. It is possible to obtain most of the antisera required for the ELISA commercially (e.g. sheep anti-mouse immunoglobulins (Eivai Bios Ltd) or the IgG fractions of rabbit anti-sheep, swine anti-rabbit, rabbit anti-human IgG, IgM, IgA (heavy-chain specific) (DAKO Ltd), and goat anti-human IgG (F(ab) fragment specific) (Northeast Biomedical Labs, Inc.). With most commercial polyvalent antisera it is advisable to separate out and use the IgG fractions so that the background 'noise' is minimized. These can be obtained quite easily using standard DEAE cellulose chromatography (see Hudson and Hay, 1980).

Absorption of antisera. It is necessary on occasion to absorb polyvalent antisera with whole bacteria to make them monospecific. We have found that a suitable method is to absorb antisera at a dilution of 1 in 10 with 1/8 volume of packed bacteria for 1 hr at 37 °C followed by an overnight incubation at 4 °C before use in the assay. Occasionally a second absorption is necessary to abolish cross-reactive antibodies completely.

Affinity-purification of antibodies. In our experiments the IgG fraction of goat anti-human IgG (F(ab) specific) was further purified by affinity chromatography on Sepharose 4B (Pharmacia) to which human IgG had been coupled according to the manufacturer's instructions. Similarly, murine antiserum to

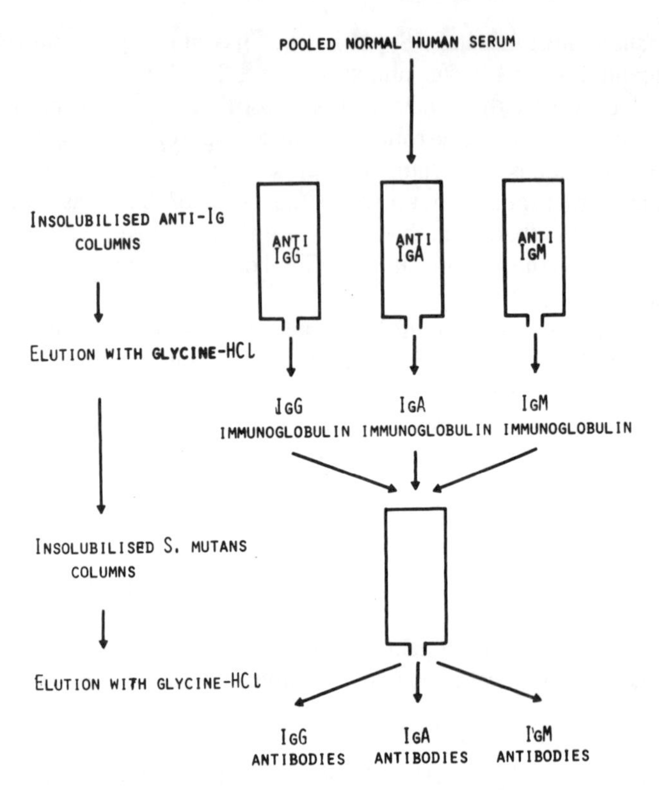

FIGURE 3 Preparation of class-specific affinity-purified antibodies to *Streptococcus mutans*. Normal human serum is separated into isotypes by passage through affinity columns containing anti-IgG, IgA, or IgM. The bound isotype is eluted, neutralized, and passed through an affinity column containing insolubilized bacteria. The bound isotype-specific antibody is eluted and neutralized

S. mutans and the other bacteria investigated were separated into IgG, IgA, and IgM fractions by affinity chromatography using insolubilized sheep anti-murine immunoglobulins and then passing the separated fractions through a column containing formalin-fixed cells of the appropriate bacteria (Figure 3). In this way isotype-specific affinity-purified antibodes can be obtained. An advantage of achieving this type of preparation is that once the protein concentration is known, and on the assumption that all the antibodies contained in the preparation are functional, then antibodies can be expressed as μg per ml or other quantitative values. Antibodies in samples of serum or secretions can also be expressed in quantitative terms by reference to a standard curve constructed from affinity-purified antibodies of known protein concentration (Figure 4). In this instance the usual assumptions of equivalent binding and parallelism between samples and standards would be made.

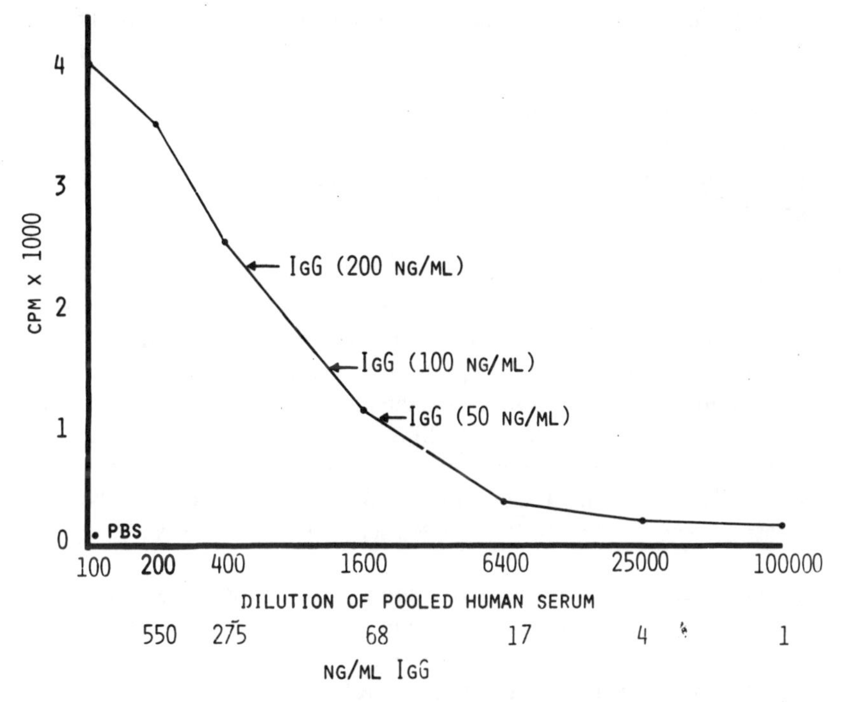

FIGURE 4 Standardization of normal human serum for isotype-specific antibody. Specific IgG antibody, affinity purified for class and antigen of known concentrations, is assayed and binding compared with the normal serum. The method assumes equivalent binding in the purified sample and the human serum

Preparation of the bacterial solid phase

We have investigated a variety of disposable tubes and microplates. We have found that for RIAs the optimal solid phase is polystyrene as wells in strips of eight (Removastrips, Dynatech Ltd), which were held in expanded polystyrene plates of 96 wells. This form of support makes it very easy to separate and count while retaining the benefits of a plate in terms of washing, etc. The type of flexible plates which are then cut up with scissors have proved less reliable and certainly more difficult to use in our hands. For ELISAs we have found polystyrene plates to be satisfactory especially if iradiated—these are discussed in more detail in Chapter 2.

Adsorption of bacteria on to the solid phase

Preliminary experiments using formaldehyde-killed or thermotreated (60 °C for 1 hr) bacteria showed that direct binding onto polystyrene wells could be achieved by incubation in PBS or alkaline buffers. However, the binding was found to be unstable and incubations of greater than 2 hr, or repeated washings, resulted in variable desorption of the bacteria from the solid phase. This considerably affected the sensitivity and the reproducibility of the assays.

The use of either an antibody-coated phase (two-step procedure) or of a methylglyoxal solution at high ionic strength (one-step procedure) resulted in stable and apparently irreversible binding of the bacteria to the solid phase, and greatly increased the sensitivity and reproducibility.

Two-step (antibody) procedure. Wells are filled with 100 μl of the IgG fraction of rabbit antiserum raised against the appropriate bacteria. The diluent is PBS, pH 7.4, containing 0.02 per cent NaN_3, and plates are incubated overnight at room temperature and then washed three times in PBS 0.1 M, pH 7.4. The amount of immunoglobulin absorbed to the plastic is only a fraction of the total applied to the well (Herrmann and Collins, 1976), and may be as little as 1 per cent with some proteins. A total of 100 μl of bacterial suspension is added to the wells and further incubated overnight at 4 °C or for shorter periods at room temperature. The incubation times are discussed in more detail below.

One-step (glyoxal) procedure. A stock solution of methylglyoxal (ICN Pharmaceuticals, Inc.) is made up as a 2 per cent solution (w/v) in distilled water and adjusted to pH 8 with 10 per cent bicarbonate solution. This can be kept at 4 °C for several months. The stock solution is diluted to 0.3 per cent with distilled water for coating. Although early experiments showed that it was possible to treat plates with glyoxal and then attach bacteria, effective attachment can be achieved by incubating the bacteria in the glyoxal solution. The latter involves one incubation step instead of two and is just as effective.

A total of 100 μl of the bacteria suspension, freshly adjusted to the apropriate concentration in 0.3 per cent glyoxal, is added to the wells and is incubated at 37°C. An incubation period of 90 min was found to be optimal (Figure 5). Control wells were filled with methylglyoxal alone.

Incubation time

In the two adsorption procedure, incubation times ranging from 6 to 24 hr were examined at temperatures of 37 °C, 4 °C, or room temperature for both steps (binding of IgG antibodies and subsequent adsorption of bacteria). Incubation periods of 8 hr or longer resulted in a maximal uptake of the IgG antibodies and also subsequently of the bacteria, and incubation at room temperature was found to be optimal. Thus, for convenience, two overnight incubations at room temperature were selected as standard conditions.

In the one-step procedure, maximal binding was obtained after 1 hr at 37 °C in a 0.3 per cent solution of methylglyoxal containing either *S. mutans*, *A. viscosus*, or *V. alcalescens* (Figure 5).

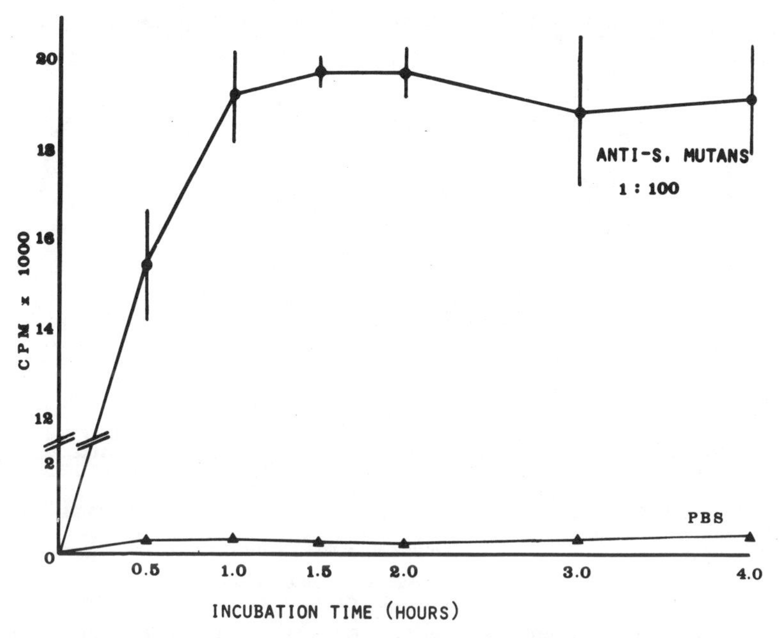

FIGURE 5 Kinetics of coating polypropylene plates with *Streptococcus mutans* cells in glyoxal. A standard serum was used to assay the effect of different coating times. No significant increase in binding was found after 1 hr, with either an RIA or ELISA

Bacterial concentration

The number of bacteria added to the polystyrene wells was found to be an important variable (Figure 6). Experiments undertaken in the standard conditions described, but with variable concentrations of bacteria, showed identical results of both procedures, although the mechanisms of attachment were different. Thus, concentrations of 2×10^9 bacteria/ml, for *V. alcalescens* or *S. mutans* and 5×10^9/ml for *A. viscosus* (Figure 6) were found to be optimal using either technique. At higher concentrations the sensitivity of the assays decreased, usually because non-specific binding increased (Figure 6).

The results of a representative experiment using *S. mutans* as the bacterial solid state are summarized in Table 1. Both procedures of bacterial attachment appeared to be reproducible and discriminative (ratio +ve/−ve). Initial attachment of bacteria to the solid phase could be achieved with PBS alone, but the variation was always greater than with the two linkage methods described, and desorption of bacteria was evident with longer incubation periods. Similar results were obtained with *V. alcalescens* and *A. viscosus* using either mouse or human sera. The one-step coating procedure was selected as the reference method for other experiments.

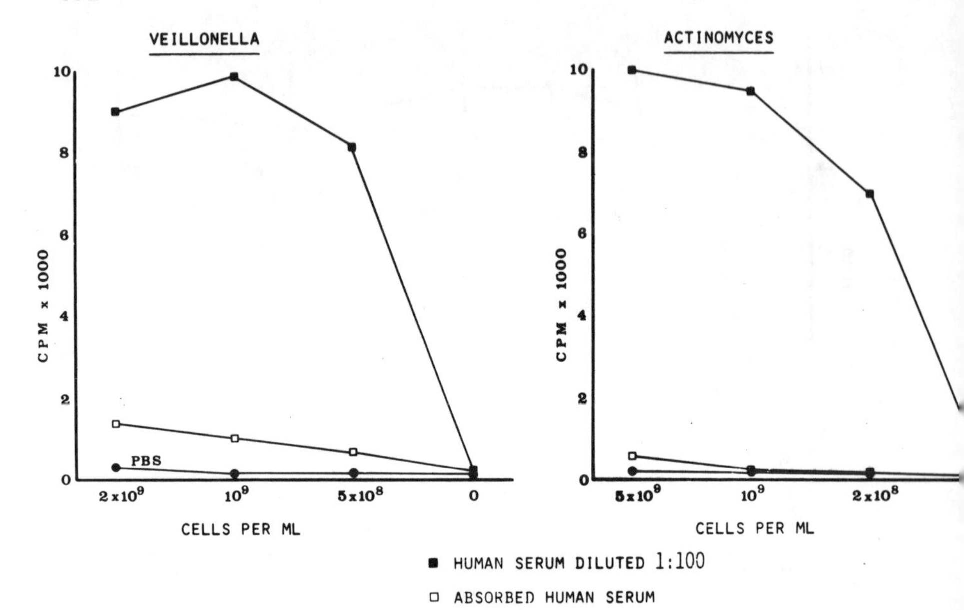

FIGURE 6 Optimal concentrations of bacteria used for coating the solid phase (for details see test). Most bacteria examined have an optimal coating concentration of between 5×10^8 and 5×10^9 in both RIA and ELISA

Blocking stage. After thorough washing with PBS, the wells are filled with PBS containing 0.5 per cent (w/v) bovine serum albumin, 0.05 per cent (v/v) polysorbate 20 (Tween 20), and 0.02 per cent NaN_3, and can be incubated for 1 hr at 37 °C or overnight at 4 °C, whichever is more convenient.

TABLE 1 Comparison of coating methods of bacteria. The methods are detailed in the text. Murine anti-*Streptococcus mutans* whole serum was used. The coefficient of variation was calculated from the mean cpm ± SD of 51 × paired samples assayed together for IgG antibodies

Method	Immune mouse serum	CV (%)	Non-immune serum	CV (%)
Antibody method (two step)	833 ± 38	4.6	10.7 ± 3.6	6.9
Glyoxal method (one step)	874 ± 46	5.3	17.6 ± 5.7	6.5
Cells alone	976 ± 81	8.3	16.8 ± 12.4	14.8

ELISA units read against standard curve.

Assay stage

Plates prepared by any method should be washed with PBS containing 0.05 per cent polysorbate 20, which is used as the diluent throughout the remainder of the assay. Samples of serum are added to the wells in volumes of 100 μl. The initial dilution of the serum samples can be 1/100 (requiring 1 μl of sample per well), or 1:1000, or even greater depending on the antibody concentration. For assay of antibodies in secretions the starting dilution is usually 1:50. We have now adopted the system of assaying a single well of sample at each of four dilutions. Each is read from the standard curve, adjusted for the dilution factor. If the dilutions have been accurate, then all four dilutions should give approximately the same value. The value for the test sample is the mean of these four dilutions providing all values fall within the range of the standards. Control wells are coated with bacteria and treated with PBS. After a standard incubation time of 90 min at 37 °C (see below for optimal times), the plates are washed four times and treated with radiolabelled or enzyme-labelled antisera (direct assay) or with an intermediate step for isotype specificity (indirect assay) (Figure 2).

Incubation times for sample

We have examined the optimal incubation times for binding of each antibody isotype; serially diluted sera were incubated for various periods of time at 37 °C in the bacteria-coated wells. Significant binding of specific antibodies of each isotype could be detected after 5 min and reached maximum values after approximately 90 min (Figure 7), although 90 per cent of IgM and IgG had bound by 30 min. A 90-min incubation of the serum sample was selected as optimal. Similar results have been found with soluble antigens. Titrations should be performed not only for different antigens but also for different fluids such as secretions.

Direct immunoassay — concentration of antiglobulin

Antiglobulin reagents should be titrated against bacterial coated wells alone (background) and against a serum sample. The dilution of antiglobulin which yielded the greatest activity (cpm or optimal density), while still producing low backgrounds, should be chosen for subsequent assays. Concentrations in the region of 2 μg/ml are often optimal for rabbit anti-mouse Ig and sheep anti-mouse Ig, whereas lower concentrations (0.1 μg/ml) are required for affinity-purified reagents. In our experience a 2 hr incubation at 37 °C is adequate for the antiglobulin reaction, since longer periods of incubation do not increase the sensitivity of the assay (Figure 7). Shorter incubation times are feasible if maximum sensitivity or binding is not required.

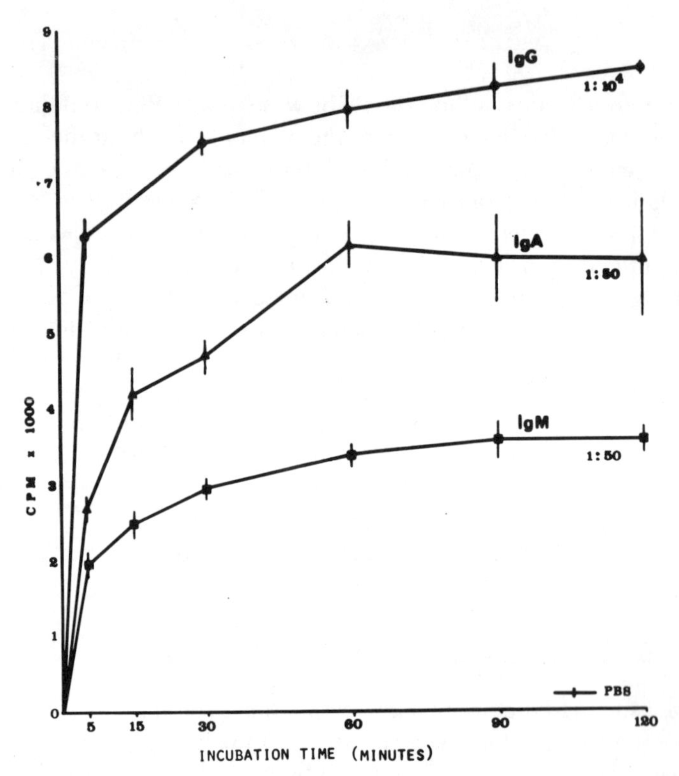

FIGURE 7 Comparison of binding of antibodies of different isotypes to *Streptococcus mutans* in relation to time. Significant binding is found by 5 min, though not at maximal levels until 90 min

Dilutions of anti-heavy-chain-specific antisera and anti-heterospecies antiglobulin reagents used in the indirect immunoassay usually fall in the range of 1–4 μg/ml.

Sensitivity

The use of isotype-specific affinity-purified antibodies of known concentration allows an estimate to be made of the sensitivty of the assay in quantitative terms. As an example, when affinity-purified antibodies to *S. mutans* of each isotype were prepared from mouse sera and assayed, antibodies of all three isotypes were detectable at 0.1 ng/ml or less. This is similar to that found with soluble antigens (Challacombe, Czerkinsky, and Rees, 1983).

Specificity of assays using whole bacterial cells

A potential problem of assays using whole bacterial cells is cross-reactivity. The objective of using cells is the indirect diagnosis of disease by looking for

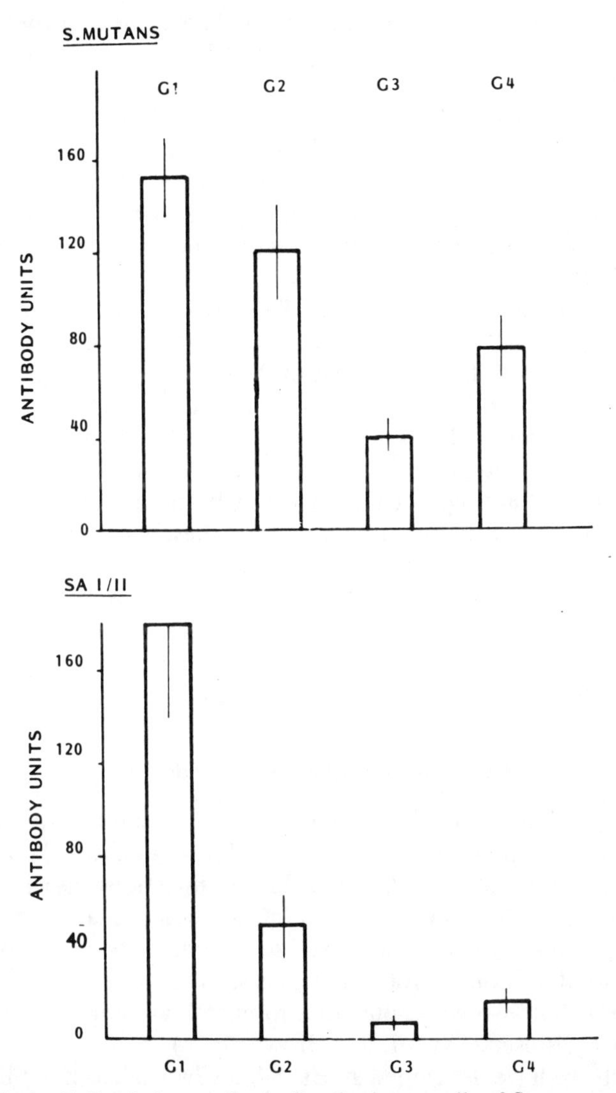

FIGURE 8 Relative IgG subclass antibody distribution to cells of *Streptococcus mutans* and streptococcal antigen I/II in 40 human sera (mean ‡SD). Values in individual sera were read from a curve of standard serum given arbitrary value of 1000 IgG antibody units, and the optical density values of IgG 1–4 antibodies read off the standard curve

changes in specific antibodies. This objective is lost if specificity to the bacteria cannot be shown. It should be borne in mind that natural antibodies induced by exposure of the host to antigens in the course of a disease may differ from antibodies induced in response to injection of antigen with adjuvant in animal models.

Essentially, specificity can be shown in two separate ways. The first is to absorb the fluid with a variety of bacteria and assay the absorbed fluid against the original organism. Thus, if a panel of ten bacteria has been used to absorb the antibodies and this is achieved significantly only with the homologous organism, it would indicate a degree of specificity. However, this specificity has only been demonstrated in terms of the organisms examined. Often, closely related species do not absorb out the antibodies whereas the homologous organism does. For example, in experiments reported by Czerkinsky *et al.* (1983), serotype specificity of antibodies directed against whole cells of *Streptococcus mutans* was found. Some serotypes of *S. mutans* did not absorb out antibodies directed against serotype c *S. mutans* and neither did other closely related species such as *S. sanguis*. Thus a high degree of specificity was indicated not only to the species but to the serotype. Serotype specificity has also been shown with *Legionella pneumophilia* cells as antigen (Barka, Thomasi, and Stadtsbaeder, 1986).

An alternative method is to attach a variety of bacteria to the solid phase and assay the same fluid against the panels of bacteria. As with absorption, if antibodies are found to bind only to one of these bacteria, then a degree of specificity has been demonstrated. This method is often complementary to the absorption method and in experiments reported with *S. mutans* serotypes the two methods correlated almost exactly (Czerkinsky *et al.*, 1983).

It is an important principle that with assays reported using whole cells or complex antigens some indication of specificity should be given.

ELISA detection of human IgG subclass antibodies

A recent interesting development in assay of antibodies to antigens is the analysis of the response in IgG subclasses. Little is known about the role and function of the subclasses of human IgG in infectious disease. However, a number of reports have suggested that IgG antibodies against glycoprotein and polysaccharide antigens may be restricted to one or two subclasses. This has been reported for non-microbial antigens such as rhesus antigens (Devey and Voak, 1974) but also with antibodies to microbial antigens such as group A streptococci (Riesen, Skvaril, and Brown, 1976).

Recently we have developed an ELISA for IgG subclass antibodies against whole cells of *S. mutans* and to a purified protein antigen derived from this bacteria designated streptococcal antigen I/II. In this assay, bacterial cells were bound to the solid phase using glyoxal and mouse monoclonal antisera against IgG and each human IgG subclass were used to detect antibodies. It was found that natural antibodies to *S. mutans* were principally of the IgG_1 and IgG_2 subclasses, though IgG_3 and IgG_4 antibodies were detectable in most subjects examined and indeed were the majority response in a few subjects. Antibodies to the Strep. antigen I/II were mainly of the IgG_1 subclass with virtually no activity detectable in the IgG_3 and IgG_4 subclasses (Figure 8).

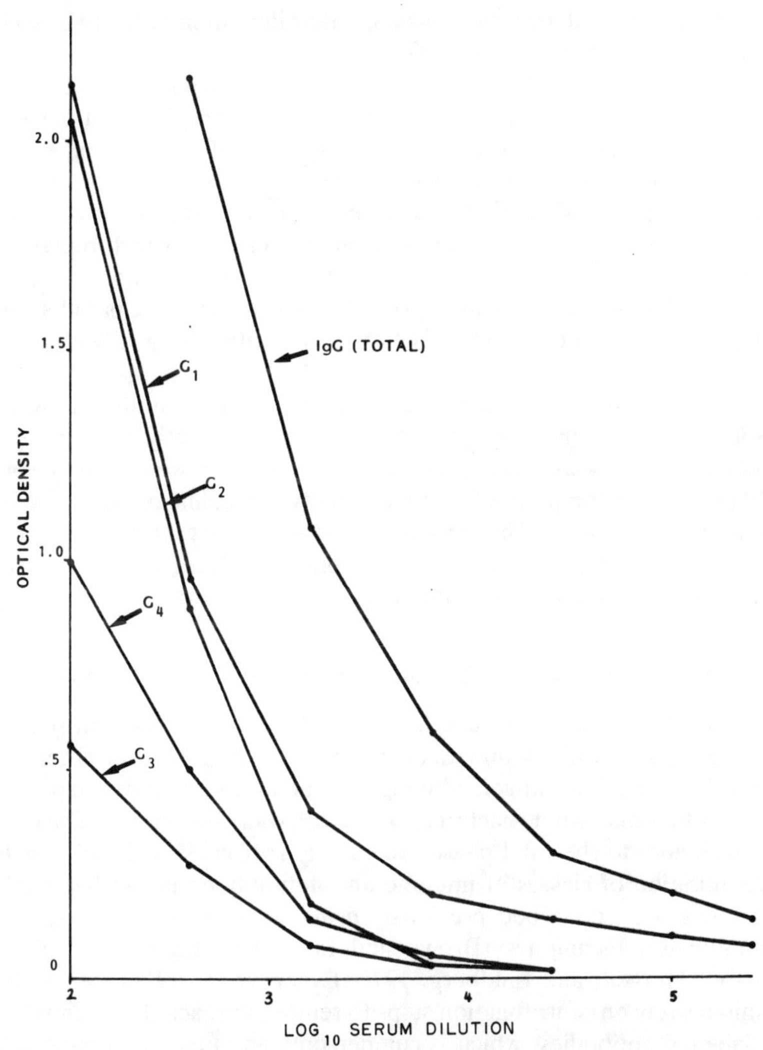

FIGURE 9 Standard curves of IgG subclass antibodies to *Streptococcus mutans* using a pooled human serum diluted in tenfold steps. Antibodies were assayed by ELISA using mouse monoclonal anti-human IgG 1–4 and alkaline-phosphatase-conjugated rabbit anti-mouse IgG. Absorbance was read at 405 nm

Whole bacterial cells can be regarded as a complex of several different antigens. In absorption studies with *S. mutans* using purified antigens the results suggested that some restriction of IgG subclasses to specific bacterial antigens was present in human sera since the SA I/II and c polysaccharide purified antigens could inhibit antibody binding of all subclasses to the whole cells of *S. mutans* whereas other purified antigens such as glucosyltransferase, lipoteichoic acid, and dextran showed greatest inhibition of the IgG_3 and IgG_4 subclasses (Challacombe *et al.*, 1986).

The methodology for detection of IgG subclasses is similar to that detailed in this chapter for the indirect method of antibody detection and depends on carefully characterized mouse anti-human IgG_1 to IgG_4 monclonal antibodies and detection with enzyme-labelled (alkaline phosphatase) rabbit anti-mouse at a dilution of approximately 1 in 500. The optimal concentration of the mouse monoclonals was found to be approximately 1 in 1000, which represented a protein concentration of about 2 μg per ml. The standard curves obtained for each of the IgG subclasses using a pooled human serum were parallel with the total IgG antibody curve (Figure 9) with the exception of IgG2 with showed a different slope.

It should be noted, however, that not all monoclonal antibodies would be acceptable for use in assay. It has been shown recently that monoclonal antibodies which were avid, potent, and specific for well-defined epitopes could partially or completely lose their activity depending on the assay system in which they were used. Thus specificity and activity in each assay system must be shown using positive and negative controls (Hussain *et al.* 1986), and optimal conditions carefully evaluated.

ASSAY OF BACTERIAL ANTIBODIES — CONCLUSIONS

The solid phase assay described is suitable for both soluble and particulate bacterial antigens and is summarized in Figure 10. Whole bacterial cells can be used as the particulate antigen, although the method is applicable to particulate antigens other than whole bacteria, such as cell walls and viruses. The methods for attachment to the solid phase are equally applicable to ELISA or RIA.

Quantitation of classes of immune and naturally occurring bacterial antibodies has been described previously using solid phase radio- or enzyme-immunological techniques (Brown and Lee, 1973, Granfors *et al.*, 1978, Carlsson, Hurvell, and Lindberg, 1976, Eggert *et al.*, 1979). Most of these techniques rely on centrifugation steps to remove unreacted immunoglobulins and labelled antibodies, which is cumbersome and time consuming and can generate bacterial losses (Brown and Lee, 1973). Attachment to the solid phase for ELISA techniques has proved unstable. Thus, the accuracy and the suitability for large-scale studies of such technical approaches remain questionable.

A key prerequisite for a sensitive and reproducible assay is stable attachment of antigen, whether soluble or particulate, to the solid phase. Adhesion, and more importantly retention, of bacteria to a polystyrene surface can be achieved by at least two different approaches. The first involves the binding of IgG antibodies to the plastic surface followed by the immuno-adsorption of the

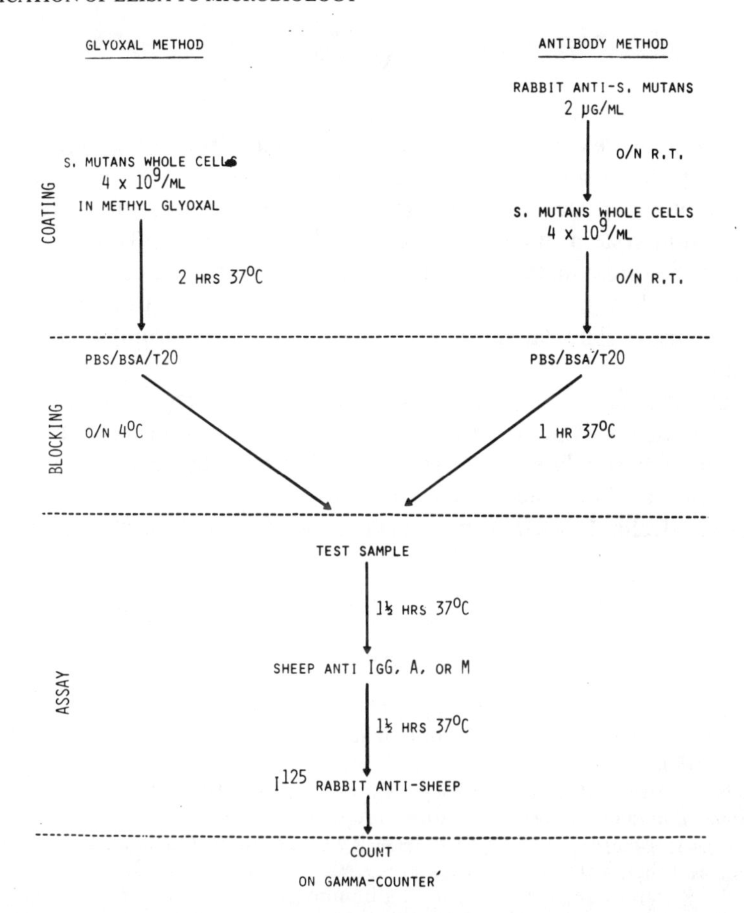

FIGURE 10 Outline of two methods of attachment of bacteria to the solid phase and assay of antibodies

corresponding micro-organism, whereas the other uses a polyaldehyde derivative (methylglyoxal) to link the bacteria directly to the solid phase. Both techniques result in apparently irreversible binding of bacteria, but the latter is preferred for its simplicity.

Although the mechanism of action of methylglyoxal is unknown, one could at least assume that it interacts as other polyaldehyde derivatives through cross-linking proteins exposed at the bacterial surface (Peters and Richards, 1977) or through electrostatic action with polystyrene.

The use of the solid phase for bacterial antibody assays is technically easy, since no centrifugation steps are necessary and the method can be adapted for the detection of antibodies in a variety of species.

SUMMARY

Solid phase assays have two distinct applications in microbiology. The direct diagnosis of disease can be made by the detection of bacterial antigen in body fluids. Both direct and indirect assays have been used. The principle is to 'capture' antigen in fluids by exposure to antibody attached to the solid phase. The detection systems may be by radiolabelling or enzyme labelling of antibodies. In the direct method the same antibody as attached to the solid phase may be used, but in the indirect method antiglobulin of a different species should be used. The latter method has proved very useful and versatile clinically.

Indirect diagnosis of disease may be achieved by detection of antibodies to microbial antigens. A major problem in this type of assay has been the reliable attachment of particulate antigens to the solid phase. This may be overcome by a variety of techniques including methylglyoxal, other fixatives, and antibody attachment. Reliable assays have been described for a wide variety of microbial diseases.

REFERENCES

Barka, N., Tomasi, J.P., and Stadtsbaeder (1985). Use of whole *Streptococcus pneumoniae* cells as a solid phase sorbent for C-reactive protein measurement by ELISA. *J. Imm. Methods*, **82**, 57–63.

Barka, N. Tomasi, J.P., and Stadtsbaeder, S. (1986). Elisa using whole *Legionella pneumophilia* cells or antigen. *J. Imm. Methods*, **93**, 77–81.

Bowden, G.H., Hardie, J.M., and Fillery, E.D. (1976). Antigens from actinomyces species and their value in identification. *J. Dent. Res.*, **55**, A192.

Brown, W.R., and Lee, E.M. (1973). Radioimmunologic measurements of naturally occurring bacterial antibodies. I. Human serum antibodies reactive with *Escherichia coli* in gastrointestinal and immunologic disorders. *J. Lab. Clin. Med.*, **82**, 125.

Bullock, S.L., and Walls, K.W. (1977). Evaluation of some of the parameters of the enzyme-linked immunospecific assay. *J. Infect. Dis.*, (Suppl.) **136**, S279–S285.

Carlsson, H.B., Hurvell, B., and Lindberg, A.A. (1976). Enzyme linked immunosorbent assay (ELISA) for titration of antibodies against *Brucella abortus* and *Yersinia enterocolitica*. *Acta Pathol. Microbiol. Scand. Sect. C. Immunol.*, **84**, 168.

Catt, K.J., and Tregear, G.W. (1967). Solid phase radioimmunoassay in antibody coated tubes. *Science*, **158**, 1570.

Challacombe, S.J., Czerkinsky, C., and Rees, A.S. (1983). Solid phase radioimmunoassay of class-specific antibodies to whole bacteria and soluble antigens. In: *Recent Developments in RAST and other Solid Phase Immunoassay Systems*. Eds Kemeny D.M. and Lessof M.H. Excerpta Medica, pp. 55–66.

Challacombe, S.J., Biggerstaff, M., Greenall, C., and Kemeny, D.M. (1986). ELISA detection of human IgG subclass antibodies to *Streptococcus mutans*. *J. Imm. Methods*, **87**, 95.

Czerkinsky, C.C., and Svennerholm, A.M. (1983). Ganglioside GMI enzyme-linked immunospot assay for sample identification of heat labile exterotoxin-producing *Escherichia coli*. *J. Clin Microbiol.*, **17**, 965–9.

Czerkinsky, C., Rees, A., Bergmeier, L., and Challacombe, S.J. (1983). Detection and specificity of naturally induced class specific antibodies to whole bacterial cells using a solid phase radio-immunoassay. *Clin. Exp. Immunol.*, **53**, 192–200.

Devey, M.E., and Voak, D. (1974). A critical study of the IgG subclasses of Rh anti-D antibodies formed in pregnancy and in immunized volunteers. *Immunology*, **27**, 1073–80.

Drow, D., Maki, D.G., and Manning, D. D. (1979). Indirect sandwich enzyme-linked immunosorbent assay for rapid detections of *Haemophilus influenzae* type b infections. *J. Clin. Microbiol.*, **10**, 442–450.

Eggert, F.M., Ebedo, L.B., Curner, B.W., and Coombes, R.R.A. (1979). A simplified procedure for measuring the class of antibacterial antibodies by mixed reverse passive haemagglutination (MRPAH). *J. Imm. Methods*, **26**, 125.

Engvall, E., and Perlmann, P. (1972). Enzyme linked immunosorbent assay. ELISA. III. Quantitation of specific antibodies by enzyme-linked anti-immunoglobulin in antigen coated tubes. *J. Immunol.*, **109**, 129–35.

Felber, J.P. (1978). Radioimmunoassay in the clinical chemistry laboratory. *Adv. Clin. Chem.*, **20**, 129–79.

Fulton, R.E., and Middleton, P.J. (1974). Comparison of immunofluorescence and isolation technique in rapid diagnosis of respiratory viral infections of children. *Infect. Immun.*, **10**, 92–101.

Granfors, K., Viljanen, M.K., Ahvonen, P., and Toivanen, P. (1978). Measurement of IgM and IgG antibodies to Yersinia by solid phase radioimmunoassay. *J. Infect. Dis.*, **138**, 232.

Hancock, K., and Tsang, V.C.W. (1986). Development and optimisation of the FAST-ELISA for detecting antibodies to *Schistosoma mansoni*. *J. Imm. Methods*, **92**, 167–76.

Harding, S.A., Scheld, W.M., McGowan, M.D., and Sande, M.A. (1979). Enzyme-linked immunosorbent assay for detection of *Streptococcus pneumoniae* antigen. *J. Clin. Microbiol.*, **10**, 339–42.

Hermann, J.E., and Collins, M.F. (1976). Quantitation of immunoglobulin adsorption to plastics. *J. Imm. Methods*, **10**, 363–6.

Hudson, L., and Hay, F.C. (1980). *Practical Immunology*, 2nd edn. Blackwell Scientific Publications.

Hussain, R., Pointdexter, R.W., Wistar, R., and Reimer, C.B. (1986). Use of monoclonal antibodies to quantify subclasses of human IgG — development of two-site immuno-enzymometric assays for total IgG subclass determinations. *J. Imm. Methods*, **93**, 89–96.

Kato, K., Umeda, U., Suzuki, F., Hayashi, D., and Kosaka, A. (1979). Use of gelatin to remove interference by serum with the solid phase enzyme-linked sandwich immunoassay of insulin. *FEBS Lett.*, **99**, 172–4.

McIntosh, K., Wilfert, C., Chernesky, M., Plotkin, S., and Mattheis, M. J. (1978). Summary of a workshop on new and useful methods in viral diagnosis. *J. Infect. Dis.*, **138**, 414–19.

Merson, M.H., Sack, R.B., Islam, S., Saklayen, G., Huda, N., Hug, I., Zulich, A.W., Yolken, R.H., and Kapikian, A.Z. (1980). Disease due to enterotoxigenic *Escherichia coli* in Bangladeshi adults: Clinical aspects and a controlled trial of tetracycline. *J. Infect. Dis.*, **141**:, 702–11.

Pepple, J., Moxon, E.R., and Yolken, R.H. (1980). Indirect enzyme-linked immunosorbent assay for the quantitation of the type specific antigen of *Hemophilus influenzae* b: a preliminary report. *J. Pediatr.*, **97**, 233–7.

Peters, K., and Richards, F.M. (1977). Chemical cross-linking: reagents and problems in studies of membrance structure. *Ann. Rev. Biochem.*, **46**, 523.

Prince, A.M. (1968). An antigen detected in the blood during the incubation period of serum hepatitis. *Proc. Natl. Acad. Sci. USA*, **60**, 814–21.

Riesen, W.F., Skvaril, F., and Brown, D. G., (1976). Natural infection of man with group A streptococci. *Scand. J. Immunol.*, **5**, 383.

Russell, M.W. (1985). Protein antigens of *Streptococcus mutans*. In: *Molecular Microbiology and Immunobiology of Streptococcus mutans*. (Ed. S. Hammond). Elsevier, Amsterdam, p. 51.

Thirumoorthi, M.C., and Dajani, A.S. (1979). Comparison of staphylococcal coagglutination, latex agglutination, and counterimmunoelectrophoresis for bacterial antigen detection. *J. Clin. Microbiol*, **9**, 28–32.

Voller, A., Bartlett, A., and Bidwell, D.E. (1976). Enzyme immunoassays for parasitic diseases. *Trans. R. Soc. Trop. Med. Hyg.*, **70**, 98–106.

Ward, J.I., Siber, G.R., Scheifele, D.W., and Smith, D.H. (1978). Rapid diagnosis of *Haemophilus influenzae* type b infections by latex particle agglutination and counter immunoelectrophoresis. *J. Pediatr.*, **93**, 37–42.

Warren, R.C., Bartlett, A., Bidwell, D.E., Richardson, M.D., Voller, A., and White, L.O. (1977). Diagnosis of invasive candidosis by enzyme immunoassay of serum antigen. *Brit. Med. J.*, **1**, 1183–5.

Winkelstein, J.A. (1970). The influence of partial treatment with penicillin on the diagnosis of bacterial meningitis. *J. Pediatr.*, **77**, 619–24.

Yolken, R.H. (1978). Use of ELISA in clinical medicine. *Hosp. Pract.*, **13**, 31.

Yolken, R.H. (1982) *Enzyme Immunoassays for the Detection of Infectious Antigens in Body Fluids: Current Limitations and Future Prospects*. National Institute of Allergy and Infectious Diseases, Thrasher Research Fund, Salt Lake City, Utah.

Yolken, R.H., and Leister, F.J. (1981). Investigation of enzyme immunoassays time courses: development of rapid assay systems. *J. Clin. Microbiol.*, **13**, 738–41.

Yolken, R.H., Greenberg, H.B., Merson, M.H., Sack, R.B., and Kapikian, A.Z. (1977). Enzyme-linked immunosorbent assay for detection of *Escherichia coli* heat-labile enterotoxin. *J. Clin. Microbiol.*, **6**, 439–44.

ELISA and Other Solid Phase Immunoassays
Edited by D.M. Kemeny and S.J. Challacombe

Appendix

D. Richards
UMDS, Guy's Hospital, London

BUFFERS

0.1 M Bicarbonate coating buffer — pH 9.6

4.24 g Na_2CO_3 (0.04 M) (BDH)
5.04 g $NaHCO_3$ (0.06 M) (BDH)

Make up to 1 litre with distilled H_2O and check the pH (approx pH 9.6). Store at 4 °C.

Citrate–phosphate buffer — pH 5 for OPD substrate

21.01 g $C_6H_8O_7.H_2O$ (0.1 M) (BDH)
28.40 g Na_2HPO_4 (0.2 M) (BDH)

Make up both reagents separately to 1 litre with distilled H_2O and add 48.5 ml of 0.1 M citric acid to 51.5 ml of 0.2 M Na_2HPO_4 producing a pH of approximately 5. Check and adjust as necessary.

0.1 M Diethanolamine buffer — pH 9.8 (for alkaline phosphatase)

101 mg $MgCl_2$. $5H_2O$ (50 mM) (BDH)
97 ml diethanolamine (BDH)
Concentrated HCl (BDH)

Dissolve $MgCl_2$ in 800 ml of distilled water. When dissolved add diethanolamine and mix thoroughly, then adjust the pH to 9.8 with concentrated HCl. Finally make up to 1 litre with distilled water and store in the dark at 4 °C.

Phosphate buffered saline (PBS) — pH 7.4 (isotonic saline)

16.7 g Na_2HPO_4 (0.012 M) (BDH)
5.7 g NaH_2PO_4 (0.003 M) (BDH)
85 g NaCl (0.15 M) (BDH)
100 mg NaN_3 (1.5 mM) (BDH)

Dissolve in distilled H_2O, make up to 10 litres, and check pH. Store at room temperature. Can be made up as a ×10 concentrated stock solution.

0.5 M Tris/HCl — pH 8.0

60.5 g Tris (hydroxymethyl) methylamine (BDH)
1 M HCl

Dissolve the Tris in 800 ml and adjust the pH to 8.0 using 1 M HCl.
Make up to 1 litre and store at 4 °C.

Coloured additives

40 mg Phenol red
60 mg Amido black

Make up both solutions separately in 100 ml of distilled H_2O and add 1 ml of each to 100 ml of buffer. These additives are optional and serve two purposes: they make it visually easier to add solutions to ELISA plates and they provide a rough guide to the pH of the buffer.

PBS/0.5% animal serum/ 0.5% Tween 20

99 ml PBS (0.05 M)
0.5 ml Animal serum
0.5 ml Tween 20

The particular animal serum used will depend on the reagents and antibodies used in the assay. Horse serum or rabbit serum is commonly used but cross-reactivity between serum and reagent antibodies may make it necessary to use another serum such as goat.

SUBSTRATES

***O*-Phenylenediamine (OPD)**

0.4 mg/ml OPD (Sigma P-1526)

0.4 μl/ml H_2O_2 (BDH)
Citrate–phosphate buffer

The substrate should be made up no earlier than 10 min before use and hydrogen peroxide added just prior to use. Incubation time of substrate is about 20 min after which the reaction is stopped by the addition of 50 μl of 2 M H_2SO_4. The absorbance must be recorded (at 492 nm) within 40 min as the substrate product is unstable.

Phosphatase substrate

1 mg/ml Sigma 104 phosphate substrate (104–105) (substrate comes as 5 mg tablets)
Diethanolamine buffer

Add tablets to diethanolamine buffer (one tablet for each 5 ml) and allow them to dissolve by mixing for 10 min. The substrate is stable at 4 °C for some time. Substrate incubation time is longer than for OPD with up to 2 hr at 37 °C. The reaction is stopped using 50 μl of 3 M NaOH. The coloured substrate product is very stable at 4 °C and plates can be stored for several days in the dark. Absorbance should be read at 405 nm.

BASIC ELISA PROTOCOLS

Indirect ELISA

Materials

Antigen
Positive standard
Quality controls
Patients' sera
Enzyme-labelled anti-human IgG
Substrate
PBS/0.5% animal serum/0.5% Tween 20
Bicarbonate buffer
Nunc Immuno-1 plate (96F)
Spectrophotometer (Titertek multiskan)

Method

(1) Coat the plate with antigen at a concentration between 100 μg/ml and 1 μg/ml, the optimum having been determined by experimentation. The antigen should be made up in bicarbonate coating buffer, pH 9.6, and

incubated with the plate at 100 μl/well overnight at 4 °C.
(2) The plate is washed three times with 300 μl of PBS/0.05% Tween 20.
(3) Positive standards, quality controls, and patients' sera at a suitable dilution are added to the plate (100 μl/well) which is then incubated at 4 °C for 2 hr.
(4) The plate is washed three times with PBS/0.05% Tween 20.
(5) Enzyme-conjugated anti-human IgG is added to the plate at 100 μl/well. The optimum concentration is determined by experiment and is usually between 1/100 and 1/500 (typically 5 μg/ml). The incubation time is 1 hr at 4 °C.

For alkaline phosphatase substrates:

(6) Plate washed twice with PBS/0.05% Tween 20 and then once with distilled water.
(7) Make up substrate in diethanolamine buffer, pH 9.8, as described above. Add 100 μl to each well and incubate at 37 °C for 1–2 hr.
(8) Stop enzyme reaction by adding 50 μl/well of 3 M NaOH.
(9) Measure absorbance at 405 nm.

For peroxidase substrates:

(6) Wash plate three times with PBS/0.05% Tween 20.
(7) Make up OPD substrate as described, add 100 μl/well and incubate at 37 °C for 40 min.
(8) Stop enzyme reaction by adding 50 μl/well of 2 M H_2SO_4.
(9) Measure absorbance at 492 nm.

Two-site ELISA for the detection of total IgE

Materials

Monoclonal anti-human IgE (Clone 7.12, kind gift Professor A. Saxon, UCLA, USA)
IgE standards
Test sera
Rabbit anti-human IgE alkaline phosphatase
Bicarbonate coating buffer, pH 9.6
Assay diluent (PBS/animal serum/Tween 20)
Diethanolamine buffer, pH 9.8
Substrate tablets (Sigma 104–105)
3 M NaOH
Washing buffer (PBS/0.05% Tween 20)
Nunc Immuno-1 plate (96F)

Spectrophotometer (Titertek multiskan)

Method

(1) Coat the plates with 100 μl/well of anti-human IgE antibody at a concentration of 1 μg/ml, made up in bicarbonate buffer. Incubate overnight at 4 °C.
(2) Wash the plate three times with 300 μl of washing buffer.
(3) Add standard, controls, and test sera at a suitable range of dilutions (1/10, 1/20, 1/50, etc.) made up in assay diluent. Add 100 μl/well and incubate at 4 °C for 3 hr.
(4) Wash the plate as before.
(5) Add the enzyme-conjugated anti-human IgE at an optimum concentration (between 1/100 and 1/500) made up in assay diluent. Incubate at 4 °C for 3 hr.
(6) Wash the plate twice with washing buffer and then once with distilled water.
(7) Add substrate and incubate for 1 hr at 37 °C or at 4 °C overnight.

Total IgG subclass ELISA

Materials

Oxoid mouse monoclonals:

Specificity	*Clone number*
IgG	8a4
IgG_1	NL16
IgG_2	6014
IgG_3	ZG4
IgG_4	RJ4

Rabbit anti-mouse immunoglobulin (DAKO cat. no. Z259)
Rabbit anti-human IgG alkaline phosphatase conjugated (DAKO cat. no. D336) (pre-absorbed against normal mouse serum as described below)
Positive standard
Quality controls
PBS/O.05% Tween 20
PBS/0.5% mouse serum/0.5% Tween 20
Bicarbonate coating buffer pH 9.6
Diethanolamine buffer pH 9.8
Substrate tablets (Sigma 104–105)
3 M NaOH
Spectrophotometer (Titertek multiskan)
Washing buffer (PBS/0.05% Tween 20)
Nunc-Immuno plate 1 (96F)

All incubation volumes are 100 μl/well.
Spectrophotometer (Titertek multiskan)

(1) Plate is coated overnight at 4 °C with a 1/3000 dilution of rabbit anti-mouse immunoglobulin in bicarbonate buffer.
(2) Wash three times with 300 μl of washing buffer.
(3) The different mouse anti-human IgG subclass monoclonals are added at a 1/1000 dilution in PBS/0.5% Tween 20 for 1 hr at 4 °C.
(4) Wash three times with 300 μl of washing buffer.
(5) The positive standard curve is diluted in PBS/ mouse serum/Tween 20 between $1/10^{-4}$ and $1/10^{-6}$ for all subclass estimations except IgG_4 ($1/10^{-3}$ to $1/10^{-6}$) and is added to the plate and incubated for 2 hr at 4 °C.
    The quality control and test sera should be used at three dilutions: 1×10^{-4}, 2×10^{-4}, 5×10^{-4} for all estimations except IgG_4 where they should be used at 1×10^{-3}, 2×10^{-3}, 5×10^{-3}.
(6) Wash three times with 300 μl of washing buffer.
(7) Enzyme labelled anti-human IgG diluted 1/300 in PBS/mouse serum/ Tween 20 is then added and incubated for 2 hr at 4 °C.
(8) Wash the plate twice with washing buffer and once with 300 μl of distilled water.
(9) Make up substrate as previously described and incubate at 37 °C for 40 min.
(10) Stop the reaction by adding 50 Ml of 3 M NaOH.
(11) Measure the absorbance at 405 nm.

Enzyme labelling of proteins

(1) One-step glutaraldehyde method

(for alkaline phosphatase)
(Avrameas, 1969)

Materials

Affinity-purified antibody
50% glutaraldehyde (Emscope Lab. Ltd)
Alkaline phosphatase (Sigma type VII-S P-5521)
0.5 M Tris pH 8
Phosphate buffered saline (PBS)
Bovine serum albumin (BSA)
NaN_3
Dialysis tubing-visking size 1, 8/32″ (Solmedia)

Method

(1) Antibody (2 mg/ml) and 5 mg of alkaline phosphatase activity (>1000 U/ml) are dialysed together against PBS, overnight at 4 °C with several changes of PBS, to remove all the ammonium sulphate present in the enzyme extracted.
(2) The concentrate is removed from the visking tubing and the volume measured. The amount of glutaraldehyde required to make a final concentration of 2% v/v is calculated. Glutaraldehyde is added to the mixture and incubated, continuously mixing, for 2 hr at 4 °C.
(3) The mixture is transferred to dialysis tubing and dialysed against PBS overnight at 4 °C with two changes of PBS.
(4) The dialysis bag is transferred to 0.5 M Tris, pH 8, and dialysed overnight at 4 °C with two changes of Tris buffer.
(5) The mixture is removed from the dialysis bag and diluted to 4 ml with Tris buffer containing 1% BSA and 0.2% NaN_3.
(6) Store in the dark at 4 °C.

(2) Two-step glutaraldehyde method
(for peroxidase or alkaline phosphatase)
(Avrameas and Ternynck, 1972)

Materials

Horseradish peroxidase (HRP) (Sigma type IV P-8375)
Phosphate buffered saline (PBS), 0.1 M, pH 6.8
50% glutaraldehyde (Emscope Lab. Ltd.)
Normal saline
Sephadex G100 (Pharmacia) or AcA 34 (LKB)
Affinity-purified antibody
Carbonate-bicarbonate buffer, 1 M, pH 9.5
Lysine, solution, 0.2 M
Saturated ammonium sulphate
Bovine serum albumin (BSA)
Millipore filter 0.22 μm (Flow Laboratories)
Glycerol

Method

(1) Horseradish peroxidase (10 mg) is dissolved in 200 μl of PBS (0.1 M pH 6.8) containing 1.25% v/v glutaraldehyde. The solution is then left mixing overnight at room temperature.
(2) The unbound glutaraldehyde is then removed either by dialysis overnight

against normal saline (two changes), or by passing the mixture down a Sephadex G100 or an LKB ACA gel filtration column, monitoring the absorbance of fractions at 280 nm and pooling the 'activated' HRP peak.

(3) The affinity-purified antibody is made up to a 5 mg/ml solution with normal saline.

(4) Mix 1 ml of activated HRP with 1 ml of antibody solution. 0.1 ml of 1 M carbonate-bicarbonate buffer pH 9.6 is then added and incubated with it at 4 °C overnight.

(5) Add 0.1 ml of 0.2 M lysine and mix at room temperature for 2 hr.

(6) The mixture is dialysed again PBS, pH 7.2, overnight at 4 °C.

(7) The HRP-conjugated antibody is then precipitated by adding an equal volume of saturated ammonium sulphate solution. The precipitate is spun down and supernatant removed. The precipitate is then washed twice and spun down using half saturated ammonium sulphate. The final precipitate is then resuspended in 1 ml of PBS.

(8) The conjugate is dialysed extensively against PBS overnight at 4 °C.

(9) The conjugate is then spun at 10 000 × *g* for 30 min to remove any sediment. The supernatant is removed and BSA or HSA (1% w/v) is added.

(10) The conjugated antibody is filtered through a 0.22 μm millipore.

(11) The conjugate is stored either at −20 °C or, if made up to a 50% glycerol concentration using equal volumes of glycerol and conjugate, it can be stored 4 °C.

(3) Periodate method
(suitable for alkaline phosphatase or peroxidase)
(Nakane and Kawaoi, 1974)

Materials

Anti-IgG immunoglobulin
Alkaline phosphatase (Sigma type VIII-S P-5521)
Horseradish peroxidase (Sigma type IV P-8375)
Na_2CO_3 buffer, 0.3 M pH 8.1
1-Fluoro-2,4-dinitrobenzene (Searle)
Ethanol
Sodium periodate, 0.08 M (dissolved in distilled, deionized water)
Ethylene glycol, 0.16 M (dissolved in distilled, deionized water)
Na_2CO_3 buffer, 0.01 M, pH 9.5
Phosphate buffered saline (PBS), pH 7.2
Sodium borohydride
Sephadex G200 85 × 1.5 cm column equilibrated with PBS, pH 7.2 (Pharmacia)

Method

(1) Dialyse 5 ml alkaline phosphatase or HRP against 0.3 M Na_2CO_3, pH 8.1, to remove all $(NH_4)_2SO_4$ and stabilizers. Add 100 μl of a 1% solution of 1-fluoro-2,4-dinitrobenzene, dissolved in absolute ethanol, to the dialysed suspenion and mix gently for 1 hr at room temperature.
(2) Then add 1 ml of 0.08 M sodium periodate (freshly made up) and gently mix for 30 min at room temperature.
(3) Add 1 ml of 0.15 M ethylene glycol and mix gently for a further hour at room temperature.
(4) Dialyse the enzyme–aldehyde solution against 0.01 M Na_2CO_3, pH 9.5, for 24 hr at 4 °C with at least three changes of buffer.
(5) The anti-IgG immunoglobulin is prepared by dialysing 5 mg against 0.01 M Na_2CO_3, pH 9.5, overnight at 4 °C.
(6) Add the dialysed immunoglobulin to the 3 ml solution of enzyme–aldehyde and mix for 2–3 hr at room temperature.
(7) NaBH (5 mg) is then added, dissolved, and then left to stand at 4 °C for 3 hr.
(8) The conjugate is then dialysed against PBS, pH 7.2, at 4 °C for 24 hr. Any precipitate that forms should be spun down and removed.
(9) The remaining conjugate should then be applied to the 85 × 1.5 cm Sephadex G-200 column and eluted using PBS, pH 7.2 at a flow rate of 5 ml/hr. Collect 2 ml fractions and monitor the absorbance at 280 nm, the first peak should contain the conjugate.
(10) Store conjugate at −20 °C in 1% BSA. Do not thaw and refreeze.

(4) Maleimide method
(for β-D-galactosidase)
(Kato *et al.*, 1976)

Materials

Affinity purified antibody or Fab′ fragments
β-D-galactosidase EC 3.2.1.23 (Sigma)
N, *N*′-*o*-phenylenedimaleimide (Aldrich)
Bovine serum albumin (BSA)
$MgCl_2$
NaOH 1 M
Sodium phosphate buffer, 0.01 M pH 7.0
Sodium acetate buffer, 0.01 M pH 5.0
Sephadex G25 (1 × 40 cm) column (Pharmacia)
Sepharose 6B (1.5 × 40 cm) column (Pharmacia)

Method

(1) The antibody should be prepared as a 3 mg/ml solution in 0.1 M sodium acetate buffer pH 5.0
(2) Two millilitres of the antibody solution is added dropwise into 1 ml of a saturated solution of *N*,*N'*-*o*-phenylenedimaleimide in 0.1 M sodium acetate pH 5.0 at 4 °C. The mixture is then incubated at 30 °C for 20 minutes.
(3) The solution is then separated on a Sephadex G-25 column (1.0 × 40 cm) pre-equilibrated with 0.1 M sodium acetate buffer pH 5.0 and the antibody containing fractions pooled.
(4) The pH of the isolated dimaleimide-treated antibody is adjusted to 6.5 using 0.25 M sodium phosphate buffer, pH 7.5.
(5) The β-D-galactosidase is dialysed against 0.01 M sodium phosphate buffer, pH 7.0, to remove the $(NH_4)_2SO_4$ present. After dialysis the pH of the mixture is adjusted to pH 7.0 and stored with final concentrations of 0.5% BSA and 1 mM $MgCl_2$.
(6) The dimaleimide treated antibody (0.5–1.0 mg) is incubated with 0.5 mg of β-D-galactosidase at 30 °C for 20 minutes in a final volume of 1 ml sodium acetate buffer pH 5.0.
(7) The pH is adjusted to 7.0 and the preparation applied to a column of Sepharose 6B (1.5 × 40 cm), which had previously been equilibrated with 0.01 M sodium phosphate buffer containing 0.1 M NaCl, 1 mM $MgCl_2$, 0.1% NaN_3 and 0.01% BSA.

Using the same buffer, 0.7 ml fractions are collected and the absorbance of each fraction is measured at 280 nm. Each fraction is then diluted 50–300 fold and the enzyme activity of 10 μl measured.

From these, the ratio of antibody to enzyme in the complex can be determined.

Absorption of antisera to remove unwanted cross-reactivity

Many comercially available antisera are either supplied as whole serum or as a partially purified immunoglobulin fraction, e.g. an ammonium sulphate cut. Even affinity purified antisera may cross-react with other angtigens and can cause short circuits in the assay.

This commonly occurs when an antiserum raised against immunoglobulin of one species cross-reacts with another species. To improve specificity, antisera can be pre-absorbed against the cross-reactive antigen, linked to agarose, prior to use in an ELISA or before enzyme labelling. Recovery of antisera after adsorption is greater if this is done prior to any affinity purification or conjugation.

Method

Several serum-agarose preparations are available commercially alternatively these can be prepared by the same method as described for linking antigen to agarose for affinity purification

(1) A sufficient quantity of serum agarose should be added to the antiserum (typically 10 ml of antiserum) and mixed together overnight at 4 °C.
(2) The mixture should then be applied to a small column or sintaglass funnel and the unbound antiserum drained from the agarose.
(3) The volume of antiserum before and after incubation with the antigen–agarose conjugate should be noted and the dilution factor taken into consideration when preparing working dilutions.

Affinity purification of antibodies

Materials

CNBr-activated Sepharose 4B (Pharmacia)
Purified antigen
Rabbit serum containing antibody to above antigen
Sodium bicarbonate 0.1 M, pH 8.0
Phosphate buffered saline (PBS)
Tris/HCl 0.1 M, pH 8.0
Ethanolamine buffer 0.1 M, pH 8.6
1 cm × 10 cm column
HCl 0.01 M
Glycine HCl 0.2 M, pH 2.8
NaCl 0.5 M
Proprionic acid, 1 M
Tris 1 M

Method

Preparation of Sepharose:

(1) Add 3 g (dry weight) of Sepharose 4B to 100 ml of 0.001 M HCl and allow to swell up for 20 min. Spin down at 2000 rpm for 5 min and remove the supernatant.
(2) Wash the Sepharose twice with PBS, spinning down the Sepharose and removing the supernatant as above. Check the pH (should be 7–8).
(3) Dissolve 10 mg of antigen in 10 ml of 0.1 M sodium bicarbonate, pH 8, measure the optical density at 280 nm and add to the Sepharaose. Mix end-over overnight at 4 °C.

(4) Spin down the Sepharose and remove the supernatant. Check its optical density and calculate the amount bound.
(5) Wash the Sepharose twice with sodium bicarbonate, as above.
(6) Wash the Sepharose twice (for 30 min each time) with 10 ml 0.1 M ethanolamine buffer, pH 8.6. This blocks any remaining CNBr-activated sites.
(7) Wash three times (5 min each) with 10 ml of PBS.
(8) Wash once with 0.01 M HCl (for 5 min) to remove any loosely bound protein.
(9) Wash twice with PBS (5 min each) and check pH is back to pH 7.
(10) Store in 0.5 M NaCl at 4°C.

Batch procedure

(1) Add 15 ml of serum to 10 ml antigen–Sepharose in a universal container. Mix at 4 °C overnight.
(2) Centrifuge at 2000 *g* for 10 min at 4 °C and remove supernatant. This should be retained.
(3) Add Sepharose–antigen–antibody complex to sintaglass 3 funnel seated on a side arm flask which should be attached to vacuum pump.
(4) Wash with PBS until OD 0.005 or less.
(5) Add 5 ml of proprionic acid to the Sepharose and mix for 1 min, draw eluate into side arm flask containing 15 ml of 1 M Tris/HCl.
(6) Repeat step 5 and measure OD of eluate at 280 nm to determine protein yield.
(7) Dialyse against PBS overnight at 4 °C, store at −20 °C.

Column method

(1) Pack column with 10 ml of antigen coated Sepharose and equilibrate with 30–40 ml of PBS.
(2) Apply 20 ml of rabbit serum by continually circulating over the column for 24 hr at 4 °C.
(3) Remove excess serum and test by double gel diffusion for unbound antibody.
(4) Wash the column thoroughly with bicarbonate buffer at 30 ml/hr, measuring the absorbance of the eluate at 280 nm. When the absorbance is less than 0.005 continue with step 5.
(5) Apply 10 ml of 0.01 M HCl or 0.2 M glycine/HCl at a flow rate of 30 ml/hr. Collect 2 min fractions into 1 ml of 0.1 M Tris/HCl pH 8. Measure the absorbance of the fractions at 280 nm and determine fractions containing antibody.

(6) Pool fractions containing proteins and dialyse against PBS or other buffer as required.
(7) Store at −20 °C.

A good affinity purification should yield between 6 and 18 mg per 20 ml of serum. The antibody should be tested in an ELISA for specificity and cross-reactivity. Antibody purified by this method can be used directly for enzyme labelling without further preparation.

REFERENCES

Avrameas, S. (1969). Coupling of enzymes to proteins with glutaraldehyde. *Immunochemistry*, **6,** 43.
Avrameas, S., and Ternynck, T. (1971). Peroxidase labelled antibody and Fab′conjugates with enhanced intracellular penetration. *Immunochemistry*, **8,** 1175.
Kato, K., Fukui, H., Hamaguchi, Y., and Ishikawa, E. (1976). Enzyme-linked immunoassay: conjugation of the Fab′ fragment of rabbit IgG with β-D-galactosidase from *E. coli* and its use in immunoassay. *J. Immunol.*, **116,** 1554.
Nakane, P.K., and Kawaoi, A. (1974). Peroxidase-labelled antibody; a new method of conjugation. *J. Histochem. Cytochem.*, **22,** 1084.

Index